Dog Behaviour, Physiology and Welfare

Dog Behaviour, Physiology and Welfare

Editors

Angelo Gazzano
Chiara Mariti

MDPI • Basel • Beijing • Wuhan • Barcelona • Belgrade • Manchester • Tokyo • Cluj • Tianjin

Editors

Angelo Gazzano
Università di Pisa
Italy

Chiara Mariti
Università di Pisa
Italy

Editorial Office
MDPI
St. Alban-Anlage 66
4052 Basel, Switzerland

This is a reprint of articles from the Special Issue published online in the open access journal *Animals* (ISSN 2076-2615) (available at: https://www.mdpi.com/journal/animals/special_issues/Dog_Behavior).

For citation purposes, cite each article independently as indicated on the article page online and as indicated below:

LastName, A.A.; LastName, B.B.; LastName, C.C. Article Title. *Journal Name* **Year**, *Article Number*, Page Range.

ISBN 978-3-03943-044-4 (Hbk)
ISBN 978-3-03943-045-1 (PDF)

Cover image courtesy of Chiara Mariti.

Contents

About the Editors

Angelo Gazzano (Dr.) graduated in Veterinary Medicine, with a PhD in physiology. Angelo Gazzano is currently Associate Professor at the University of Pisa (Italy). His research is mainly focused on veterinary physiology and behavioural medicine. He is a diplomate of the European College of Animal Welfare and Behavioural Medicine.

Chiara Mariti (Dr.) graduated in Veterinary Medicine, with a PhD in equine physiology, Chiara Mariti is currently a senior researcher at the University of Pisa (Italy). Her research is mainly focused on pet behaviour and welfare and anthrozoology.

Preface to "Dog Behaviour, Physiology and Welfare"

Dogs were the first animal species to be domesticated. Large amounts of evidence and science-based knowledge support the unique role that domestic dogs play in human life. Dogs can play many roles, ranging from companionship to very specialized roles. Dogs, though descended from the wolf, show noticeable differences when compared to their ancestor, leading to the need for studies focused on the species, as demonstrated by several comparative studies highlighting differences between wolves and domestic dogs. One main difference concerns inter-specific relationships. The kinds of relationship domestic dogs can establish with humans greatly vary and this is a relatively new field of research whose results can provide relevant inputs for both humans and dog welfare. In the last years, research on dog behaviour and welfare is also increasing, with studies ranging from those having an ethological approach to those more focused on applied ethology, physiology, endocrinology, anthrozoology and behavioural medicine. The objective of this Special Issue and book is to publish research papers dealing with dog behaviour, physiology and welfare and their interrelations with the dog–human relationship, to strengthen the knowledge of our "best friend". Thanks to the open access policy, we hope to ensure these data are available to all stakeholders, ranging from people working in the field of dog training, canine behavioural medicine, shelters, etc. to researchers and, possibly, even dog owners.

Angelo Gazzano, Chiara Mariti
Editors

animals

MDPI

Article

A Retrospective Analysis of Complaints to RSPCA Queensland, Australia, about Dog Welfare

Hao Yu Shih [1,*], Mandy B. A. Paterson [2] and Clive J. C. Phillips [1]

[1] Centre for Animal Welfare and Ethics, University of Queensland, White House Building (8134), Gatton Campus, Gatton, QLD 4343, Australia; c.phillips@uq.edu.au

[2] Royal Society for the Prevention of Cruelty to Animals Queensland, Brisbane, QLD 4076, Australia; mpaterson@rspcaqld.org.au

* Correspondence: haoyu.shih@uqconnect.edu.au

Received: 29 April 2019; Accepted: 22 May 2019; Published: 27 May 2019

Simple Summary: Animal neglect and cruelty are important welfare and social issues, and dogs are one of the most commonly reported species to have experienced both. Most previous studies related to canine cruelty and welfare focused on animal abuse and dog fighting. However, literature dealing with the milder but more common forms of animal welfare concerns is limited. Therefore, this retrospective study aimed to understand the epidemiology of different types of canine welfare complaints in Queensland in the past decade and also to identify risk factors and their roles in different types of welfare complaints. The number of complaints received each year increased by 6.2% annually. The majority of complaints were neglect-related rather than related to deliberate cruelty, with the most common complaints being that dogs had poor body conformation, insufficient food and/or water, and receiving inadequate exercise. Poor living conditions and leaving dogs in a hot vehicle unattended were more commonly reported in recent years, potentially due to higher public awareness. Adult dogs that were reported were more likely to be alleged to have been poisoned, left unattended in a hot car, abandoned, and to have had inadequate exercise and shelter, compared with puppies. Puppies that were reported were more likely to be alleged to have experienced cruelty, lack of veterinary support, overcrowding, poor living and health conditions, and inappropriate surgery. Recognising which dogs are at most risk of cruelty will inform strategies to address this serious welfare problem.

Abstract: Animal neglect and cruelty are important welfare and social issues. We conducted an epidemiological study of dog welfare complaints and identified risk factors. The retrospective study included 107,597 dog welfare complaints received by RSPCA Queensland from July 2008 to June 2018. The risk factors considered were the age of dogs and the year of being reported. The number of complaints received each year increased by 6.2% per year. The most common complaints were poor dog body conformation, insufficient food and/or water, dogs receiving inadequate exercise, and dogs being confined or tethered. Increasing numbers were most evident for poor living conditions and leaving dogs in a hot vehicle unattended, both of which may have resulted from increasing public awareness. The majority of complaints were neglect-related rather than related to deliberate cruelty. Compared with puppies, adult dogs were more likely to be reported to have been poisoned, left unattended in a hot car or abandoned, as well as to have had inadequate exercise and shelter. Reported puppies were more likely to be alleged to have experienced cruelty, lack of veterinary support, overcrowding, poor living and health conditions, and inappropriate surgery. In conclusion, animal neglect was the most commonly reported welfare concern in dogs. Due to an assumed increasing public awareness of some types of cruelty, the trends of reported concerns differed. Adult dogs and puppies were reported to be involved in different types of welfare concerns. Strategies to address cruelty to dogs can be informed by an understanding of risk factors and trends in types of cruelty.

Keywords: dog; canine welfare; canine cruelty; neglect; RSPCA; age

1. Introduction

Animal cruelty involves all human behaviours towards animals that are morally and/or legally unacceptable, causing them to be inflicted with unnecessary and unjustifiable physiological, psychological, and behavioural discomfort or pain [1,2]. It is a complex issue implicating animal welfare, moral concerns, criminal activity, and violence [2–4]. It is regulated by state and territory law in Australia; for example, in Queensland by the Animal Care and Protection Act 2001 (ACPA [1]). This state-based legislation empowers the State to appoint inspectors, some of whom are employed by the Royal Society for the Prevention of Cruelty to Animals, Queensland (RSPCA Qld), to investigate potential breaches of the Act and enforce compliance with the Act [1]. There are two main offences under the ACPA, one is failure to fulfil duty of care responsibilities and the other is cruelty. There are a number of other specified offences. The Act recognises that a person who has charge of an animal owes that animal a duty of care. Failure to provide such care potentially constitutes a 'breach of duty of care' offence. This offence covers such actions as not providing sufficient food, water, exercise, veterinary care, and suitable living conditions. It is not only the owner that has a duty of care towards an animal. Anyone who even temporarily is in charge of an animal has a duty of care. The second major offence is 'animal cruelty' and the Act describes what it sees as cruelty in Section 18. A cruel act towards an animal can be committed by anyone, whether it is their own animal, another domestic animal, or even a wild animal [1]. It is important to note, that under the ACPA, the intention of a person to be cruel is not a prerequisite for committing the offence of cruelty. If an action carried out by a person causes pain and suffering and the action was intentional (that is not accidental), the person may be charged with cruelty. The intention to carry out the action must be proved but not the intention to be cruel. If a lack of action deprives an animal of its fundamental needs then they may be charged with a breach of their duty of care or even cruelty, depending on the circumstances. Motivation may be considered during sentencing [1]. Other offences under the Act include unreasonable abandonment or release, the carrying out of prohibited surgical procedures (e.g., tail docking, ear cropping, debarking, etc.), being involved in, or having items used for, a prohibited event such as dog or cock fighting, and allowing an animal to injure or kill another animal [1].

Potential cases are reported to RSPCA through various means. RSPCA Qld has a 'Cruelty Complaints' telephone number manned 24 h a day, seven days a week; complaints also come in through emails. Complaints can be made by members of the public but also by veterinarians and veterinary nurses, council officers, and other government and non-government employees visiting a location as part of their duties. Animals surrendered to the RSPCA or that come in as strays may be investigated if cruelty is suspected. They are considered by RSPCA Qld inspectors and further investigated if necessary.

According to the annual statistics of RSPCA Qld, there were 15,102 animal welfare complaints reported by the general public in 2011 [5], which had increased to 17,929 by 2017 [6]. Of all species falling victim to animal welfare concerns, dogs (*Canis familiaris*) are one of the most commonly reported species [7].

Various risk factors have been identified as contributing to an unsuccessful dog–owner relationship, which potentially results in neglect or abuse. These include the age of the dog [8,9], dog behaviour [8,10–12], physical attributes of the dog [9,13], the owner's motivation to care for the dog [14,15], the owner's attachment to the dog [12,16], costs of keeping the dog [16,17], and the owner's socioeconomic status [18,19]. In relation to actions carried out by third parties, most studies have focused on organised industries such as dog coursing [20] and fighting [21]. There has also been research into the origin of 'noxious abuses', e.g., cruelty involving intentional abuse, such as beating, shooting, and burning, that lead to severe physical injuries to the animals [7,15]. Literature dealing with the milder but more common forms of animal welfare concerns is limited. One report considers neglect, such as exposing dogs to poor nutrition, keeping dogs in a backyard for hours without a shelter, and failing to meet exercise needs [2]. Most studies [20,22,23] stress the moral, legal, and social aspects of animal cruelty, and few explore the epidemiological dimension of this topic. This study addresses the epidemiology of diverse animal welfare concerns reported by the general public, instead of actual neglect or cruelty cases in a typical Western society. It also aims to identify the age of dogs as a risk factor for different forms of canine welfare complaints. Other risk factors, breed and socioeconomic status of the complainant, will be the subject of future papers.

2. Materials and Methods

2.1. Materials

From July 2008 to June 2018, RSPCA Qld received 129,036 canine welfare complaints. Some involving more than one dog were recorded as multiple complaints sharing the same case number, while others were recorded as one complaint with multiple animals. To avoid sample bias due to multiple entries, we only retained the first complaint of case numbers with multiple entries, discarding 21,439 entries as a result. There remained 107,597 canine welfare complaints for this retrospective study. Complaints that fell within the zone of responsibility of RSPCA Qld (determined by a Memorandum of Understanding between RSPCA Qld and Biosecurity Queensland, the Government Department tasked with the administration of ACPA) were investigated by RSPCA Qld inspectors. All other complaints were referred to Biosecurity Queensland to be investigated by their inspectors.

Complaints were recorded in Shelter Buddy®, the RSPCA Qld database. The following information was requested from the reporter of each incident by the inspector at the time of taking the complaint: the number of dogs involved (n = 106,104), their age (n = 107,597), their breed(s) (n = 92,021), the coded complaint type(s) (n = 106,983), the suburb (n = 107,413), and the postcode (n = 107,270); in addition, the date was recorded (n = 107,597). Dogs' ages were dichotomised into adult dog and puppy, based on reporters' interpretation. It is important to recognise that the information recorded from the complainant may have been inaccurate or inaccurately interpreted, e.g., a small dog is commonly referred to as a puppy. Records regarding breed and the number of dogs involved were based on either complainants' initial reports or comments from trained inspectors, again recognising inaccuracies with identification of the breed and the number of dogs involved. The 'complaint code' was selected by the staff member receiving the call or email from a drop-down menu of 18 possible complaints (Appendix A, Table A1). Multiple 'complaint codes' were able to be selected for each case, according to the description of what was alleged to have happened to the dog(s), and each was treated as a separate code for analysis.

2.2. Statistical Analysis

Data were analysed using the statistical package Minitab® 17.3.1. Descriptive analysis was used to investigate the distribution of complaint codes. Polynomial regression analysis and simple linear regression analyses were used to model the prevalence of different complaint codes from 2008 to 2018. The model chosen was that with the highest R-sq value, after ensuring that all components in the model were significant ($p < 0.05$). In 2008 and 2018, only data from July to December, and January to June were available, respectively. Therefore, data in 2009 and 2017 were used to test for within year variation in code citation rates for 2008 and 2018, respectively. Specifically, chi-squared analyses were conducted to compare whether the reported prevalence of each complaint code from January to June were different from those in July to December in 2009 and 2017. If there was no significant ($p < 0.05$) difference between the two six-month periods in that complaint code in 2009 and/or 2017, then the prevalence of the particular complaint code in the six-month period in 2008 and/or 2018 was/were assumed to be partially representative of the entire year(s). However, if there was a significant difference between the two six-month periods in that complaint code in 2009 and/or 2017, the data of the specific complaint code in 2008 and/or 2018 were excluded from the polynomial regression analyses of year effects. After that, a Grubbs' test was used to identify outliers of each complaint code, which were excluded from polynomial and simple linear regression analyses. In polynomial regression analyses and simple linear regression analyses, years were entered as input variables and the prevalence of the complaint code as the output. The models were chosen on the basis of significant p values and the greatest R-sq values yielded. Three complaint codes, Causing captive animal to be injured/killed by a dog (N = 29), Keeping or using animal for blooding/coursing a dog (N = 18), and Emergency relief (N = 8) were not included in polynomial and simple linear regression analyses because the number of reported cases in the past decade was too few. Eighteen stepwise forward binary logistic regression models were constructed to understand how dogs' ages correlated with each complaint code. To determine the effect of age on complaint codes, age was entered (in dichotomous data form) into a binary logistic regression model as a fixed factor, using a logit function, with an alpha value to enter of 0.15. Complaint codes were entered into the model as outcomes. Separate models were constructed for each complaint code with the same input variable.

3. Results

3.1. Complaint Codes and Dogs' Ages

There were 18 complaint codes in total (Appendix A Table A1). On average, each case involved 1.76 (SEM = 0.003) codes. The distribution of complaint codes is presented in Figure 1. The most common codes, listed in declining order, were Poor dog condition (n = 29,982, 27.9%), Insufficient food and/or water (n = 28,265, 26.3%), No exercise/confined/tethered (n = 27,913, 25.9%), and Abandonment (n = 21,626, 20.1%). Overall, 93.67% (N = 100,791) of reported cases involved reported adult dogs and 6.33% (N = 6806) of reported cases involved reported puppies.

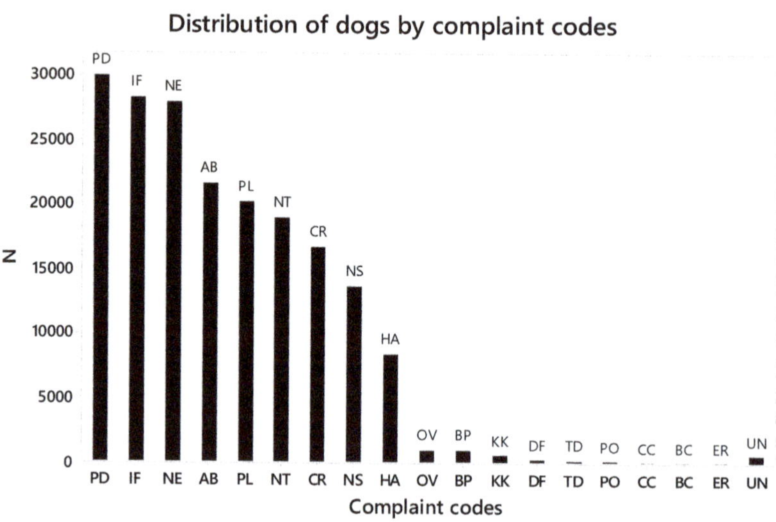

Figure 1. Distribution of dogs by complaint code. PD-Poor dog condition (27.9%, N = 29,982); IF-Insufficient food and/or water (26.3%, N = 28,265); NE-No exercise/confined/tethered (25.9%, N = 27,913); AB-Abandonment (20.1%, N = 21,626); PL-Poor living condition (18.7%, N = 20,162); NT-No treatment (17.6%, N = 18,963); CR-Cruelty (15.5%, N = 16,661); NS-No shelter (12.7%, N = 13,682);

HA-Hot animal in car (7.8%, N = 8384); OV-Overcrowding (0.9%, N = 978); BP-Baiting/poisoning (0.9%, N = 974); KK-Knowingly allowing an animal to kill/injure another (0.6%, N = 600); DF-Dog fighting or other prohibited offence (0.3%, N = 277); TD-Tail docking or other surgical procedure (0.2%, N = 214); PO-Prohibition order breached (0.1%, N = 133); CC-Causing captive animal to be injured/killed by a dog (0.03%, N = 29); BC-Keeping or using animal for blooding/coursing a dog (0.02%, N = 18); ER [a]-Emergency relief (0.01%, N = 8); UN-Unknown (0.6%, N = 614). [a] Emergency relief, as opposed to emergency rescuing which occurred when an animal encountered an urgent situation not related to domestic violence, was provided based on the ACPA, Section 123 [1].

3.2. *Trends of Complaint Types*

The number of complaints received annually increased by 6.2% per year, and the incidence of most complaint codes changed over the ten years. Results of the Chi-squared analyses showed that the prevalence from January to June and from July to December was significantly ($p < 0.05$) different for Poor living condition and Baiting/poisoning in 2009, No treatment and Poor dog condition in 2017, and No exercise/confined/tethered and Hot animal in car in both 2009 and 2017 (Appendix A, Table A2). Therefore, the data for Poor living condition and Baiting/poisoning in 2008, No treatment and Poor dog condition in 2018, and No exercise/confined/tethered and Hot animal in car in both 2008 and 2018 were excluded from the analyses of year effects. The prevalence of Poor dog condition in 2008 was an outlier ($p = 0.029$), and therefore was excluded as well. Figure 2 demonstrates the trends and the equations used for polynomial regression or simple linear analysis of each complaint code. These trends can be classified into five patterns: negative linear, positive linear, concave, monotonic, and irregular. Negative linear models included No exercise/confined/tethered, overcrowding, and Tail docking or other surgical procedure. Positive linear models included Poor living conditions, Hot animal in car, and Prohibition order breached. A concave pattern, indicating that the prevalence increased to a peak and then slowly decreased, was observed for No treatment, Abandonment, No shelter, and Knowingly allowing an animal to kill/injure another, for which codes the prevalence reached a peak in 2015, 2014, 2015, and 2011, respectively. In monotonic patterns, the trend was to generally increase, but not consistently, e.g., the prevalence of Poor dog condition generally increased, except for 2011–2016. Finally, some complaint codes had irregular changes over time. Cruelty, Insufficient food and/or water, Baiting/poisoning, and Dog fighting or other prohibited offence could not be modelled as they were reported sporadically over the ten years.

<table>
<tr><th>Complaint Code</th><th>Figure</th><th>Pattern</th><th>p Value/R-Sq</th></tr>
<tr><td>No exercise/confined/tethered</td><td>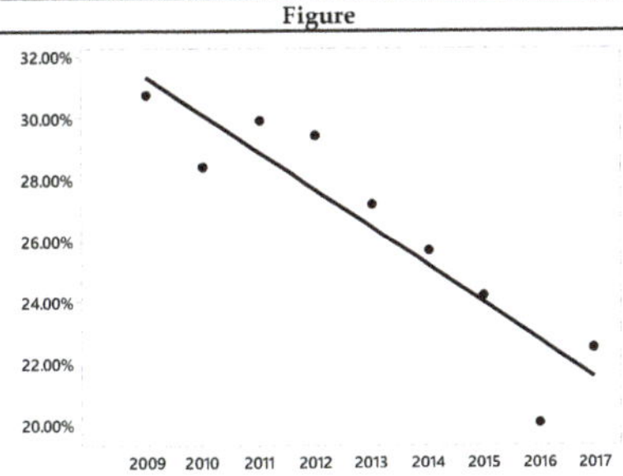
Y = 24.85 − 0.01221 X</td><td>Negative linear</td><td><0.001/84.4%</td></tr>
</table>

Figure 2. *Cont.*

Overcrowding

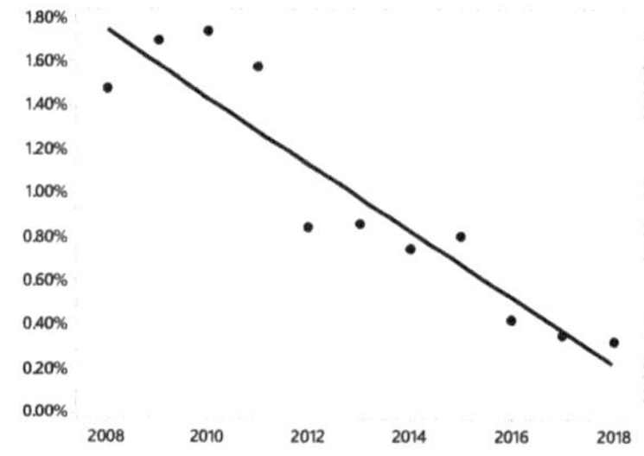

Y = 3.081 − 0.001526 X

Negative linear — <0.001/86.5%

Tail docking or other surgical procedure

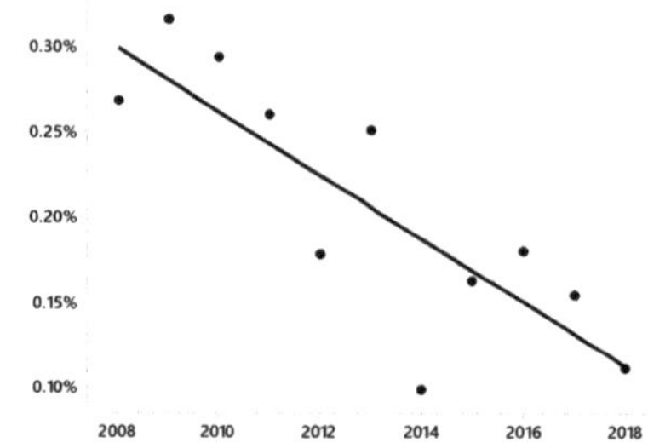

Y = 0.3768 − 0.000186 X

Negative linear — 0.002/69.2%

Poor living condition

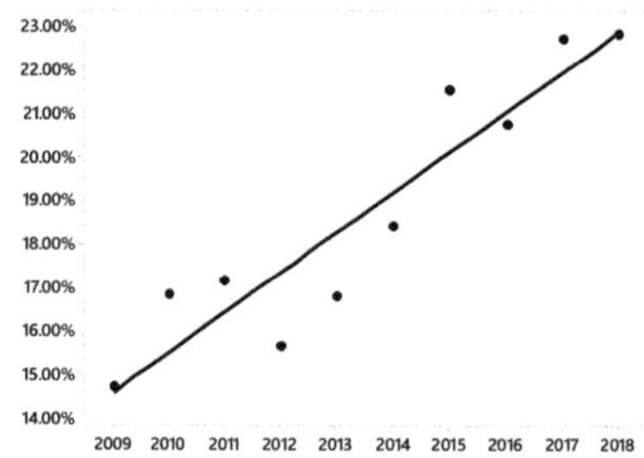

Y = −18.34 + 0.009202 X

Positive linear — <0.001/86.7%

Hot animal in car

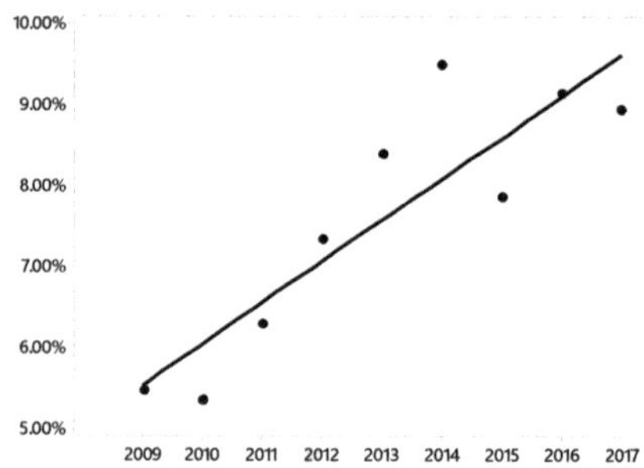

Y = −10.17 + 0.005089 X

Positive linear — 0.001/78.5%

Figure 2. *Cont.*

Prohibition order breached

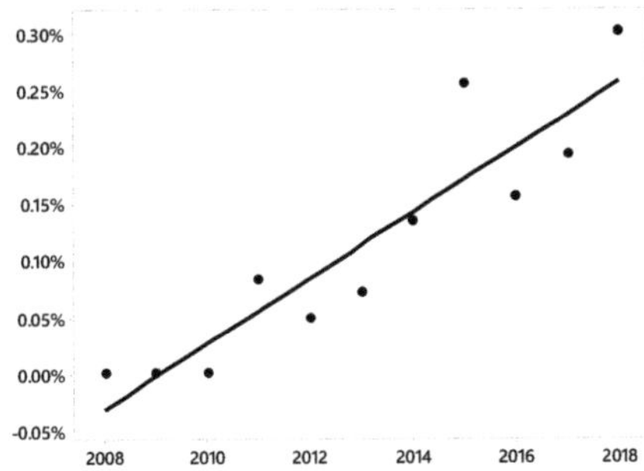

$Y = -0.5797 + 0.000289\ X$

Positive linear <0.001/83.8%

No treatment

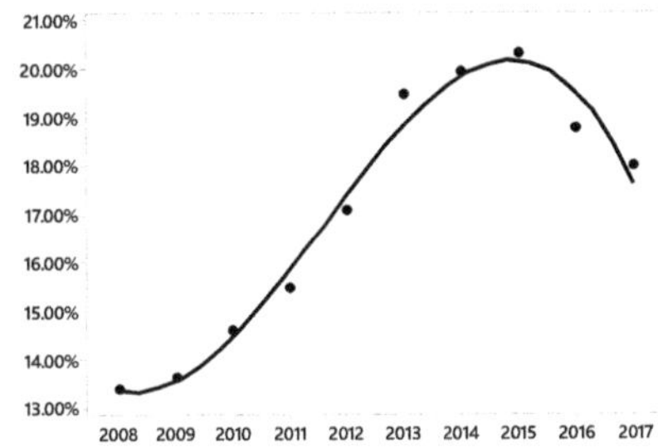

$Y = 3889846 - 5801\ X + 2.884\ X^2 - 0.000478\ X^3$

Concave <0.001/97.7%

Abandonment

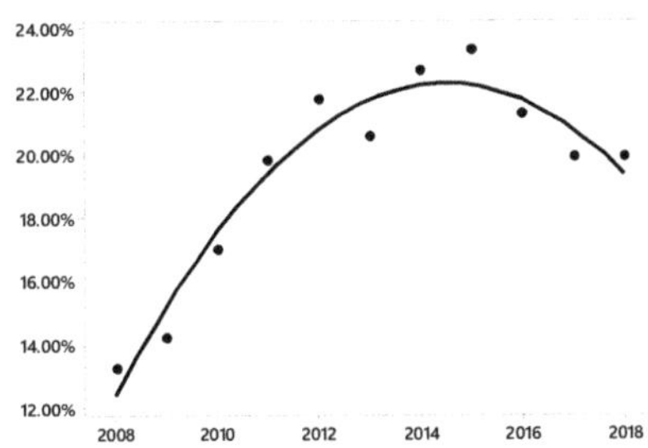

$Y = -9487 + 9.419\ X - 0.002338\ X^2$

Concave <0.001/93.4%

No shelter

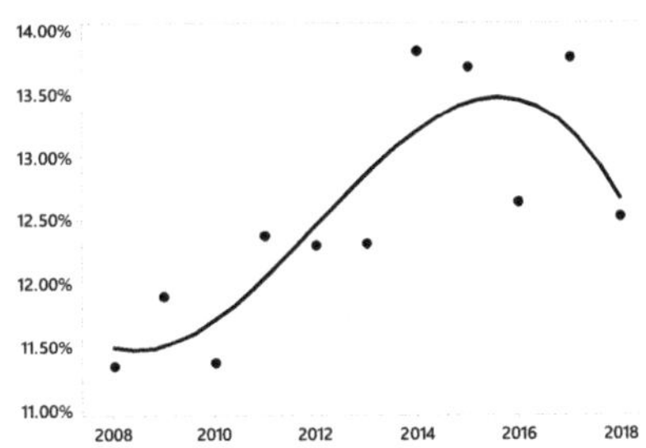

$Y = 886771 - 1322\ X + 0.6571\ X^2 - 0.000109\ X^3$

Concave 0.023/72.2%

Figure 2. *Cont.*

Knowingly allowing an animal to kill/injure another	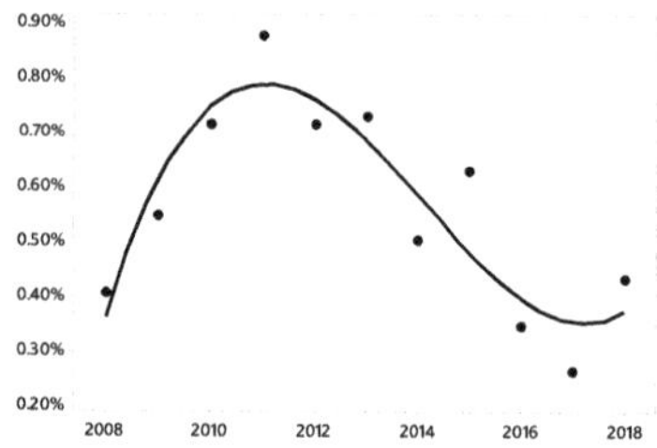	Concave	0.005/82.9%
	$Y = -304963 + 454.2\,X - 0.2255\,X^2 + 0.000037\,X^3$		
Poor dog condition	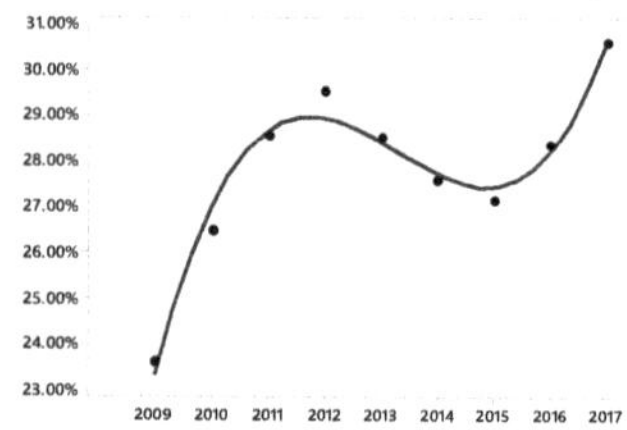	Monotonic	<0.001/97.2%
	$Y = -8311541 + 12385\,X - 6.152\,X^2 + 0.001018\,X^3$		
Cruelty	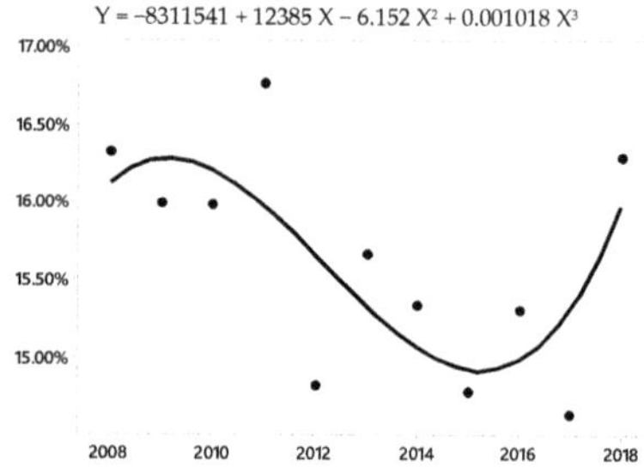	Irregular	0.132/53.0%
	$Y = -960945 + 1433\,X - 0.7120\,X^2 + 0.000118\,X^3$		
Insufficient food and/or water	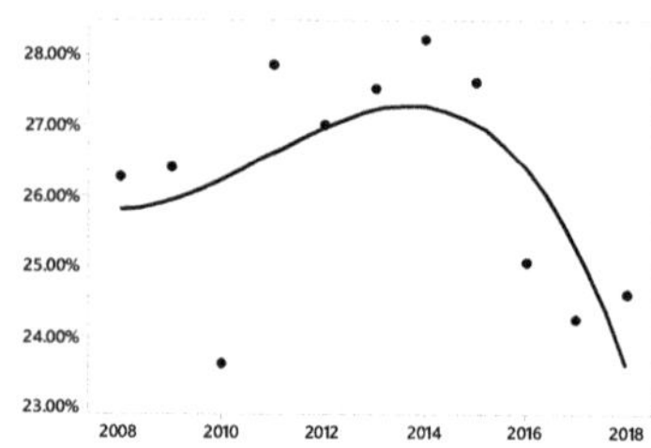	Irregular	0.217/44.9%
	$Y = 1258720 - 1878\,X + 0.934\,X^2 - 0.000155\,X^3$		
Baiting/ poisoning	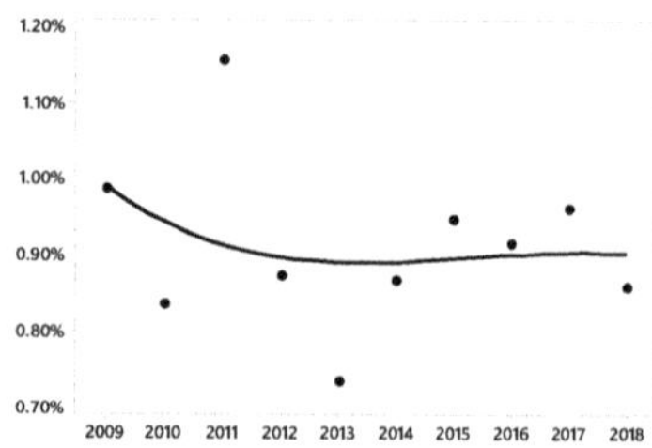	Irregular	0.917/7.6%

Figure 2. *Cont.*

Dog fighting or other prohibited offense	$Y = 40993 - 61.0\ X + 0.0303\ X^2 - 0.000005\ X^3$ 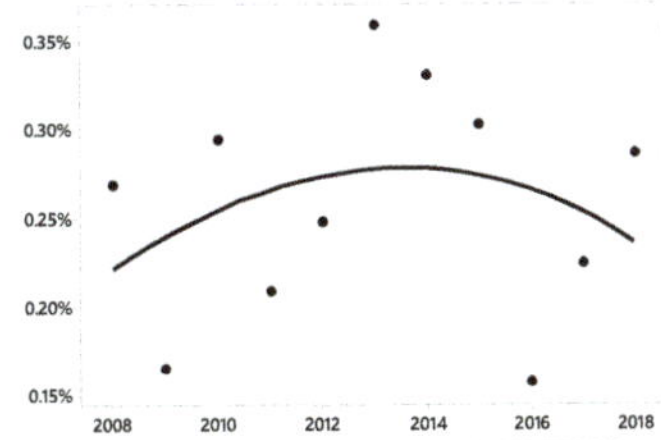 $Y = 1981 - 3.0\ X + 0.00151\ X^2 - 0.000000\ X^3$	Irregular	0.883/8.4%

Figure 2. Polynomial regression of each complaint code. The X axis represents the year, and the Y axis represents the prevalence of the complaint code.

3.3. *Risk Factors for Different Complaint Codes*

We considered age as a risk factor. The relationships between dogs' age and complaint codes are displayed in Table 1 and Figure 3. Compared to adult dogs, puppies were more likely to be reported for alleged Tail docking or other surgical procedure (OR = 9.87, $p < 0.001$), Overcrowding (OR = 4.44, $p < 0.001$), Poor living condition (OR = 1.45, $p < 0.001$), No treatment (OR = 1.33, $p < 0.001$), Cruelty (OR = 1.27, $p = 0.001$), and Poor dog condition (OR = 1.23, $p < 0.001$). Adult dogs were significantly more likely to be reported as an alleged case of a Hot dog in car (OR = 0.41, $p < 0.001$), Baiting/poisoning (OR = 0.42, $p < 0.001$), Abandonment (OR = 0.53, $p < 0.001$), No exercise/confined/tethered (OR = 0.64, $p < 0.001$) and No shelter (OR = 0.91, $p = 0.037$).

Table 1. Odds ratio of each variable in the logistic regression model of complaint codes. The outputs of these models were different complaint codes. The input variable was dog age (puppy or dog).

Complaint Code	Puppy/Dog OR (CI) [(a)]	*p* Value
Tail docking or other surgical procedure	9.87 (7.30, 13.34)	<0.001
Overcrowding	4.44 (3.70, 5.32)	<0.001
Poor living condition	1.45 (1.35, 1.55)	<0.001
No treatment	1.33 (1.24, 1.42)	<0.001
Cruelty	1.27 (1.18, 1.37)	0.001
Poor dog condition	1.23 (1.16, 1.30)	<0.001
No shelter	0.91 (0.84, 1.00)	0.037
No exercise/confined/tethered	0.64 (0.60, 0.69)	<0.001
Abandonment	0.53 (0.49, 0.58)	<0.001
Baiting/poisoning	0.42 (0.27, 0.66)	<0.001
Hot animal in car	0.41 (0.35, 0.47)	<0.001
Causing captive animal to be injured/killed by dog	_ [b]	_ [b]
Dog fighting or other prohibited offence	_ [b]	_ [b]
Emergency relief	_ [b]	_ [b]
Insufficient food and/or water	_ [b]	_ [b]
Keeping or using animal for blooding/coursing a dog	_ [b]	_ [b]
Knowingly allowing an animal to kill/injure another	_ [b]	_ [b]
Prohibition order breached	_ [b]	_ [b]

[(a)] Dog age was only classified as dog or puppy. Odds ratio refers to puppy relative to dog. [(b)] Age factor was not selected in the logistic regression model.

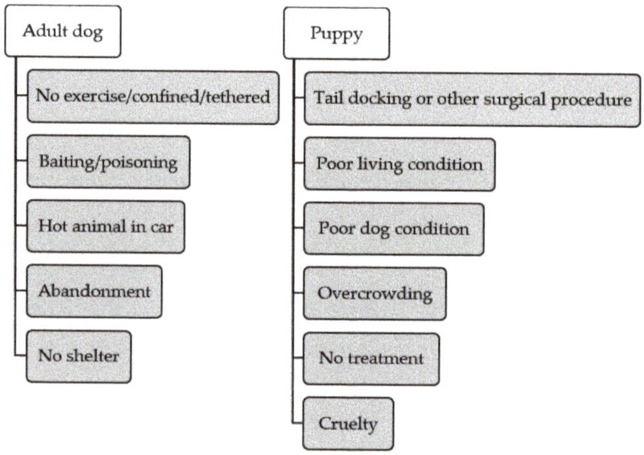

Figure 3. Positive relationships between dog age (adult dog and puppy) and complaint codes. Complaint codes listed under 'Adult dog' and 'Puppy' are complaints commonly involving adult dogs and puppies, respectively.

4. Discussion

4.1. Complaint Codes

This study reports the prevalence and progression of canine welfare complaints over the past decade. The complaints came from members of the public and may not represent all animal welfare issues or breaches of the ACPA. It must also be recognized that not all calls were found, on investigation, to represent a breach of the ACPA or even a dog suffering poor welfare. The number of complaints may be a representation of the degree of awareness in the community of animal welfare and the vigilance of many. A small number may be vexatious.

The descriptive analyses of different complaints show that the majority of alleged complaints related to neglect rather than deliberate maleficence. In this study, poor body and living conditions, insufficient food and water, and lack of the provision of exercise were the most commonly reported. This is in line with previous research that animal neglect or cruelty was most likely a result of ignorance due to lack of knowledge or forgetfulness [2,4].

In most cases, but particularly where it was decided not to continue to a prosecution, education of the dog owner was undertaken. To address neglect-related issues, people in charge of the animal were informed about food and water requirements, as well as the need for exercise, human companionship, and what represents good living conditions for a dog. For example, diets were recommended that are complete and balanced to replace homemade and all raw meat diets [24], and dog owners were likely to be informed that there were specialized products designed for specific ages of dogs [25] or dogs with specific health concerns [26]. Regular exercise is essential to promote health and quality of life [13,27,28]. Unfortunately, there are sometimes mismatches between an owner's exercise capability and the needs of their dog which inspectors recognise and advise accordingly [13,28]. The amount and frequency of exercise recommended varies according to the age, breed, size, fitness, and health of the dog. According to the ACPA, owners should ensure their dogs exercise for two hours after being continuously confined (e.g., caged or tethered) for 24 h, or for 1 h after 24 h confinement and another 1 h in the next 24 h [1]. This is the minimal time for most dogs to exercise and the majority benefit from receiving more [29]. The United Kingdom kennel club has published guidelines detailing suitable exercise amounts for each breed [29]. These are general in nature and not prescriptive.

Poor living conditions are negatively associated with a dog's quality of life, and increase the risk of diseases caused by ringworm, Giardia, Cryptosporidium, Toxocara, and Ancylostoma, as well as infestation with ectoparasites such as fleas [30–33]. Some of these agents are zoonotic and can cause health problems in humans.

4.2. Trends Over Time

Overall, the number of complaints received each year increased by 6.2% per year in the past decade, which may be contributed by the increasing population in Queensland in parallel with growing dog ownership in Queensland, and people's rising awareness of animal welfare. The population growth rate in Queensland was around 2% per year from 2008 to 2017 [34,35], and the pet population breakdown across the states and territories mirrored the country's population distribution [36]; it is reasonable that dog ownership in Queensland has also been increasing at a similar rate. Given the mismatch between the growth rate of canine welfare complaints (around 6% annually) and Queensland population (around 2% annually) for the same period of time, we believe that increasing public awareness and propensity to report animal cruelty may be another contributing factor [37]. The issue of animal welfare is becoming popular with the Australian public, in terms of their knowledge, concerns, willingness to participate, and legislation [38,39]. This is reflected by the reported increased concern for animal welfare among the general public [40–42], animal protection movements from advocacy groups [43], and public media [44], as well as the development of a closer relationship between humans and animals [45] and people's inclination to report cases involving animal harm to the police and RSPCA [37]. Such an increase in public awareness may be associated with the connections between animal abuse and human violence [23,46], and animal and human health [30,47], as well as more general dog ethical issues, such as dog consumption in Asia [48]. Moreover, people increasingly acknowledge the importance of good nutrition [49], canine behaviour [50,51] and emotions [52], and the goal of achieving a no-kill policy and improving the live release rate for roaming or sheltering dogs [39,53]. At RSPCA Qld, the number of volunteers has also grown from 2000 in 2011 to over 5000 in 2018 [6], suggesting that more people are actively concerned about the welfare of dogs.

The trends seen with the different complaints are also important to consider. Dogs were commonly reported as being in poor condition, living in poor environments, or being left in a heated vehicle. The increasing frequency of these three complaints may also be the result of a growing awareness of animal welfare among the public. However, the importance of animal welfare law enforcement and public education should not be underestimated. Humans' changing lifestyle may also influence the prevalence of certain kinds of animal welfare concerns. For instance, more people nowadays own a car and travel with their dogs frequently [54]. Consequently, dogs are at greater risk of being left unattended in a vehicle. This is particularly hazardous in Queensland, where median summer maximum temperatures of around 30 °C can lead to dangerously high temperatures inside cars [55]. Apart from the increasing trends for the types of complaints mentioned above, some complaint categories, particularly those involving food or water insufficiency, cruelty, dog poisoning, and dog fighting demonstrated irregular patterns, suggesting that they have occurred, or have been reported, inconsistently over the ten years [56]. These complaints also have the potential to jeopardize the dogs' life, and may also be related to violence and crimes [4,21,57]; therefore, these should be closely monitored. Finally, in Queensland, tail docking and other inappropriate surgical procedures (e.g., ear cropping and declawing) have been banned under the ACPA since 2001, unless they are undertaken by and under the recommendation of a veterinary surgeon [1]. Our results demonstrate a steady decline in the number of such cases, which indicates that the law is being adhered to.

4.3. Adult Dogs and Puppies

Apart from gaining an overall understanding of the prevalence and trends of different complaints, we also identified dog age as a risk factor in our dataset. Figure 3 summarizes the positively significant correlations between dog age and complaint codes. Adult dogs were more likely to be alleged to be subjected to activity-related welfare issues—for example being confined. They were also more likely to be reported as abandoned or left in a heated vehicle. The higher abandonment potential is supported by an anthropomorphic theory that states that as adult dogs possess fewer infant-like traits, and thus are regarded as less attractive, they are less likely to evoke our nurturing instinct, creating a weaker bond with humans [12,58,59]. Adult dogs may have a greater chance of being left in a hot car because they are more difficult to carry and are usually not allowed to enter public places. On the other hand, puppies need more care and are easier to carry with a portable confinement box, therefore, it is less likely that they will be transported in a car and left when the driver is absent. Drivers may also imagine adult dogs to be more robust than puppies and more able to cope with high temperatures. This problem may be solved by promoting more pet-friendly environments (e.g., shops and restaurants) where dogs can stay with their owners and not be left in a vehicle alone. Increased information and signage in car parks reminding drivers of the danger to their dog of hot cars should also be encouraged. As for puppies, tail docking is generally conducted on three to five-day old puppies [60,61]; therefore, it is not surprising that this complaint category was more commonly reported with puppies than with adults. As the ban on tail docking was introduced in Queensland in 2001 [1], it is likely that some adult dogs were not reported because they were docked before the ban was introduced.

Additionally, in this study, puppies were alleged to be more likely to suffer from cruelty, which may result from the fact that puppies are submissive, and therefore tend to satisfy the controlling motivation for animal cruelty [15]. This finding is partially supported by a previous study that focused on non-accidental injuries of dogs and cats, and which revealed that dogs less than two years of age were more susceptible to intentional abuse [62]. In addition to the submissive nature of puppies, the authors suggest that young dogs are less manageable and thus may provoke owners with aggressive potential [62]. Another inconclusive result was found in a research trial investigating animal cruelty and domestic violence; in that study, authors did not find dogs' age to be a predisposing factor of being targeted for abuse [46]. Finally, access to appropriate medical support and suitable living conditions such as a good environment and enough space is another concern for puppies. These welfare concerns are probably indicators for animal hoarding or puppy farms [63,64].

4.4. Strengths and Limitations

One strength of this study is that it describes the trends in dog welfare complaints in Queensland over the last ten years and represents a large database, which allowed trends to be determined. It also correlates dog age with specific complaints. However, the current study also has its limitations. First, dogs were only classified as puppy or dog by complaints, which may hide important details related to age [9]. Second, this dataset covers only coastal, highly populated parts of Queensland, and thus generalization should be made with caution. Finally, the complaint code choices made by Call Centre staff were made on the basis of information obtained from the public, which indicates the potential of inaccurate reports. However, the study presents an analysis of what was reported and reflects changes in public awareness and motivation to act. Future studies could assess the accuracy of what people report and include more risk factors; for instance, breeds, behavioural issues of dogs, socioeconomic levels, and the history of an unsuccessful ownership have been reported to negatively influence dog ownership [8]. Dealing with behavioural problems and preventing people with a history of poor dog ownership from acquiring a new dog could reduce the risk of similar incidents being repeated. These factors are useful for addressing canine welfare issues and associated crimes and thus should be considered in future studies.

5. Conclusions

This study identified prevalence, trends, and the age of dogs as a risks factor for different types of complaints. Breed of the dog and socioeconomic status of the complainant will be the subject of future papers. Some neglect-related complaints, such as offering insufficient food and water, providing poor living conditions, and leaving a dog unattended in a heated vehicle apparently became more prevalent in recent years, probably indicating greater public awareness rather than an increase in neglectful behaviour. However, some serious complaints have been consistently reported over the past decade, including those involving animal abuse or severe injuries, and consequently should be closely monitored. The age of dogs was correlated with complaints about abandonment, neglect-related mistreatment, cruelty, and inappropriate surgery. Adult dogs were more likely to be reported as receiving inadequate exercising and shelter, having been abandoned, and having been left unattended in a hot vehicle; puppies were more likely to be reported as having poor living and health conditions, having undergone inappropriate surgery, and having suffered abuse. Recognising which dogs are at most risk of cruelty will inform strategies to address this serious welfare problem. Furthermore, the local or state government can direct specific attention to the most common and growing types of neglect and cruelty.

Author Contributions: Conceptualization, C.J.C.P. and M.B.A.P.; methodology, H.Y.S., C.J.C.P. and M.B.A.P.; software, H.Y.S. and C.J.C.P.; validation, H.Y.S., M.B.A.P. and C.J.C.P.; formal analysis, H.Y.S.; investigation, H.Y.S.; resources, M.B.A.P. and C.J.C.P.; data curation, H.Y.S., M.B.A.P. and C.J.C.P.; writing—Original draft preparation, H.Y.S.; writing—Review and editing, C.J.C.P. and M.B.A.P.; visualization, H.Y.S.; supervision, C.J.C.P. and M.B.A.P.; project administration, C.J.C.P.

Funding: This research received no external funding.

Acknowledgments: We thank RSPCA, QLD for providing the database, and RSPCA inspectors for consultation.

Conflicts of Interest: Mandy B.A. Paterson is employed as the principal scientist by RSPCA, Qld. None of the authors receive any interest or financial support from people or organizations who can bias the research.

Appendix A

Table A1. Description of each complaint code alleging a welfare issue.

Complaint Code	Description
Abandonment	An animal was abandoned/left by the owner either at their abode or somewhere else such as in the bush.
Baiting/Poisoning	An animal was poisoned or planned to be poisoned.
Causing captive animal to be injured/killed by dog	A person let a captive animal be injured/killed by a dog.
Cruelty	A person was reported to have abused an animal.
Dog fighting or other prohibited offence	A person was reported for allowing dogs to fight or conducting other specifically prohibited acts.
Emergency relief	Emergency relief is required for an animal left unattended because its owner experienced an emergency (e.g., flood or being hit by a car).
Hot animal in car	An animal was left unattended in a car during hot weather.
Insufficient food and/or water	An animal has insufficient food and/or water.
Keeping or using animal for blooding/coursing a dog	A person used a live bait for blooding/coursing a dog.
Knowingly allowing an animal to kill/injure another	A person allowed one animal to kill/injuring another one, and did nothing to stop them.
No exercise/confined/tethered	An animal is confined or tethered and not given a suitable amount of exercise.
No shelter	An animal is not provided with suitable shelter provisions.
No treatment	An animal did not receive appropriate medical treatment when needed.
Overcrowding	The number of animals is too high for the living space provided.
Poor dog condition	The general condition of an animal is poor. (e.g., messy/matted coat, pussy eyes, etc.)
Poor living condition	The living environment of the animal is poor.
Prohibition order breached	An owner violated a prohibition order [(a)].
Tail docking or other surgical procedure	Tail docking or other surgical procedure (e.g., declaw removal, etc.) was conducted on an animal.
Unknown	Unknown

[(a)] Prohibition order—A prohibition order is given by the court when a person convicted of an animal welfare offense must not possess any or a specific animal for a prescribed period of time [1].

Table A2. Chi-squared tests comparing prevalence of different complaint codes in two six-month periods, from January to June and from July to December, in 2009 and 2017.

Complaint Code	Prevalence % (N) January–June, 2009	Prevalence % (N) July–December, 2009	*p*-Value	Prevalence % (N) January–June, 2017	Prevalence % (N) July–December, 2017	*p*-Value
No Treatment	13.2% (500/3787)	13.9% (576/4134)	0.343	18.9% (1203/6369)	17.1% (1135/6632)	0.008
Abandonment	13.8% (523/3787)	14.5% (601/4134)	0.354	20.4% (1298/6369)	19.4% (1288/6632)	0.171
Cruelty	15.2% (575/3787)	16.7% (692/4134)	0.059	14.6% (933/6369)	14.6% (970/6632)	0.970
Insufficient food and/or water	26.9% (1017/3787)	25.9% (1073/4134)	0.364	24.4% (1551/6369)	24.1% (1599/6632)	0.748
No shelter	11.3% (428/3787)	12.4% (513/4134)	0.128	13.2% (843/6369)	14.3% (948/6632)	0.080
No exercise/confined/tethered	31.8% (1206/3787)	29.7% (1226/4134)	0.035	21.4% (1366/6369)	23.4% (1554/6632)	0.007
Hot Animal in Car	4.7% (177/3787)	6.2% (256/4134)	0.003	8.3% (530/6369)	9.5% (630/6632)	0.018
Poor living conditions	13.8% (523/3787)	15.5% (640/4134)	0.036	23.0% (1462/6369)	22.5% (1490/6632)	0.507
Baiting/poisoning	0.7% (27/3787)	1.2% (51/4134)	0.018	0.8% (52/6369)	1.1% (73/6632)	0.097
Poor dog condition	24.1% (914/3787)	23.1% (956/4134)	0.291	31.9% (2033/6369)	29.4% (1948/6632)	0.002
Overcrowding	1.6% (61/3787)	1.8% (73/4134)	0.593	0.4% (25/6369)	0.3% (20/6632)	0.377
Knowingly allowing an animal to kill/injure another	0.5% (20/3787)	0.6% (23/4134)	0.864	0.3% (20/6369)	0.2% (14/6632)	0.250
Tail docking or other surgical procedure	0.3% (13/3787)	0.3% (12/4134)	0.675	0.1% (9/6369)	0.2% (11/6632)	0.721
Dog fighting or other prohibited offence	0.2% (6/3787)	0.2% (7/4134)	0.905	0.2% (10/6369)	0.3% (19/6632)	0.118
Prohibition order breached	0% (0/3787)	0% (0/4134)	–	0.2% (15/6369)	0.2% (10/6632)	0.270

– Unable to conduct Chi-squared test because no cases were reported in 2009 and 2017.

References

1. Queensland Government–Department of Agriculture and Fisheries. Animal Care and Protection Act 2001. Available online: https://www.legislation.qld.gov.au/view/pdf/inforce/current/act-2001-064 (accessed on 22 April 2019).
2. Mogbo, T.C.; Oduah, F.N.; Okeke, J.J.; Ufele, A.N.; Nwankwo, O.D. Animal cruelty: A review. *J. Nat. Sci. Res.* **2013**, *3*, 94–98.
3. Ascione, F.R. Children who are cruel to animals: A review of research and implications for developmental psychopathology. *Anthrozoös* **1993**, *6*, 226–247. [CrossRef]
4. Lockwood, R.; Arkow, P. Animal abuse and interpersonal violence: The cruelty connection and its implications for veterinary pathology. *Vet. Pathol.* **2016**, *53*, 910–918. [CrossRef]
5. The Royal Society for the Prevention of Cruelty to Animals Queensland (RSPCA Qld). Annual Report 2011/2012. Available online: https://www.rspcaqld.org.au/who-we-are/annual-report (accessed on 15 November 2018).
6. The Royal Society for the Prevention of Cruelty to Animals Queensland (RSPCA Qld). Annual Report 2017–2018. Available online: https://www.rspcaqld.org.au/who-we-are/annual-report (accessed on 15 November 2018).
7. Tallichet, S.E.; Hensley, C. Rural and urban differences in the commission of animal cruelty. *Int. J. Offender Ther.* **2005**, *49*, 711–726. [CrossRef] [PubMed]
8. Weng, H.Y.; Kass, P.H.; Hart, L.A.; Chomel, B.B. Risk factors for unsuccessful dog ownership: An epidemiologic study in Taiwan. *Prev. Vet. Med.* **2006**, *77*, 82–95. [CrossRef] [PubMed]
9. New, J.C.J.; Salman, M.D.; King, M.; Scarlett, J.M.; Kass, P.H.; Hutchison, J.M. Characteristics of shelter-relinquished animals and their owners compared with animals and their owners in U.S. pet-owning households. *J. Appl. Ani. Welf. Sci.* **2000**, *3*, 179–201. [CrossRef]
10. Westgarth, C.; Christian, H.E.; Christley, R.M. Factors associated with daily walking of dogs. *BMC Vet. Res.* **2015**, *11*, 116–128. [CrossRef]
11. Kim, Y.M.; Kim, S.A.; Lee, S.M.; Choi, Y.J.; Kim, B.J.; Shin, N.S. Canine behavioral problems and their effect on relinquishment of the Jindo dog. *J. Vet. Sci.* **2010**, *11*, 345–350. [CrossRef]
12. Kwan, J.Y.; Bain, M.J. Owner attachment and problem behaviors related to relinquishment and training techniques of dogs. *J. Appl. Ani. Welf. Sci.* **2013**, *16*, 168–183. [CrossRef]
13. Pickup, E.; German, A.J.; Blackwell, E.; Evans, M.; Westgarth, C. Variation in activity levels amongst dogs of different breeds: Results of a large online survey of dog owners from the UK. *J. Nutr. Sci.* **2017**, *6*, e10. [CrossRef] [PubMed]
14. Lim, C.; Rhodes, R.E. Sizing up physical activity: The relationships between dog characteristics, dog owners' motivations, and dog walking. *Psychol. Sp. Exerc.* **2016**, *24*, 65–71. [CrossRef]
15. Newberry, M. Associations between different motivations for animal cruelty, methods of animal cruelty and facets of impulsivity. *Psychol. Crime Law* **2018**, *24*, 500–526. [CrossRef]
16. Hart, L.A.; Weng, H.Y. Impact of the economic recession on companion animal relinquishment, adoption, and euthanasia: A Chicago animal shelter's experience. *J. Appl. Ani. Welf. Sci.* **2012**, *15*, 80–90. [CrossRef]
17. Dolan, E.D.; Scotto, J.; Slater, M.; Weiss, E. Risk factors for dog relinquishment to a Los Angeles municipal animal shelter. *Animals* **2015**, *5*, 1311–1328. [CrossRef] [PubMed]
18. Kasperbauer, T.J. Animals and the expanding moral circle. In *Subhuman: The Moral Psychology of Human Attitudes to Animals*; Oxford University Press: New York, NY, USA, 2018; pp. 152–158. ISBN 978-019-069-581-1.
19. Haidt, J.; Koller, S.H.; Dias, M.G. Affect, culture, and morality, or is it wrong to eat your dog? *J. Personal. Soc. Psychol.* **1993**, *65*, 613–628. [CrossRef]
20. McEwan, A.; Skandakumar, K. The welfare of greyhounds in Australian racing: Has the industry run its course? *Aust. Anim. Prot. Law J.* **2011**, *6*, 52–74.
21. Kalof, L.; Taylor, C. The discourse of dog fighting. *Humanit. Soc.* **2007**, *31*, 319–333. [CrossRef]
22. Burley, S. My dog's the champ: An analysis of young people in urban settings and fighting dog breeds. *Anthropol. Matter. J.* **2008**, *10*, 1–18.
23. Becker, F.; French, L. Making the links: Child abuse, animal cruelty and domestic violence. *Child Abuse Rev.* **2004**, *13*, 399–414. [CrossRef]

24. Laflamme, D.P.; Abood, S.K.; Fascetti, A.J.; Fleeman, L.M.; Freeman, L.M.; Michel, K.E.; Bauer, C.; Kemp, B.L.E.; Doren, J.R.V.; Willoughby, K.N. Pet feeding practices of dog and cat owners in the United States and Australia. *J. Am. Vet. Med. Assoc.* **2008**, *232*, 687–694. [CrossRef] [PubMed]
25. Frantz, N.Z.; Yamka, R.M.; Friesen, K.G. The effect of dietary protein on body composition and renal function in geriatric dogs. *Intern. J. Appl. Res. Vet. Med.* **2007**, *5*, 57–64.
26. Elliott, D.A. Nutritional management of chronic renal disease in dogs and cats. *Vet. Clin. Small Anim. Pract.* **2006**, *36*, 1377–1384. [CrossRef] [PubMed]
27. Cutt, H.; Corti, B.G.; Knuiman, M.; Burke, V. Dog ownership, health and physical activity: A critical review of the literature. *Health Place* **2007**, *13*, 261–272. [CrossRef]
28. Degeling, C.; Burton, L.; McCormack, G.R. An investigation of the association between socio-demographic factors, dog-exercise requirements, and the amount of walking dogs receive. *Can. J. Vet. Res.* **2012**, *76*, 235–240.
29. The Kennel Club. Breed Information Centre. Available online: https://www.thekennelclub.org.uk/services/public/breed/Default.aspx (accessed on 15 November 2018).
30. Robertson, I.D.; Irwin, P.J.; Lymbery, A.J.; Thompson, R.C.A. The role of companion animals in the emergence of parasitic zoonoses. *Int. J. Parasitiol.* **2000**, *30*, 1369–1377. [CrossRef]
31. Palmer, C.S.; Traub, R.J.; Robertson, I.D.; Devlin, G.; Rees, R.; Thompson, R.C.A. Determining the zoonotic significance of Giardia and cryptosporidium in Australian dogs and cats. *Vet. Parasitol.* **2008**, *154*, 142–147. [CrossRef] [PubMed]
32. Drouot, S.; Mignon, B.; Fratti, M.; Roosje, P.; Monod, M. Pets as the main source of two zoonotic species of the Trichophyton mentagrophytes complex in Switzerland, Arthroderma vanbreuseghemii and Arthroderma benhamiae. *Vet. Dermatol.* **2009**, *20*, 13–18. [CrossRef]
33. Henn, J.B.; Gabriel, M.W.; Kasten, R.W.; Brown, R.N.; Theis, J.H.; Foley, J.E.; Chomel, B.B. Gray foxes (*Urocyon cinereoargenteus*) as a potential reservoir of a *Bartonella clarridgeiae*-like bacterium and domestic dogs as part of a sentinel system for surveillance of zoonotic arthropod-borne pathogens in Northern California. *J. Clin. Microbiol.* **2007**, *45*, 2411–2418. [CrossRef]
34. Queensland Government Statistician's Office. Population Growth Highlights and Trends, Queensland, 2018 Edition. Available online: http://www.qgso.qld.gov.au/products/reports/pop-growth-highlights-trends-qld/pop-growth-highlights-trends-qld-2018-edn.pdf (accessed on 15 November 2018).
35. Australian Bureau of Statistics. Regional Population Growth, Australia, 2008–2009. Available online: http://www.abs.gov.au/ausstats/abs@.nsf/Products/3218.0~{}2008-09~{}Main+Features~{}Queensland (accessed on 26 March 2018).
36. Animal Medicines Australia. Pet Ownership in Australia|2016. Available online: http://animalmedicinesaustralia.org.au/wp-content/uploads/2016/11/AMA_Pet-Ownership-in-Australia-2016-Report_sml.pdf (accessed on 26 March 2018).
37. Taylor, N.; Signal, T.D. Community demographics and the propensity to report animal cruelty. *J. Appl. Ani. Welf. Sci.* **2006**, *9*, 201–210. [CrossRef] [PubMed]
38. Taylor, N.; Signal, T.D. Willingness to pay: Australian consumers and "on the farm" welfare. *J. Appl. Ani. Welf. Sci.* **2009**, *12*, 345–359. [CrossRef]
39. Srinivasan, K. The biopolitics of animal being and welfare: Dog control and care in the UK and India. *Trans. Inst. Brit. Geogr.* **2013**, *38*, 106–119. [CrossRef]
40. Neumann, S.L. Animal welfare volunteers: Who are they and why do they do what they do? *Anthrozoös* **2010**, *23*, 351–364. [CrossRef]
41. Ferrari, J.R.; Loftus, M.M.; Pasek, J. Young and older caregivers at homeless animal and human shelters: Selfish and selfless motives in helping others. *J. Soc. Distress Homel.* **1999**, *8*, 37–49. [CrossRef]
42. Main, D.C.J. Evolution of animal-welfare education for veterinary students. *J. Vet. Med. Educ.* **2010**, *37*, 30–35. [CrossRef] [PubMed]
43. Greenebaum, J. "I'm not an activist!": Animal rights vs. animal welfare in the purebred dog rescue movement. *Soc. Anim.* **2009**, *17*, 289–304. [CrossRef]
44. Tiplady, C.M.; Walsh, D.-A.B.; Phillips, C.J.C. Public response to media coverage of animal cruelty. *J. Agri. Environ. Ethic* **2013**, *26*, 869–885. [CrossRef]
45. McGreevy, P.D.; Starling, M.; Branson, N.J.; Cobbc, M.L.; Calnon, D. An overview of the dog–human dyad and ethograms within it. *J. Vet. Behav.* **2012**, *7*, 103–117. [CrossRef]

46. Tiplady, C.M.; Walsh, D.B.; Phillips, C.J.C. Intimate partner violence and companion animal welfare. *Aust. Vet. J.* **2012**, *90*, 48–53. [CrossRef]
47. Degeling, C.; Kerridge, I.; Rock, M. What to think of canine obesity? Emerging challenges to our understanding of human–animal health relationships. *Soc. Epistemol.* **2013**, *27*, 90–104. [CrossRef]
48. Serpell, J.A. Having our dogs and eating them too: Why animals are a social issue. *J. Soc. Issues* **2009**, *65*, 633–644. [CrossRef]
49. Carter, R.A.; Bauer, J.E.; Kersey, J.H.; Buff, P.R. Awareness and evaluation of natural pet food products in the United States. *J. Am. Vet. Med. Assoc.* **2014**, *245*, 1241–1248. [CrossRef]
50. Cooper, J.J.; Cracknell, N.; Hardiman, J.; Wright, H.; Mills, D. The welfare consequences and efficacy of training pet dogs with remote electronic training collars in comparison to reward based training. *PLoS ONE* **2014**, *9*, e102722. [CrossRef] [PubMed]
51. Notari, L.; Gallicchio, B. Owners' perceptions of behavior problems and behavior therapists in Italy: A preliminary study. *J. Vet. Behav.* **2008**, *3*, 52–58. [CrossRef]
52. Zupan, M.; Buskas, J.; Altimiras, J.; Keeling, L.J. Assessing positive emotional states in dogs using heart rate and heart rate variability. *Physiol. Behav.* **2016**, *155*, 102–111. [CrossRef]
53. Weiss, E.; Patronek, G.; Slater, M.; Garrison, L.; Medicus, K. Community partnering as a tool for improving live release rate in animal shelters in the United States. *J. Appl. Ani. Welf. Sci.* **2013**, *16*, 221–238. [CrossRef] [PubMed]
54. Dotson, M.J.; Hyatt, E.M.; Clark, J.D. Traveling with the family dog: Targeting an emerging segment. *J. Hosp. Mark. Manag.* **2011**, *20*, 1–23. [CrossRef]
55. Australian Government Bureau of Meteorology. Climate Outlooks—Monthly and Seasonal. Available online: http://www.bom.gov.au/climate/outlooks/#/temperature/maximum/median/seasonal/0 (accessed on 7 March 2019).
56. Yilmaz, O. Dog fighting in some European countries. *Int. J. Livest. Res.* **2016**, *6*, 20–25. [CrossRef]
57. Tarver, E.C. The dangerous individual('s) dog: Race, criminality and the 'Pit bull'. *Cult. Theory Crit.* **2014**, *55*, 273–285. [CrossRef]
58. Archer, J.; Monton, S. Preferences for infant facial features in pet dogs and cats. *Ethology* **2010**, *117*, 217–226. [CrossRef]
59. Golle, J.; Lisibach, S.; Mast, F.W.; Lobmaier, J.S. Sweet puppies and cute babies: Perceptual adaptation to babyfacedness transfers across species. *PLoS ONE* **2013**, *8*, e58248. [CrossRef]
60. Bennett, P.C.; Perini, E. Tail docking in dogs: A review of the issues. *Aust. Vet. J.* **2003**, *81*, 208–218. [CrossRef] [PubMed]
61. Noonan, G.J.; Rand, J.S.; Blackshaw, J.K.; Priest, J. Behavioural observations of puppies undergoing tail docking. *Appl. Ani. Behav. Sci.* **1996**, *49*, 335–342. [CrossRef]
62. Murno, H.M.C.; Thrusfield, M.V. "Battered pets": Non-accidental physical injuries found in dogs and cats. *J. Small Ani. Pract.* **2001**, *42*, 279–290. [CrossRef]
63. Lockwood, R. Cruelty toward cats: Changing perspectives. In *The State of the Animals III: 2005*, 1st ed.; Salem, D.J., Rowan, A.N., Eds.; Humane Society Press: Washington, DC, USA, 2005; pp. 15–26. ISBN 978-097-484-005-5.
64. The Kennel Club. Puppy Farming. Available online: https://www.thekennelclub.org.uk/our-resources/kennel-club-campaigns/puppy-farming/ (accessed on 27 February 2019).

animals

MDPI

Article

Feeding Enrichment in a Captive Pack of European Wolves (*Canis Lupus Lupus*): Assessing the Effects on Welfare and on a Zoo's Recreational, Educational and Conservational Role

Giacomo Riggio [1], Chiara Mariti [2], Chiara Boncompagni [1], Simone Corosaniti [1], Massimiliano Di Giovanni [3], Asahi Ogi [2], Angelo Gazzano [2,*] and Robert Thomas [4]

1 Vethos, Via E. Besta 46, 00167 Rome, Italy; giacomoriggio@gmail.com (G.R.); boncompagnichiara@gmail.com (C.B.); scorosaniti@hotmail.it (S.C.)

2 Department of Veterinary Sciences, University of Pisa, Viale delle Piagge 2, 56124 Pisa, Italy; chiara.mariti@unipi.it (C.M.); asahi.ogi@vet.unipi.it (A.O.)

3 Research Department, Fondazione Bioparco di Roma, Viale del Giardino Zoologico 20, 00197 Rome, Italy; massimiliano.digiovanni@bioparco.it

4 Deanery of Biomedical Sciences, University of Edinburgh, Teviot Place, Edinburgh EH8 9AG, UK; rob.thomas@ed.ac.uk

* Correspondence: angelo.gazzano@unipi.it

Received: 11 April 2019; Accepted: 4 June 2019; Published: 8 June 2019

Simple Summary: Feeding enrichment is widely used to improve the welfare of zoo animals, but it may also affect zoo visitors' experience and perception of the animals. The objective of this study was to assess the effects of a naturalistic and a non-naturalistic feeding enrichment program, on both wolf behaviour and visitors' interest in the exhibit. A questionnaire was administered to visitors with the aim of assessing whether our feeding enrichment programs might affect their perception of captive wolf welfare as well as their attitude towards wolf conservation issues. Our findings suggest that, although wolves seemed to benefit from enrichment, their behavioural responses were highly variable among individuals. Visitors' interest in the exhibit and perception of captive wolf welfare improved by observing the wolves interacting with food, especially when novel feeding objects were provided. Finally, their attitude towards wolf conservation issues did not change in relation to enrichment, but improved when they observed the wolves performing feeding-related behaviours. These findings may help zoos implement enrichment programs that are effective for enhancing their wolves' welfare as well as their recreational and educational role.

Abstract: This study investigated the effects of two feeding enrichment programs on the behaviour of a captive pack of European wolves (*Canis lupus lupus*) and their correlation with both zoo visitors' interest towards the exhibit and their overall perception of the species. Behavioural data (exploration, stereotypies, social interactions, activity/inactivity rates) were collected on four male wolves during four two-week long phases: initial control, hidden food, novel object, final control. Three observation sessions were performed daily: before, during and after feeding. Number of visitors and their permanence in front of the exhibit were recorded. After watching the wolves, visitors were asked to fill out a brief questionnaire in order to investigate their perception of captive wolf welfare, as well as their attitude towards wolf conservation issues. Despite the high inter-individual variability in their behavioural response, all wolves seemed to benefit from feeding enrichment. With regard to visitors, interest in the exhibit increased when enrichment was provided. Visitors' perception of the level of welfare of wolves improved if they attended a feeding session, especially during the novel object phase. Visitors' attitude towards wolf conservation issues also improved during feeding sessions, regardless of enrichment provision.

Keywords: zoo; wolf behaviour; animal welfare; visitor; conservation; education

1. Introduction

Wolf conservation has always been a very controversial topic because of the impact of wolf predatory activity on livestock [1]. If on one hand, wolves held in zoos may help raise public awareness on the need to preserve the species, on the other they may have their welfare compromised by the impossibility to perform natural behaviours, including predation itself [2,3].

In the Carnivora, predatory motivation influences the individual's behaviour from the early stages of its psycho-physical development, even in apparently extraneous contexts than those related to food consumption (i.e., play) [4]. Furthermore, in social species like the wolf, group hunting may also have a cohesive function among members of the pack [5,6].

For management and ethical reasons, captive carnivores are rarely given the opportunity to perform predatory behaviour, as food is often served ready to be consumed [7–9]. Constantly preventing animals from appeasing their predatory motivation may lead to chronic frustration and stress [10]. Abnormal behaviours may develop as coping strategies in response to the stressful environment [11].

Feeding enrichment programs aim to promote the expression of natural behaviours and the interactions with a more adequately stimulating environment [12,13]. In captive carnivores, such programs are often aimed to stimulate at least some phases of predatory behaviour [14]. The ultimate goal is, of course, that of enhancing the animals' level of welfare [15,16]. Several behavioural indicators of welfare have been used to assess the effects of feeding enrichment programs. In carnivores, feeding enrichment has been shown to decrease stereotypies [14,17], improve behavioural diversity [18,19], increase exploration [20] and activity levels [21].

Improving animal welfare may be beneficial not just for the animals, but for zoo visitors [22,23]. For instance, animals that interact with their environment have been shown to increase visitor interest in the exhibit [22,24–26]. Furthermore, sleeping or pacing animals objectively provide less information about the behavioural repertoire of the species and may fail to meet visitors' expectations of the animals' behaviours [27]. Visitors who have a negative experience may leave the zoo with wrong or little, if at all, new knowledge on the species and a sense of frustration for not having their expectations met [8,28]. On the opposite, more active and naturally behaving animals have been shown to improve visitors' perception of animal welfare and, as a consequence, their perception of the educational importance of zoos [8,29,30].

This study aimed to implement two distinct feeding enrichment procedures in a captive pack of European wolves (*Canis lupus lupus*) in order to (1) simultaneously assess their effects on both captive wolves' behaviour and visitors' interest towards the exhibit, (2) assess the possible correlation between enrichment and visitors' perception of captive wolf welfare, as well as their attitude towards wolf conservation.

2. Materials and Methods

2.1. Time and Setting

The study was carried out at the Bioparco of Rome (Italy) for two consecutive months, between March and May 2017. The wolf enclosure is one of the largest exhibits of the park. It is a 200 m^2 area, surrounded by a 3 m high wooden palisade on one long side (the visitor side). A central 10 m^2 trapezoidal recess virtually divides the enclosure in a right and a left area. Windows in the two diagonal sides of the recess allow visitors to observe the animals in both areas. When necessary, metal sliding doors are used to physically divide the enclosure in three smaller areas (Figure 1).

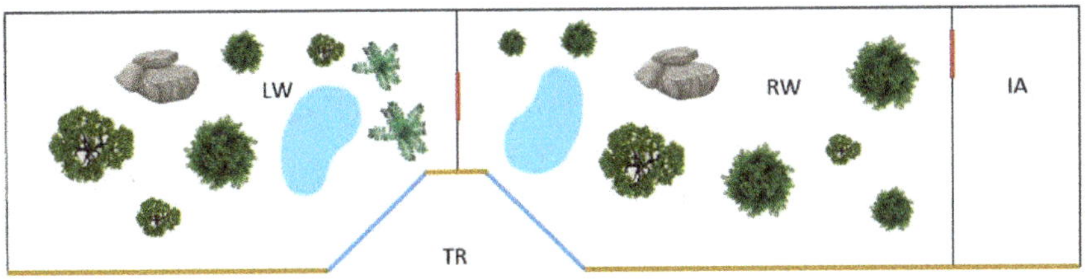

Figure 1. Simplified map of the wolf exhibit. LW = left wing, RW = right wing, IA = isolation area, TR = trapezoidal recess. Red lines = sliding doors, blue lines = windows, yellow lines = wooden palisade.

2.2. Animals

Experimental subjects were four European wolves (*Canis lupus lupus*). All of them were un-neutered male siblings: wolf 1, 2 and 3 were 8 years old at the time of the study and belonged to the same litter, while wolf 4 was 1 year younger. All of them had always lived together and they were moved together from another zoo where they were born and raised. They were fed once a day for 6 days a week and their diet comprised either chunks of buffalo or cow meat, or entire rabbits or quails.

2.3. Experimental Design

The experimental protocol was scheduled into four different phases: initial control (ICp), hidden food (HFp), novel object (NOp), final control (FCp). Each phase lasted 2 weeks. ICp, in which no enrichment was administered, served as a baseline control. Food was administered as usual, in the small area at the edge of the right wing of the enclosure. During HFp, food was divided into 12 pieces, regardless of its nature. Four of them were singly buried 10 to 15 cm deep in holes purposely dug by the keepers. Four were suspended 2–3 m-high on tree branches. The remaining pieces were singly inserted deep in the middle of four woodpiles previously built in each wing of the enclosure. These piles had been introduced into the enclosure 2 months before the beginning of the study in order to allow the wolves to habituate to the new elements. During NOp food had to be wrapped in eight canvas bags in order to make it more difficult for the wolves to reach for it. In addition, a hard-plastic feeding ball (Aussiedog®) filled with smaller chunks of meat had to be randomly left around the area. Again, during FCp, no enrichment was provided. During this phase, which served as final double control, food was administered using the same procedure as in ICp. Feeding procedures were standardized across the four phases. Keepers would alternately lock the wolves in the left or right area, by using the central metal sliding door. Hence, they would enter the other area in order to clean it and place food. At the end of these procedures they would step out, lock the cage, and free the wolves. Once the sliding door was opened, keepers had 2 min to walk out of sight to the wolves. The study was observational in nature. The animals involved in this study were housed at the zoo and were part of a program of enrichment promoted by the zoo itself, i.e., the procedures were thus not carried out for research purposes. No animal care license nor approval of ethical committees were therefore needed.

2.4. Wolf Data Collection

Behavioural observations were carried out daily, 5 days a week, from Tuesday to Saturday. Every day, three observation sessions were performed before, during and after feeding. Each session lasted 1 h, plus a 1-h interval between consecutive sessions, for a total of 3 h per day. In order to meet keepers' working and break hours and zoo opening hours, but still avoid predictability by wolves on feeding times, observations could start at three different times of the day: 9.30, 11.00, or 12.15. Daily observation starting times were randomly set. Behavioural data were collected through instantaneous scan sampling technique [31]. During each hour of observation, the behaviour of each wolf was recorded every 1 min. The ethogram comprised 16 behavioural categories (Table 1) and was obtained by integrating behaviours described in scientific literature [5,18,32–34] with direct behavioural observations over a 7-day period, prior to the study. Since behavioural data from the first observation session on 29 March 2017 had not been recorded, data from the entire day were eliminated.

Table 1. Description of wolf behavioural categories grouped in macrocategories used for statistical purposes. Social interactions were categorized as either positive or negative. All behaviours, except stereotypies, were categorized as either active or inactive.

Behavioural Category	Description
Stereotypy	An apparently functionless fixed behavioural pattern that is repeatedly performed for at least three consecutive times
Inactive	
Sleeping	To lie down with closed eyes
Inactive	To sit or lie down with open eyes
Active	
Solo active	To stand motionless or move without interacting with other wolves
Solo play	To tug, chase, pull to pieces, jump on objects. To run alone, chase own tail, lie supine and roll or squirm in a playful manner
Exploratory	To sniff ground, objects, trees and plants, when not aimed at the acquisition of food
Foraging	To move food around, cache, dig and sniff, when aimed at the acquisition of food
Feeding	To take food into the mouth and swallow it
Drinking	
Marking	To urinate with raised leg
Self-directed	To nip, lick or scratch its own fur or skin, rub against a tree, stretch
Negative social interactions	
Dominant	To stand tall with rigid posture and tail, stand over another wolf with tail high, grab the muzzle of another wolf while maintaining a rigid posture
Submissive	To lie on the back with tail between legs, crouch
Aggressive	To show bare teeth, growl, snap, bite, attack, fight
Positive social interactions	
Affiliative	To sniff another wolf, lick another wolf, rub the muzzle against a wolf, rub the muzzle or body one another, greet, stand or lie close while wagging tails, put a paw or the head on another wolf while keeping tail down or wagging, stand over another wolf with tail down, chorus howl
Play	Two or more subjects that engage in motor patterns such as chase, run around one another, kick, jump and even jaw spar, snap or bite with not enough pressure to cause injury, play invitation

2.5. Visitor Data Collection

Data from visitors were collected by an additional observer using the same observation schedule used for the wolves. Both the number of visitors and the duration of their permanence in front of the exhibit were recorded [35]. The number of visitors in front of the enclosure was recorded by instantaneous scan sampling with 1-min intervals for the entire hour of observation for each session. Children up to 16 years were excluded from the count. The amount of time (seconds) spent in front of the exhibit by every third visitor (excluding children) was also recorded [28,29,36,37]. Visitor permanence recording began when the selected subject walked across the virtual line that delimited the wolf exhibit observation zone (trapezoidal recess) and ended when they crossed it the opposite way.

2.6. Questionnaire

When visitors exited the observation zone, a researcher asked them (not children) to fill in a brief questionnaire on wolves. Visitors selected for filling the questionnaire and those selected for recording their permanence in front of the exhibit were not necessarily the same. All visitors were asked to fill the questionnaire although not all of them would accept (respondents $n = 630$). The questionnaire consisted of two sections. The first section investigated on demographic information, which are summarized in Table 2. The second section consisted of 11, 1–5 Likert-scale items partially based on surveys used in previous studies [28,38] (Table 3). Visitors could respond by marking with an X a number from 1 to 5 depending respectively on the degree of disagreement/agreement with the corresponding statement, with 1 meaning "strongly disagree" and 5 meaning "strongly agree." Items' positivity and negativity was intended in relation to the quality of visitors' perception towards wolves.

Table 2. Demographic information for questionnaire respondents.

Item/Response	Frequency	Percentage (%)
Gender		
Female	344	54.6
Male	280	44.4
Age		
<20	59	9.4
20–29	184	29.2
30–39	176	27.9
40–49	133	21.1
50–59	43	6.8
>60	32	5.1
Ever been in a zoo before		
No	47	7.5
Yes, more than once	476	75.6
Yes, once	106	16.8
Nationality		
Italian	440	69.8
British	30	4.8
American	20	3.2
French	13	2.1
Dutch	11	1.7
Romanian	11	1.7
Bulgarian	9	1.4
German	9	1.4
Russian	9	1.4
Irish	6	1.0

Nationalities below 1% were not reported.

Table 3. Likert-scale items and scoring system.

Items	Score (Level of Disagreement/Agreement)						
(1) I would love to spot a wolf in the wild	S. D.	1	2	3	4	5	S. A.
(2) Wolves are mean animals *	S. D.	1	2	3	4	5	S. A.
(3) Wolves are very dangerous to humans *	S. D.	1	2	3	4	5	S. A.
(4) Wolves in zoos behave like they do in documentaries	S. D.	1	2	3	4	5	S. A.
(5) Wolves in the wild are a serious threat to livestock *	S. D.	1	2	3	4	5	S. A.
(6) It is important to have wolves in zoos for education purposes	S. D.	1	2	3	4	5	S. A.
(7) Wolf reintroduction programs should be implemented in those areas from where the wolf disappeared	S. D.	1	2	3	4	5	S. A.
(8) Wild wolves that prey on cattle should be systematically eliminated *	S. D.	1	2	3	4	5	S. A.
(9) Illegal killing of wild wolves should be severely punished	S. D.	1	2	3	4	5	S. A.
(10) Wolves in zoos make me feel sad *	S. D.	1	2	3	4	5	S. A.
(11) The level of welfare of wolves in zoos is worrisome *	S. D.	1	2	3	4	5	S. A.

S.D. = strongly disagree, S.A. = strongly agree. Negative items are marked with an *.

2.7. Statistical Analysis

All the statistical analyses were run with the software SPSS Statistic 17.0 (Chicago, IL, USA).

Due to the small sample size and the possible individual differences in behavioural responses to enrichment, data were analysed separately for each wolf, which acted as its own control. For analysis purposes, all behaviours except stereotypies were grouped into broader macro-categories (Table 1). Exploratory behaviour was also analysed singularly. In order to assess potential differences in each wolf behaviour among phases and among observation sessions, the Kruskal Wallis test and then Wilcoxon signed rank test were applied ($p < 0.05$ multiple comparison corrections were performed using the Benjamini–Hochberg procedure). The Wilcoxon signed rank test was also used to compare data obtained from feeding sessions in the first and second week for both hidden food and novel objects phases, with the aim of assessing a possible effect of the decreased novelty of the enrichment.

Potential correlations between number of visitors in front of the exhibit and respectively phase, session and number of active wolves, were analysed using the Spearman test ($p < 0.05$). The Kruskal Wallis test was used to investigate potential differences in visitor stay time in front of the exhibit across observation sessions and experimental phases. When appropriate, pair-wise comparisons were carried out using the Mann–Whitney U test ($p < 0.05$) for independent variables.

As for the questionnaire, descriptive statistics was used for demographic information. Furthermore, all the negative items on the Likert-scale questionnaire were reverse-scaled to match with the score of positive items. Principal Component Analysis (PCA) was used to identify a possible underlying structure across the items; promax rotation with a correlation matrix was used.

Items were included in a component if their loading in that component was >0.50 and their loading in the other components was <0.25. The three principal components found were further analysed by using a Kruskal Wallis Test and then a Mann–Whitney U test ($p < 0.05$) in order to assess differences among phases and sessions.

3. Results

3.1. Wolf Behavioural Results

Within each enrichment phase, none of the comparisons between the first and second week of observation during feeding session revealed a statistically significant difference. Activity and inactivity rates did not significantly differ across phases, for any of the wolves. Significant results were obtained for stereotypies, social behaviour and exploration, as reported below.

A significant difference was found in the rate of stereotypic behaviours for wolf 2 and wolf 4 across phases. For wolf 2, stereotypy rates were significantly lower in the NOp when compared with the HFp ($p = 0.012$) and the FCp ($p = 0.001$). Similarly, wolf 4 stereotypy rates were significantly lower during the NOp when compared with the HFp ($p = 0.003$) and the FCp ($p = 0.001$). In addition, wolf 4 stereotypy rates tended to be higher in FCp compared to ICp ($p = 0.054$) (Figure 2).

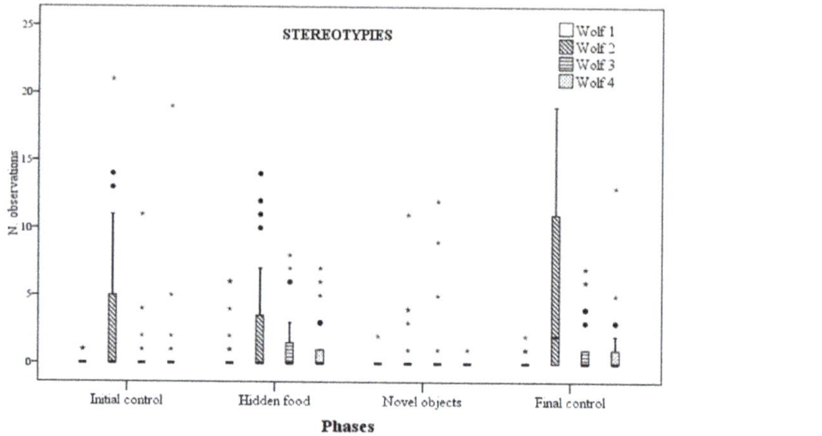

Figure 2. Number of times stereotypic behaviour was observed for each wolf across the four experimental phases.

Social behaviour rates significantly differed for wolf 1, wolf 2 and wolf 4 across phases, whilst no difference was observed for wolf 3. More specifically, for wolf 1, negative social behaviour rates were significantly lower during NOp compared to ICp ($p = 0.013$). However, they were also lower during FCp compared to ICp ($p = 0.025$). Similarly, for wolf 2, negative social behaviour rates were significantly lower during NOp, if compared with ICp ($p = 0.012$). However, they were also different between control phases, being significantly lower during the FCp ($p = 0.042$) (Figure 3). As for wolf

4, statistical differences were instead found for positive social behaviour rates, that resulted higher during HFp than ICp (p = 0.011) and NOp (p = 0.024) (Figure 4).

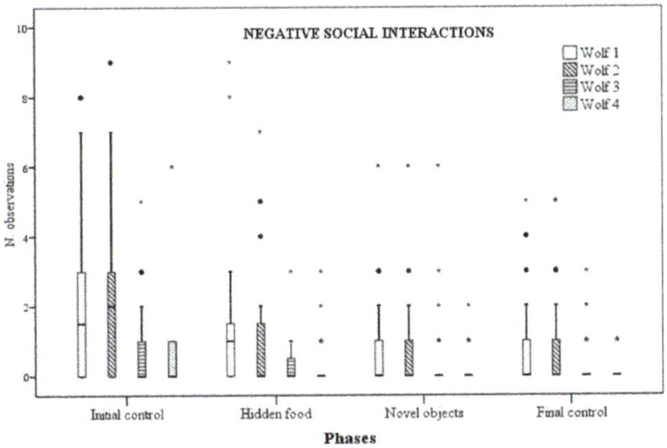

Figure 3. Number of times negative social interactions were observed for each wolf across the four experimental phases.

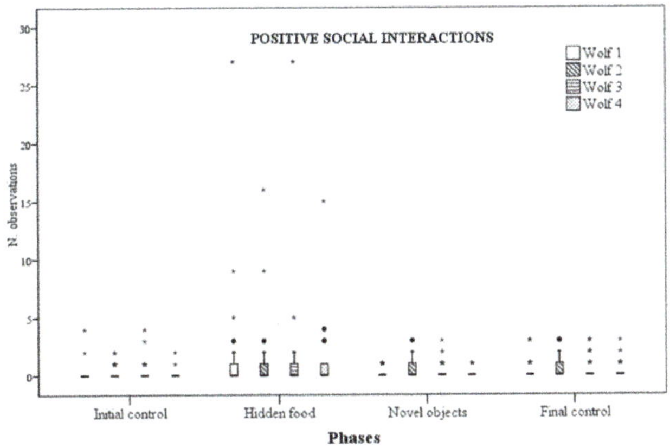

Figure 4. Number of times positive social interactions were observed for each wolf across the four experimental phases.

For all wolves there were no statistically significant differences in exploratory behaviour rates between initial and final control phases. Wolf 1 exploratory behaviour rates were significantly higher during the NOp when compared with both ICp (p = 0.008) and FCp (p = 0.029). For wolf 2, they were significantly higher in NOp when compared with any other phases (NOp versus ICp: p = 0.001; NOp versus HFp: p = 0.001; NOp versus FCp: p = 0.001), as well as in HFp when compared with ICp (p = 0.043). Additionally, for wolf 3 exploration was significantly higher in NOp when compared with any other phases (NOp versus ICp: p = 0.001; NOp versus HFp: p = 0.002; NOp versus FCp: p = 0.001). Finally, for wolf 4, exploratory behaviour rates were significantly higher in both the enrichment phases when compared with the control phases (HFp versus ICp: p = 0.001; HFp versus FCp: p = 0.001; NOp versus ICp: p = 0.005; NOp versus FCp: p = 0.003) (Figure 5).

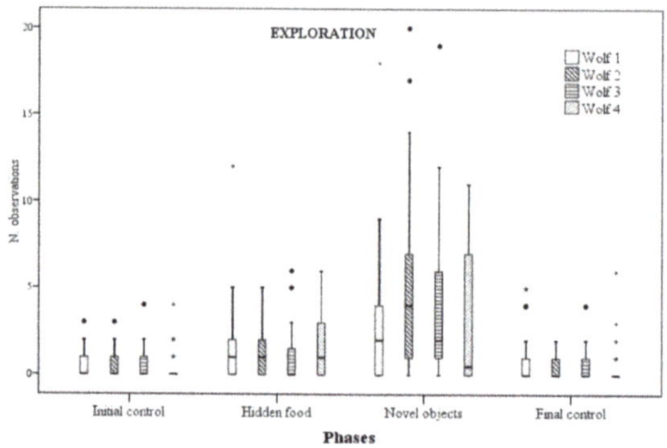

Figure 5. Number of times exploratory behaviour was observed for each wolf across the four experimental phases.

3.2. *Visitors' Interest Results*

A positive correlation was found between mean number of visitors in front of the exhibit and experimental phase (Rho value > 0.001), as well as mean number of visitors and observation session (Rho value > 0.001). The mean number of visitors was the highest during NOp ($p < 0.001$). Unexpectedly, it was also higher during FCp phase than HFp ($p < 0.001$) (Figure 6). As for the sessions, the highest mean number of visitors was found during feeding sessions ($p < 0.001$), regardless of the use of enrichment (Figure 7). Moreover, a positive correlation between the number of active wolves and the number of visitors in front of the exhibit was found ($p < 0.001$).

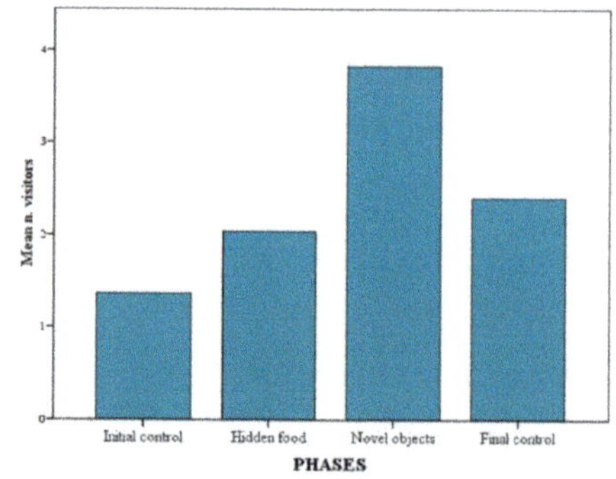

Figure 6. Mean number of visitors in front of the exhibit across the four experimental phases.

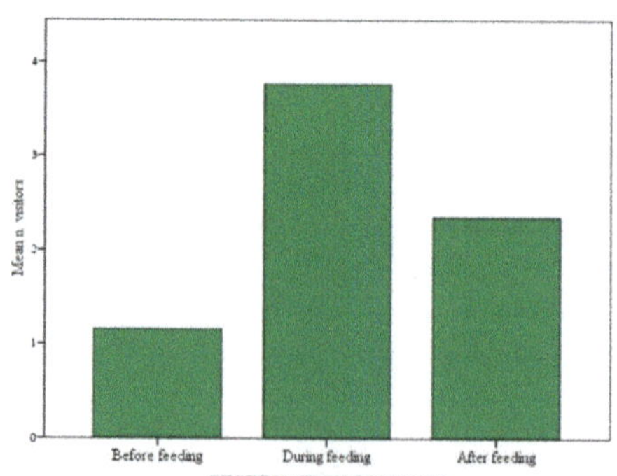

Figure 7. Mean number of visitors in front of the exhibit across the three daily observation sessions.

Time spent by visitors in front of the exhibit differed significantly across phases. It was longer during NOp than during any other phase ($p < 0.001$). It was also longer during HFp when compared with both control phases ($p < 0.001$). No significant difference was found between control phases. Furthermore, visitors' permanence in front of the exhibit also differed across observation sessions, being the longest during feeding sessions ($p < 0.001$) and longer before feeding than after feeding ($p = 0.016$).

3.3. *Questionnaire Results*

The 11, 1–5 Likert-scale items from the questionnaire were analysed using a Principal Component Analysis (PCA) with promax rotation ($p < 0.05$). The component correlation matrix (Table 4) did not show a correlation between the three components. The PCA model was adequate as verified by the Keiser Meyer Olkin (KMO) test (KMO = 0.694; $p < 0.001$). Three principal components (PC1, PC2 and PC3) were identified that explained 49.895% of the variation (Table 5). PC1 was characterized by those items relating to the respondent's perception of wolf welfare in zoos. PC2 was characterized by those items relating to the respondent's perception of wolves as a threat to humans and human activities. Finally, PC3 comprised those items relating to respondent's attitude towards wild wolf population management measures. No significant differences in questionnaire responses were found when data from the three daily sessions were compared across phases. On the contrary, differences were found for PC1 and PC3 across sessions. In particular, respondents scored significantly higher for PC1 and PC3 when questionnaires were administered during feeding when compared with sessions before feeding (PC1: $p = 0.016$; PC3: $p = 0.010$), suggesting that their perception of captive wolves' welfare and their attitude towards wild population management measures improved by watching the animals interacting with food, regardless of the use of enrichment. However, by comparing questionnaire responses during the sole feeding sessions across the four experimental phases, we found that respondents scored higher for PC1 when they observed the wolves get fed during NOp and HFp rather than during ICp.

Table 4. Component correlation matrix obtained for principal component analysis with promax rotation on questionnaire items.

Component	1	2	3
1	1.000	−0.158	−0.138
2	−0.158	1.000	0.257
3	−0.138	0.257	1.000

Table 5. Results of the principal component analysis and promax rotation carried out on items of the questionnaire. For each item, the load on each component has been reported.

Item	Component		
	Zoo Wolf Welfare	Wolf as a Threat	Wolf In-Situ Conservation
Wolves in zoos make me feel sad	**0.743**	−0.196	−0.114
It is important to have wolves in zoos for education purposes	**0.724**	−0.156	−0.037
The level of welfare of wolves in zoos is worrisome	**0.693**	0.012	0.042
Wolves in zoos behave like they do in documentaries	**0.578**	0.000	−0.239
Wolves are very dangerous to humans	−0.085	**0.814**	0.144
Wolves are mean animals	−0.120	**0.772**	0.163
Wolves in the wild are a serious threat to livestock	−0.065	**0.598**	0.242
Wolves reintroduction programs should be implemented in those areas where the wolf disappeared	0.044	0.063	**0.737**
Illegal killing of wild wolves should be severely punished	−0.088	0.175	**0.732**
Wild wolves that prey on cattle should be systematically eliminated	−0.209	0.307	0.557
I would love to spot a wolf in the wild	−0.195	0.459	0.512

Loadings with an absolute value greater than 0.5 (and lower than 0.25 for another component) are in bold.

4. Discussion

4.1. Wolves

In this study, none of our significant results regarding wolf behaviour were consistent across all four experimental subjects. This suggests individual factors played a major role in determining the wolves' responses to feeding enrichment. Three behavioural categories were used to assess wolf responses to enrichment: stereotypies, quality of social interactions and exploration.

Stereotypic behaviours are probably the most used behavioural indicator of welfare in captive animal studies [39,40]. They have been defined as "repetitive, invariant behavioural patterns with no apparent goal" [41]. In this study, two distinct behaviours fit such definition: pacing and jumping against the fence. The latter was performed by only one individual, namely wolf 2. Both of them fall in the category of "locomotor stereotypies," which are the most common type of stereotypy in captive carnivores [42,43]. In order to meet the criteria of "repetitive, invariant behavioural pattern" we reported such behaviours only when they were performed for at least three consecutive times [27,42]. In this study, two out of four wolves showed a significantly lower rate of stereotypic behaviours when novel artificial feeding objects were provided. Although the underlying processes of stereotypic behaviour in zoo animals are not yet clearly understood, one of the most widely accepted theories is that it represents a copying strategy performed by an individual in response to a chronic stress condition generated by a sub-optimal and restrictive environment [2,39,42,44,45]. Therefore, a reduction in the rate of stereotypic behaviours may be suggestive of increased welfare, especially when other behavioural indicators of welfare improve concurrently [39].

Quality of social interactions was assessed classifying social behaviour in either "positive" or "negative" [32]. Social play and affiliative interactions were considered to be positive, whereas agonistic and aggressive interactions were considered to be negative [32]. Again, in the context of our study, a quite high degree of individual variability in wolves' behavioural responses to enrichment was observed. While wolf 4 increased positive social interactions when food was hidden around the enclosure, wolf 1 and wolf 2 showed a decrease in negative interactions when artificial feeding objects were provided, whereas wolf 3 showed no significant change in any enrichment programs. As by nature, social interactions occur among more individuals, the interpretation of results needs to take into consideration the possible reciprocal influence of the behaviour of the various parties involved. In this case, the decrease of negative social interactions observed in wolf 1 and 2 are likely to be the result of the decrease of negative social behaviours in one wolf; the parallel trend in the second wolf probably mirrors the first decrease, as a consequence of receiving fewer negative interactions. However, looking at the whole unit, both the direct and indirect impact of enrichment can be regarded as beneficial.

An increase in exploration when novel feeding objects were provided was the only behavioural response common to the four wolves. Nevertheless, for wolf 2 and wolf 4 exploration also increased when food was hidden around the enclosure. In accordance with previous studies [18–20,46,47], this finding suggests that the relationship between feeding enrichment and exploratory behaviour is more linear and less affected by individual variables than the other behavioural indicators we assessed.

Overall, both feeding enrichment programs seemed to be effective at positively modifying our wolves' behaviour. However, inconsistency of behavioural responses among wolves indicates that individual variables might have qualitatively and quantitatively affected the beneficial effects of enrichment. Previous studies also found a high inter-individual variability in behavioural responses to feeding enrichment [18,48]. Since all of our experimental subjects were males with equal rearing conditions, this finding cannot be attributed to gender or rearing differences, as suggested by Cummings et al. [18] in their study on maned wolves. In our case, it is most likely explained by individual temperament differences and pack social dynamics [16,48–50].

A major limitation of this study is the small size of the sample, which prevents us from drawing any conclusion at the species level. Studies that investigate enrichment programs in zoo animals are often performed on small samples [18,34,51–53] and in specific environments that are likely not representative of the entire zoo population. Being opportunistic, our study presents these limitations. Our findings should be regarded as preliminary and further studies should be performed on larger samples and different captive environments.

In addition, each enrichment program was provided for 2 weeks only. The lack of differences between behaviours displayed in the first and second week of both enrichment phases suggest that, although the novelty of the enrichment decreased, it maintained beneficial effects in the analysed period. This is relevant for zoo animal caretakers, as their investment seems to have persistent benefits. However, 2 weeks may be a sufficient amount of time to determine short-term effects, but not enough to assess a possible long-term impact. This should be investigated in future studies by extending periods of enrichment.

Lastly, for practical reasons we could not assess any physiological indicator of acute or chronic stress, which could have helped us draw a clearer picture of the internal changes that each animal underwent during the study and their link with the behaviours observed [54–56].

4.2. Visitors

The number of visitors and the duration of permanence in front of the exhibit have often been used as indicators of their interest in zoo exhibits [22,23,35]. While the latter may be a more reliable parameter, the former may be more strongly affected by other variables such as season, day of the week, time of the day and weather [57]. This may explain why visitor number was higher in the final control phase than during the first enrichment phase. Nonetheless, our overall findings suggest that visitors were more interested in the exhibit during feeding sessions and during enrichment phases, especially when novel non-naturalistic objects were involved. In the latter case both number and time of permanence in front of the exhibit increased significantly. In previous studies, greater visitor interest in the exhibit has been linked to higher levels of animals activity [24,26,58]. However, although we found visitor number to be positively correlated with the number of active wolves at a given time, no significant difference in animal activity levels across experimental phases was detected. Other variables, such as the type of enrichment used [59] or the type of behaviour elicited [28,38,59] may also affect visitors' interest in the exhibit.

Interestingly, those experimental conditions that increased visitors' interest also enhanced their perception of captive wolf welfare. In fact, among daily observation sessions, the highest scores for animal welfare-related questions were obtained from those visitors who observed the wolves during feeding times, regardless of the use of enrichment. This result reflects those of previous studies in which visitors' perception of captive wolf welfare improved by observing the animals perform active natural behaviours, such as feeding [28,29]. Furthermore, when we compared responses obtained from visitors who observed the wolves during feeding times, across all four experimental conditions, we found that higher scores for welfare-related questions were obtained when enrichment was provided, especially the non-naturalistic type. Previous studies [30,60,61] did not find any differences in the effects of naturalistic versus non-naturalistic enrichment on visitors' perception of the animals' well-being. However, some methodological differences, such as type of enrichment used [30,60,61], species involved [30,60,61], partial or total lack of complementary data on the animals' behaviour [30,60] may explain the different results.

More importantly, visitors seemed to be aware of the implications that enrichment might have had on the wolves' welfare. Price et al. [29] suggest that visitors' perception of animal welfare may in turn be influenced by their own perception of zoos' commitment at caring for their animals. In our study, artificial enrichment objects may have rendered the zoo's commitment more evident in the eyes of the visitors.

Although this assumption should be further investigated with more specific questions in future studies, it may also explain why visitors who had higher perception of the wolves' welfare also had a more positive perception of the educational role of zoos [59]. On one hand, negative emotions elicited by perceiving animals suffering may generate a deep sense of distrust in the zoo as an animal preservation institution, thus diminishing the value of its educational role [8]. On the other hand, animals stimulated to engage in natural behaviours, such as foraging and feeding, may not only be perceived as happier by visitors [8], but they may objectively provide more information about the species' behavioural repertoire, if compared with inactive or stereotyping animals [29].

Findings from previous studies indicate that visitors who observe animals perform active natural behaviours show greater appreciation for the species' biological significance and greater conservation intent [8,38,62]. According to that, our visitors showed a more positive attitude towards wolf conservation issues when they observed the animals during feeding times, regardless of the use of enrichment. However, feeding enrichment, whether naturalistic or non-naturalistic, failed to further improve visitors' attitude towards wolf conservation. Due to the increasing wild wolf population, Italy is currently going through a period of intense political debate over wolf conservation measures. Actual or spurious episodes of attacks to livestock and even citizens reported by the media are likely to affect people's attitude on the matter. This may explain why, in our study, visitors did not seem to change their perception of wolves as a threat to humans and human activities. Finally, it should be taken into account that the wolf is a historically controversial species. The cross-generational "evil wolf" myth embedded in the occidental culture may represent a significant obstacle for a mind shift on attitude towards wolf conservation [63]. Similar studies on other species or conducted in other countries may lead to different results.

5. Conclusions

Overall, both hiding food within the enclosure and providing animals with novel artificial feeding objects appeared effective at modifying the behaviour of our wolves in a way that suggests an increased level of welfare. However, inconsistency in results across experimental subjects indicates that individual variables play an important role in determining the degree and the type of behavioural response to feeding enrichment. Among the behavioural indicators assessed, an increase in exploration activity when novel feeding objects where provided was the only change in behaviour common to all the wolves.

Feeding enrichment, especially when artificial objects were used seemed to be effective at increasing visitors' interest in the exhibit. Visitors' perception of captive wolf welfare and the educational role of zoos, as well as their attitude towards wolf conservation issues were more positive when they observed the wolves during feeding times, regardless of the use of enrichment. Feeding enrichment, especially the non-naturalistic type, further improved visitors' perception of captive wolf welfare. On the contrary, it failed to modify visitors' attitude towards wolf conservation issues. In order to confirm this study's findings future research on the effects feeding enrichment on wolves and visitors should be conducted on larger animal samples and in different countries.

Author Contributions: Conceptualization, G.R., C.B. and A.O.; methodology, G.R., C.M., A.O.; formal analysis, C.M.; investigation, G.R., C.B., S.C.; resources, M.D.G.; data curation, G.R., C.M.; writing—original draft preparation, G.R.; writing—review and editing, G.R., C.M., A.G. and R.T.; supervision, A.G., R.T.; project administration, M.D., R.T.

Funding: This research received no external funding.

Acknowledgments: The authors would like to thank the Biorparco di Roma curator Yitzhak Yadid and the carnivore keepers Marco Subrizi, Mauro Bedin, Massimiliano Pellicciotta, Barbara Leoni, Francesco Pompei and Nino Pinsone for their active participation in the study; Aussiedog® for providing them with its enrichment products.

Conflicts of Interest: The authors declare no conflict of interest.

References

1. Marucco, F.; Mcintire, E.J.B. Predicting spatio-temporal recolonization of large carnivore populations and livestock depredation risk: Wolves in the Italian Alps. *J. Appl. Ecol.* **2010**, *49*, 789–798. [CrossRef]
2. Clubb, R.; Mason, G.J. Captivity effects on wide-ranging carnivores. *Nature* **2003**, *425*, 473–474. [CrossRef] [PubMed]

3. Mason, G.J. Species differences in responses to captivity: Stress, welfare and the comparative method. *Trends Ecol. Evol.* **2010**, *25*, 713–721. [CrossRef] [PubMed]
4. Hall, S.L.; Bradshaw, J.W.S.; Robinson, I.H. Object play in adult domestic cats: The roles of habituation and disinhibition. *Appl. Anim. Behav. Sci.* **2002**, *79*, 263–271. [CrossRef]
5. Mech, L.D.; Boitani, L. *Wolves: Behavior, Ecology and Conservation*; Mech, L.D., Boitani, L., Eds.; The University of Chicago Press: Chicago, IL, USA, 2010.
6. Marshall-Pescini, S.; Cafazzo, S.; Virányi, Z.; Range, F. Integrating social ecology in explanations of wolf–dog behavioral differences. *Curr. Opin. Behav. Sci.* **2017**, *16*, 80–86. [CrossRef]
7. Szokalski, M.S.; Litchfield, C.A.; Foster, W.K. Enrichment for captive tigers (Panthera tigris): Current knowledge and future directions. *Appl. Anim. Behav. Sci.* **2012**, *139*, 1–9. [CrossRef]
8. McPhee, M.E.; Carlstead, K. The importance of maintaining natural behaviors in captive mammals. In *Wild Mammals in Captivity: Principles and Techniques for Zoo Management*; Kleiman, D.G., Allen, M., Thompson, K., Eds.; University of Chicago Press: Chicago, IL, USA, 2010; pp. 303–313, ISBN 0226440117.
9. Shepherdson, D.J.; Carlstead, K.; Mellen, J.D.; Seidensticker, J. The influence of food presentation on the behavior of small cats in confined environments. *Zoo Biol.* **1993**, *12*, 203–216. [CrossRef]
10. Clubb, R.; Vickery, S.S.; Latham, N.R. Motivation, and motivational explanations for stereotypies. In *Stereotypic Animal Behaviour: Fundamentals and applications to welfare*; Mason, G., Rushen, J., Eds.; CAB International: Wallingford, UK, 2006; pp. 12–13.
11. Meehan, C.L.; Mench, J.A. The challenge of challenge: Can problem solving opportunities enhance animal welfare? *Appl. Anim. Behav. Sci.* **2007**, *102*, 246–261. [CrossRef]
12. Carlstead, K.; Shepherdson, D.J. Alleviating stress in zoo animals with environmental enrichment. In *The Biology of Animal Stress: Basic Principles and Implications for Animal Welfare*; Moberg, G.P., Mench, J.A., Eds.; CAB International: Wallingford, UK, 2000; pp. 337–354.
13. Bassett, L.; Buchanan-smith, H.M. Effects of predictability on the welfare of captive animals. *Appl. Anim. Behav. Sci.* **2007**, *102*, 223–245. [CrossRef]
14. Bashaw, M.J.; Bloomsmith, M.A.; Marr, M.; Maple, T.L. To hunt or not to hunt? A feeding enrichment experiment with captive large felids. *Zoo Biol.* **2003**, *22*, 189–198. [CrossRef]
15. Mellen, J.; Macphee, M.S. Philosophy of Environmental Enrichment: Past, Present, and Future. *Zoo Biol.* **2001**, *226*, 211–226. [CrossRef]
16. Mason, G.; Clubb, R.; Latham, N.; Vickery, S. Why and how should we use environmental enrichment to tackle stereotypic behaviour? *Appl. Anim. Behav. Sci.* **2007**, *102*, 163–188. [CrossRef]
17. McPhee, M.E. Intact carcasses as enrichment for large felids: Effects on on- and off-exhibit behaviors. *Zoo Biol.* **2002**, *21*, 37–47. [CrossRef]
18. Cummings, D.; Brown, J.L.; Rodden, M.D.; Songsasen, N. Behavioral and Physiologic Responses to Environmental Enrichment in the Maned Wolf (Chrysocyon brachyurus). *Zoo Biol.* **2007**, *26*, 331–343. [CrossRef] [PubMed]
19. Kistler, C.; Hegglin, D.; Würbel, H.; König, B. Feeding enrichment in an opportunistic carnivore: The red fox. *Appl. Anim. Behav. Sci.* **2009**, *116*, 260–265. [CrossRef]
20. Wagman, J.D.; Lukas, K.E.; Dennis, P.M.; Willis, M.A.; Carroscia, J.; Gindlesperger, C.; Schook, M.W. A work-for-food enrichment program increases exploration and decreases stereotypies in four species of bears. *Zoo Biol.* **2018**, *37*, 3–15. [CrossRef] [PubMed]
21. Andrews, N.L.P.; Ha, J.C. The Effects of Automated Scatter Feeders on Captive Grizzly Bear Activity Budgets. *J. Appl. Anim. Welf. Sci.* **2014**, *17*, 148–156. [CrossRef]
22. Altman, J.D. Animal activity and visitor learning at the zoo. *Anthrozoos* **1998**, *11*, 12–21. [CrossRef]
23. Anderson, U.S.; Kelling, A.S.; Pressley-Keough, R.; Bloomsmith, M.A.; Maple, T.L. Enhancing the zoo visitor's experience by public animal training and oral interpretation at an otter exhibit. *Environ. Behav.* **2003**, *35*, 826–841. [CrossRef]
24. Bitgood, S.; Patterson, D.; Benefield, A. *Understanding Your Visitors: Ten Factors That Influence Visitor Behavior*; Jacksonville State University, Psychology Institute: Jacksonville, AL, USA, 1986.
25. Reade, L.S.; Waran, N.K. The modern zoo: How do people perceive zoo animals? *Appl. Anim. Behav. Sci.* **1996**, *47*, 109–118. [CrossRef]
26. Margulis, S.W.; Hoyos, C.; Anderson, M. Effect of felid activity on zoo visitor interest. *Zoo Biol.* **2003**, *22*, 587–599. [CrossRef]

27. Swaisgood, R.R.; Shepherdson, D.J. Scientific Approaches to Enrichment and Stereotypies in Zoo Animals: What's Been Done and Where Should We Go Next? *Zoo Biol.* **2005**, *24*, 499–518. [CrossRef]
28. Miller, L.J. Visitor reaction to pacing behavior: Influence on the perception of animal care and interest in supporting zoological institutions. *Zoo Biol.* **2012**, *31*, 242–248. [CrossRef]
29. Price, E.C.; Ashmore, L.A.; McGivern, A.-M. Reactions of zoo visitors to free-ranging monkeys. *Zoo Biol.* **1994**, *13*, 355–373. [CrossRef]
30. McPhee, M.E.; Foster, J.S.; Sevenich, M.; Saunders, C.D. Public perceptions of behavioral enrichment: Assumptions gone awry. *Zoo Biol.* **1998**, *17*, 525–534. [CrossRef]
31. Altmann, J. Observational study of behavior: Sampling methods. *Behaviour* **1974**, *49*, 227–267. [CrossRef]
32. Frézard, A.; Le Pape, G. Contribution to the welfare of captive wolves (*Canis lupus lupus*): A behavioral comparison of six wolf packs. *Zoo Biol.* **2003**, *22*, 33–44. [CrossRef]
33. Range, F.; Ritter, C.; Virányi, Z. Testing the myth: Tolerant dogs and aggressive wolves. *Proc. R. Soc. B Biol. Sci.* **2015**, *282*, 20150220. [CrossRef]
34. Wells, D.L.; Coleman, D.; Challis, M.G. A note on the effect of auditory stimulation on the behaviour and welfare of zoo-housed gorillas. *Appl. Anim. Behav. Sci.* **2006**, *100*, 327–332. [CrossRef]
35. Fernandez, E.J.; Tamborski, M.A.; Pickens, S.R.; Timberlake, W. Animal-visitor interactions in the modern zoo: Conflicts and interventions. *Appl. Anim. Behav. Sci.* **2009**, *120*, 1–8. [CrossRef]
36. Ross, S.R.; Lukas, K.E. Zoo Visitor Behavior at an African Ape Exhibit. *Visit. Stud. Today* **2005**, *8*, 4–12.
37. Luebke, J.F.; Matiasek, J. An Exploratory Study of Zoo Visitors' Exhibit Experiences and Reactions. *Zoo Biol.* **2013**, *32*, 407–416. [CrossRef] [PubMed]
38. Hacker, C.E.; Miller, L.J. Zoo Visitor Perceptions, Attitudes, and Conservation Intent After Viewing African Elephants at the San Diego Zoo Safari Park. *Zoo Biol.* **2016**, *35*, 355–361. [CrossRef] [PubMed]
39. Mason, G.J.; Latham, N.R. Can't stop won't stop: Is stereotipy a reliable animal welfare indicator? *Anim. Welf.* **2004**, *13*, 57–69.
40. Koene, P. Behavioral Ecology of Captive Species: Using Behavioral Adaptations to Assess and Enhance Welfare of Nonhuman Zoo Animals. *J. Appl. Anim. Welf. Sci.* **2013**, *16*, 360–380. [CrossRef] [PubMed]
41. Mason, G.J. Stereotypies: A critical review. *Anim. Behav.* **1991**, *41*, 1015–1037. [CrossRef]
42. Clubb, R.; Mason, G.J. Natural behavioural biology as a risk factor in carnivore welfare: How analysing species differences could help zoos improve enclosures. *Appl. Anim. Behav. Sci.* **2007**, *102*, 303–328. [CrossRef]
43. Clubb, R.; Vickery, S.S. Locomotory stereotypies in carnivores: Does pacing stem from hunting, ranging or frustrated escape? In *Stereotypic Animal Behaviour: Fundamentals and Applications to Welfare*; Mason, G.J., Rushen, J., Eds.; CAB International: Wallingford, UK, 2006; pp. 58–85.
44. Mason, G.J. Stereotypic behaviour in captive animals: Fundamentals and implications for welfare and beyond. In *Stereotypic Animal Behaviour: Fundamentals and Applications to Welfare*; Mason, G.J., Rushen, J., Eds.; CAB International: Wallingford, UK, 2006; pp. 325–356.
45. Brummer, S.P.; Gese, E.M.; Shivik, J.A. The effect of enclosure type on the behavior and heart rate of captive coyotes. *Appl. Anim. Behav. Sci.* **2010**, *125*, 171–180. [CrossRef]
46. McGowan, R.T.S.; Robbins, C.T.; Alldredge, J.R.; Newberry, R.C. Contrafreeloading in grizzly bears: Implications for captive foraging enrichment. *Zoo Biol.* **2010**, *29*, 484–502. [CrossRef]
47. Schneider, M.; Nogge, G.; Kolter, L. Implementing unpredictability in feeding enrichment for Malayan sun bears (Helarctos malayanus). *Zoo Biol.* **2014**, *33*, 54–62. [CrossRef]
48. Vasconcellos, A.S.; Alvarenga, C.; Pizzutto, C.S.; Ades, C. Environmental enrichment for maned wolves (Chrysocyon brachyurus): Group and individual effects. *Anim. Welf.* **2009**, *18*, 289–300.
49. O'Connell, N.E.; Beattie, V.E. Influence of environmental enrichment on aggressive behaviour and dominance relationships in growing pigs. *Anim. Welf.* **1999**, *8*, 269–279.
50. Honess, P.E.; Marin, C.M. Enrichment and aggression in primates. *Neurosci. Biobehav. Rev.* **2006**, *30*, 413–436. [CrossRef] [PubMed]
51. Fischbacher, M.; Schmid, H. Feeding Enrichment and Stereotypic Behavior in Spectacled Bears. *Zoo Biol.* **1999**, *371*, 363–371. [CrossRef]
52. Hocking, D.P.; Salverson, M.; Evans, A.R. Foraging-Based Enrichment Promotes More Varied Behaviour in Captive Australian Fur Seals (Arctocephalus pusillus doriferus). *PLoS ONE* **2015**, *10*, 1–13. [CrossRef] [PubMed]

53. Renner, M.J.; Lussier, J.P. Environmental enrichment for the captive spectacled bear (Tremarctos ornatus). *Pharmacol. Biochem. Behav.* **2002**, *73*, 279–283. [CrossRef]
54. Broom, D.M. Animal welfare: Concepts and measurement. *J. Anim. Sci.* **1991**, *49*, 4167–4175. [CrossRef] [PubMed]
55. Möstl, E.; Palme, R. Hormones as indicators of stress. *Domest. Anim. Endocrinol.* **2002**, *23*, 67–74. [CrossRef]
56. Wielebnowski, N.C.; Fletchall, N.; Carlstead, K.; Busso, J.M.; Brown, J.L. Noninvasive assessment of adrenal activity associated with husbandry and behavioral factors in the North American clouded leopard population. *Zoo Biol.* **2002**, *21*, 77–98. [CrossRef]
57. Hewer, M.J.; Gough, W.A. Weather sensitivity for zoo visitation in Toronto, Canada: A quantitative analysis of historical data. *Int. J. Biometeorol.* **2016**, *60*, 1645–1660. [CrossRef]
58. Bitgood, S.; Patterson, D. Principles of Exhibit Design. *J. Visit. Behav.* **1987**, *II*, 4–6.
59. Ballantyne, R.; Packer, J. Visitors' Perceptions of the Conservation Education Role of Zoos and Aquariums: Implications for the Provision of Learning Experiences. *Visit. Stud.* **2016**, *19*, 193–210. [CrossRef]
60. Kutska, D. Variation in visitor perceptions of a polar bear enclosure based on the presence of natural vs. un-natural enrichment items. *Zoo Biol.* **2009**, *28*, 292–306. [CrossRef] [PubMed]
61. Jacobson, S.L.; Hopper, L.M.; Shender, M.A.; Ross, S.R.; Leahy, M.; Mcnernie, J. Zoo visitors' perceptions of chimpanzee welfare are not affected by the provision of artificial environmental enrichment devices in a naturalistic exhibit. *J. Zoo Aquar. Res.* **2017**, *5*, 56–61.
62. Swanagan, J.S. Factors influencing zoo visitors' conservation attitudes and behavior. *J. Environ. Educ.* **2000**, *31*, 26–31. [CrossRef]
63. Jürgens, U.M.; Hackett, P.M. The Big Bad Wolf: The Formation of a Stereotype. *Ecopsychology* **2017**, *9*, 33–43. [CrossRef]

animals

MDPI

Article

Breed Group Effects on Complaints about Canine Welfare Made to the Royal Society for the Prevention of Cruelty to Animals (RSPCA) Queensland, Australia

Hao Yu Shih [1,*], Mandy B. A. Paterson [2] and Clive J. C. Phillips [1]

1 Centre for Animal Welfare and Ethics, University of Queensland, White House Building (8134), Gatton Campus, Gatton, QLD 4343, Australia
2 Royal Society for the Prevention of Cruelty to Animals Queensland, Brisbane, QLD 4076, Australia
* Correspondence: haoyu.shih@uqconnect.edu

Received: 20 May 2019; Accepted: 24 June 2019; Published: 26 June 2019

Simple Summary: This retrospective study involves 107,597 dog welfare complaints received by the Royal Society for the Prevention of Cruelty to Animals (RSPCA) Queensland from 2008 to 2018. Results show that, compared to pure breed dogs, cross-breed dogs were more likely to be reported in welfare complaints. Poisoning, lack of veterinary support, abuse, and being left unattended in a hot vehicle were common complaints in pure breed dogs; while insufficient shelter, exercise and food/water, as well as overcrowding and abandonment, were more commonly reported in cross breed dogs. Utility breeds, terriers and working dogs were most likely to be reported, while toy, non-sporting breeds and gundogs were least likely to be reported. Common complaint types for utility dogs were: insufficient food/water, shelter and exercise, and poor living conditions; for terriers: abandonment, intentional abuses and killing or injuring another animal; for working dogs: insufficient food/water, shelter and exercise; for toy dogs: lack of veterinary care, overcrowding and staying in a hot vehicle alone; for non-sporting dogs: lack of veterinary care, being left in a hot vehicle unattended and poor body conditions; and for hounds: killing or injuring another animal, intentional abuses and poor body conditions.

Abstract: Cruelty- and neglect-related canine welfare concerns are important welfare and social issues. Dog breed has been identified as a risk factor for bad welfare, and yet its role in different types of canine welfare concerns has not been fully investigated. We conducted a retrospective study of 107,597 dog welfare complaints received by RSPCA Queensland from July 2008 to June 2018. The breed of the dog involved in the incident was either recorded as stated by the complainant or by the inspector attending the case. Dog breed was divided into groups following the Australian National Kennel Club nomenclature. Dogs of a non-recognised breed were more likely to be reported in welfare complaints than recognised breed dogs. Recognised breed dogs had a greater risk of being reported with poisoning, lack of veterinary support, abuse and being left unattended in a hot vehicle; while non-recognised breed dogs had greater risk of being reported with insufficient shelter, exercise and food/water, as well as overcrowding and abandonment. Utility breeds, terriers and working dogs were most likely to be reported, while toy, non-sporting breeds and gundogs were least likely to be reported. Common complaint types for utility dogs were: insufficient food/water, shelter and exercise, and poor living conditions; for terriers: abandonment, intentional abuses and killing or injuring another animal; for working dogs: insufficient food/water, shelter and exercise; for toy dogs: lack of veterinary care, overcrowding and staying in a hot vehicle alone; for non-sporting dogs: lack of veterinary care, being left in a hot vehicle unattended and poor body conditions; and for hounds: killing or injuring another animal, intentional abuses and poor body conditions. Breed groups rather than breeds may be the best method of breed identification in a public reporting system as they group similar breeds together, and as our research shows, they relate to types of animal welfare complaints. Understanding the relationship between breed group and canine welfare complaints may help authorities improve public education programs and inform decision-making around which breed a new owner should choose.

Keywords: canine welfare; breed; canine cruelty; neglect; RSPCA

---

1. Introduction

Animal cruelty can be defined as any socially [1] or legally [2] unacceptable behaviour causing unnecessary pain, discomfort or distress to animals. It is an important social issue, affecting not only animals but also the entire society, for example, there appears to be a strong relationship between animal cruelty and other criminal activities, such as domestic violence [3]. Animal cruelty can happen to any animal, and the dog (*Canis familiaris*) is one of the most common species reported for suspected animal cruelty [4].

For over 15,000 years, dogs have been domesticated to fit into our society with humans intentionally taming them, which facilitates the development of close bonds between humans and dogs [5,6]. According to the Australian National Kennel Club (ANKC), there are 208 recognised breeds that are categorised into seven breed groups—toys, terriers, gundogs, hounds, working dogs, utility, and non-sporting—on the basis of their physical characteristics, temperaments, behaviours, and functions [7]. For example, greyhounds were originally bred to chase and hunt in the Egyptian deserts 5000 years ago, and are categorised as hounds by the ANKC [7,8]. Border collies were selected for their sheep herding abilities in western Europe, northern England and Scotland over 100 years ago, and are classified as working dogs by the ANKC [7,9,10]. However, the original selection traits are often of little value today and their purpose has broadened to become purely aesthetic or related to entertainment (including gambling), causing them to experience some breed-specific mistreatments. For instance, greyhounds [11] and huskies [12] are used in the racing industry, and greyhounds have been widely reported to experience welfare issues associated with their training, living conditions and injuries during racing. The greyhound racing industry has been associated with "live baiting", where a live animal, such as a chicken or possum, is used as a lure to train greyhounds to race [11–13]. Border collies are popular for companionship, but their herding instincts may not be fulfilled in a home environment, which predisposes them to behavioural problems, such as bike or runner chasing, and can finally lead to unsuccessful ownerships [14]. Pit bull type dogs, such as the American pit bull terrier, American Staffordshire terrier and Staffordshire bull terrier are often thought of as aggressive [15]. These dogs may be made to participate in illegal dog fights or used for pig hunting, and animal management laws may be biased against them. In Australia, the importation of American pit bull terriers is banned [16] and the law requires all American pit bull terriers currently in Australia to be sterilised. In addition, these dogs must be kept strictly confined when at home, and when taken out be muzzled and wear easily identifiable collars [15,17]. Similarly, in America, some states (e.g., Indiana and Louisiana) require pit bull owners to obtain a special license and maintain \$100,000 to \$300,000 in liability insurance to cover any potential injuries caused by the dogs [18]. Many studies highlight the negative welfare that may be experienced by racing and fighting dogs [17,19]. The welfare issues they experience are not only breed specific, but are directly related to specific industries (e.g., dog racing and fighting) [20]. However, neglect-related issues, such as failing to provide suitable food and water, veterinary support and suitable living conditions, are more common, yet less discussed [3,21]. To our knowledge, there has been little consideration of the correlation between canine breed and different forms of animal welfare concerns, particularly those related to neglect.

Accurate breed identification is useful in many areas including in shelters, veterinary clinics, research, and even the media [22]. How dogs are identified influences the way they are perceived and how people interpret their behaviours [23–25]. For instance, dogs identified as terriers, especially American pit bull terriers and Staffordshire bull terriers are perceived as playful, curious, fearless, chase prone and aggressive; however, dogs identified as toy breeds are seen as sociable [25]. Current breed identification in the majority of facilities is based on observation, but such visual identification of breed is problematic and often inaccurate [22,26]. In a laboratory-based experiment, 986 people engaged in dog-related professions were asked to visually identify the breed of 20 dogs using video clips. The visual identification was later compared with DNA identification. The results showed that over 50% of participants failed to visually identify dog breeds that matched the DNA identification, and agreement by over 50% of participants was only found with 35% ($n = 7/20$) of dogs [22]. Another study exploring inter-observer agreement among shelter staff differentiating pit-bull-type dogs versus non-pit-bull-type dog revealed moderate reliability (76%–83%) [26]. Consequently, a better breed identification method is needed, and identifying dogs per group or type (e.g., pit-bull type), rather than by specific breed, may result in higher agreement among individuals and be more useful.

In Queensland, Australia, animals are protected by the Animal Care and Protection Act 2001 (ACPA) [2]. This state-based legislation appoints inspectors, some of whom are employed by the Royal Society for the Prevention of Cruelty to Animals, Queensland (RSPCA Qld), to investigate potential breaches of, and enforce compliance with, the Act [2]. There are two main offences under the ACPA: failure to fulfil duty of care responsibilities and cruelty. There are a number of other specified offences. The Act recognises that a person who has charge of an animal owes that animal a duty of care. Failure to provide such care is the basis of the "breach of duty of care" offence. This offence covers such actions as not providing sufficient food, water, exercise, veterinary care and suitable living conditions. It is not only the owner that has a duty of care towards an animal. Anyone who is even temporarily in charge of an animal has a duty of care. The second major offence is "animal cruelty" and according to Section 18 of the Act, cruelty describes any action that causes unjustifiable and unnecessary physical and mental discomfort to animals, inappropriate confinement or transport, unreasonable injuries and inhumane death [2]. A cruel act can be committed by anyone towards an animal, whether it is their own animal, another domestic animal or even a wild animal [2]. It is important to note, that under the ACPA, the intention of a person to be cruel is not a necessary element of a cruelty offence to be proven in Queensland. If an action carried out by a person causes pain and suffering and the action was intentional, the person may be charged with cruelty. The intention to carry out the action must be proved but not the intention to be cruel. If a lack of action deprives an animal of its fundamental needs, then the person who has a duty of care towards the animal may be charged with a breach of their duty of care or cruelty depending on the circumstances. Intention may be considered during sentencing however [2]. Other offences under the Act include unreasonable abandonment or release, the carrying out of prohibited surgical procedures (e.g., tail docking, ear cropping, debarking, etc.), being involved in, or having items used for, a prohibited event, such as dog or cock fighting, and allowing an animal to injure or kill another animal [2].

The public can report suspected welfare concerns to the RSPCA via a "Cruelty Complaints" telephone number, which operates 24 h a day, seven days a week, or by email. In addition, complaints can be made by veterinarians and veterinary nurses, council officers, and other government and non-government employees visiting a location as part of their duties. Finally, animals entering a shelter may trigger an investigation if cruelty or neglect is suspected.

In this study, we aimed to evaluate whether breed was an important factor in relation to canine welfare concerns. This report is the second in a series relating to the analysis of RSPCA Qld canine welfare complaint data [21,27]. We hypothesized that certain breeds would have a higher risk of being reported. We also hypothesized that some breeds would be at higher risk of suffering specific welfare issues than others. Other risk factors, age of the dog [21] and socioeconomic status of the complainant are the subject of other papers [21,27].

2. Materials and Methods

From July 2008 to June 2018, RSPCA Qld received 129,036 canine welfare complaints. Some involving more than one dog were recorded as multiple complaints sharing the same case number, while others were recorded as one complaint with multiple animals. To avoid sample bias due to multiple entries, we only retained the first complaint of case numbers with multiple entries, discarding 21,439 entries as a result. There remained 107,597 canine welfare complaints for this retrospective study. The data analysis was originally undertaken on the entire dataset and then redone with the reduced number. Finding the complaint distribution and demographics to be similar, we opted for the reduced dataset to avoid problems with pseudoreplication. Animal welfare complaints that fell within the geographical zone of responsibility of RSPCA Qld (determined by a Memorandum of Understanding between RSPCA Qld and Biosecurity Queensland, the Government Department tasked with the administration of ACPA) were investigated by RSPCA Qld inspectors. All other complaints were referred to Biosecurity Queensland to be investigated by their inspectors. However, all complaints coming into RSPCA Qld were included in this analysis.

All complaints were recorded in ShelterBuddy® (RSPCA, Queensland, Australia), the RSPCA Qld database. The following information was requested from the reporter of each incident at the time of taking the complaint: the number of dogs involved and their age, breed(s) (if known), the "complaint code(s)", the suburb, postcode, and in addition, the date was recorded. All cases were investigated either by RSPCA Qld inspectors (n = 100,432) or Biosecurity Qld inspectors (n = 7165). It is recognized that some of the calls did not relate to a breach of the ACPA or to a genuine welfare concern. The outcome data for these complaints was not analysed in this research. This research is focused on the complaint calls coming in to RSPCA Qld.

Dogs were classified according to two broad age ranges, being dog and puppy, based on reporters' interpretation. It is important to recognise that the information recorded from the complainant may be inaccurate or inaccurately interpreted, e.g., a small dog is commonly referred to as a puppy in Queensland. Records regarding breed and the number of dogs involved were based on either complainants' initial reports or comments from trained inspectors, again recognising inaccuracies with identification of the breed. The "complaint code" was selected by the staff member receiving the call or email from a drop-down menu of 18 possible complaints (Appendix Table A1) [21]. Multiple "complaint codes" were able to be selected for each case according to the description of what was alleged to have happened to the dog(s), and each was treated as a separate code for analysis.

2.1. Dog Breeds

The distribution of breeds was compared to the breeds of registered dogs obtained from the councils of two cities situated close to the RSPCA Qld headquarters, namely Ipswich City Council and Gold Coast City Council for the same period. Any breed in our data that was documented in any of the following kennel clubs—Australian National Kennel Council (ANKC) [7], New Zealand Kennel Club (NZKC) [28], American Kennel Club (AMKC) [29] and United Kennel Club (UKC) [30]—was considered a recognised breed (RB) and was added to our breed list (Appendix Table A2). Any breed in our data that was not recognised by at least one of the major kennel clubs listed above was classified as a non-recognised breed (N-RB), including all crossbred dogs without any identified breed. In our dataset, it was decided that if more than one dominant breed was listed, the first breed mentioned would be used. For instance, Great Dane × Bull Arab was categorized as Great Dane (Appendix Table A2).

To achieve a secondary representation of breed recognition, RB breeds were amalgamated into the following seven breed groups based on the breed inclusion categories of the ANKC: toys, terriers, gundogs, hounds, working dogs, utility, and non-sporting. Breeds not listed by the ANKC, but recognised by the NZKC, AMKC, or UKC, were categorized into one of the seven groups based on the description of each kennel club. Some breeds (e.g., Australian Koolie and Bull Arab), though listed by the council registrations and thus on the breed list (Appendix Table A2), were not recognized as breeds by any major kennel club worldwide. Therefore, these breeds were categorized as N-RB. If the breed description was left blank, the dogs' breed was considered unknown (n = 15,576/107,597), and these complaints were excluded from any data analysis related to breed factors.

2.2. Statistical Analysis

Data was analysed using the statistical package Minitab® 17.3.1. (Minitab, LLC., State College, PA, USA) Descriptive analysis was first used to investigate the distribution of RB/N-RB and the seven breed groups. Complaints reported in July 2017 and June 2018 that contained breed information provided by RSPCA inspectors (n = 95) were used to examine the agreement of breed identification between the complainant and inspectors. Apart from simple percent agreement measurements, Cohen's kappa coefficient was calculated. Cohen's kappa is a statistical method measuring agreement with qualitative assessments among different raters. It is more robust than a percentage because it considers the possibility of the agreement occurring by chance [26,31]. To examine whether RB, N-RB or certain breed groups were more likely to be reported, the study group was compared with the registration data from Gold Coast and Ipswich City councils where all owned dogs, including working dogs on farms, are required to be registered [32,33]; this was done using Pearson chi-square tests. Eighteen stepwise forward binary logistic regression models were constructed to understand how breed factors correlated with each complaint code. The binary logistic regression model is a nonlinear model using a logistic function to describe the relationship of independent variables and a dependent variable with two possible values, such as yes/no, 0/1, or healthy/sick [34]. The stepwise forward selection refers to a step-by-step method of adding the most significant dependent variable into the model [35]. To determine the effect of breed (RB/N-RB or breed group) on complaint codes, breed (RB/N-RB or breed group) was entered into the binary logistic regression model as a fixed factor, using logit models with the alpha value to enter being 0.15. Complaint codes were entered into the model as outcomes. Separate models were constructed for each complaint code with the same input variable.

3. Results

3.1. Dog Characteristics

Common breeds reported by the complainants were Staffordshire bull terrier (10.5%, n = 10/95), American Staffordshire terrier (10.5%, n = 10/95), Maltese (6.3%, n = 6/95), and Bullmastiff (5.3%, n = 5/95). Overall, the agreement between complainants and inspectors of breed identification was 23.2% (Cohen's Kappa = 0.074, indicating a slight agreement [31]), and the agreement of breed group identification was 77.8% (Cohen's Kappa = 0.69, indicating a substantial agreement [31]). Therefore, breed groups were used for further analyses.

In the study group, 32.7% (n = 35,178) of dogs were N-RB, while only 1.7% (n = 1733) of dogs in the Gold Coast and Ipswich councils' data were listed as N-RB. Around 53% (n = 56,843) of dogs in our data and 98.3% (n = 99,266) of dogs in the councils' data were of RB (Table 1). The remaining dogs (14.5%, n = 15,576) in our database were unspecified. Thus, there was an over-representation of N-RB and an under-representation of RB in our dataset. The most common breed group to be reported for canine welfare concerns in our dataset were terriers (28.2%, n = 16,030), followed by working dogs (24.8%, n = 14,085), utility dogs (15.6%, n = 8,857), toy dogs (9.2%, n = 5223), non-sporting dogs (8.9%, n = 5071), gundogs (7.8%, n=4417), and hounds (5.6%, n = 3160) (Table 2). The most common breed group registered by the city councils were also terriers (22.2%, n = 22,056), but followed by toy dogs (21.0%, n = 20,796), working dogs (17.8%, n = 17,637), non-sporting dogs (14.0%, n = 13,915), gundogs (11.6%, n = 11,504), utility dogs (7.8%, n = 7770), and hounds (5.6%, n = 5581) (Table 2).

Table 1. Distribution of RB and N-RB in our study group, and in the Ipswich City Council and Gold Coast City Council registrations, with the overrepresentation coefficient (both RB and N-RB were significantly different (chi-square $p < 0.001$).

Breed	Study Group	Ipswich and Gold Coast	Overrepresentation Coefficient [a]
N-RB	35,178 (32.7%)	1733 (1.7%)	19.24
RB	56,843 (52.8%)	99,266 (98.3%)	0.54
Unknown	15,576 (14.5%)	0 (0%)	_ [b]
Total	107,597 (100%)	100,999 (100%)	

[a] Percent of breeds in our study/percent of breeds in the councils' registrations. 1.00 signifies equal representation in our database, and Ipswich and Gold Coast City Councils registrations. [b] Unable to calculate the overrepresentation coefficient because there was no dog with an unknown breed in the councils' data.

Table 2. Distribution of each breed group in our study, and in Ipswich City Council and Gold Coast City Council registrations, with the overrepresentation coefficient.

Breed Groups	RSPCA	Ipswich and Gold Coast	Overrepresentation Coefficient [a]	p-Value (Chi-Square Tests)
Terriers	16,030 (28.2%)	22,056 (22.2%)	1.27	<0.001
Working Dogs	14,085 (24.8%)	17,637 (17.8%)	1.39	<0.001
Utility	8857 (15.6%)	7770 (7.8%)	1.99	<0.001
Toys	5223 (9.2%)	20,796 (21.0%)	0.44	<0.001
Non-sporting	5071 (8.9%)	13,915 (14.0%)	0.64	<0.001
Gundogs	4417 (7.8%)	11,504 (11.6%)	0.67	<0.001
Hounds	3160 (5.6%)	5581 (5.6%)	0.99	0.602
Total	56,843 (100%)	99,266 (100%)		

[a] Percent of breeds in our study/percent of breeds in the councils' registrations. 1.00 signifies equal representation in our database, and Ipswich and Gold Coast City Councils registrations.

3.2. Predispositions of RB/N-RB to Welfare Complaints

Table 1 summarizes the numbers and percentages of RB and N-RB in our data and the councils' data, along with the overrepresentation coefficients. This coefficient is a simple method to compare the two percentages (see explanation below Table 1). Our results indicate that N-RB were at a greater risk of being reported than RB ($p < 0.001$).

We further explored the association between RB/N-RB and different complaint codes. A logistic regression model was generated (Table 3). In the model, there were significant correlations between RB/N-RB and nine ($n = 9/18$) complaint codes. RB had significantly greater risk of being reported with the following complaint codes, listed in increasing order of odds ratio (OR): baiting/poisoning (OR = 0.36, $p < 0.001$), no treatment (OR = 0.59, $p < 0.001$), cruelty (OR = 0.95, $p = 0.004$), and hot animal in car (OR = 0.95, $p = 0.043$). Meanwhile, N-RB had significantly greater risks of experiencing the following complaint codes, listed in declining order of odds ratio (OR): no exercise/confined/tethered (OR = 1.32, $p < 0.001$), overcrowding (OR = 1.32, $p < 0.001$), abandonment (OR = 1.25, $p < 0.001$), no shelter (OR = 1.19, $p < 0.001$), and insufficient food and/or water (OR = 1.08, $p < 0.001$).

Table 3. Odds ratio of each variable in the logistic regression model of complaint codes. The outputs of these models were different complaint codes. The input variables were N-RB/RB and breed groups.

Complaint Code	N-RB/RB p-Value/or (CI) [a]	Breed Group p-Value
Emergency relief	– [c]	– [c]
Hot animal in car	0.043/0.95 (0.90, 1.00)	<0.001
Keeping or using animal for blooding/coursing a dog	– [c]	– [c]
Prohibition order breached	– [c]	– [c]
Causing captive animal to be injured/killed by dog	– [c]	– [c]
Poor dog condition	– [c]	<0.001
Overcrowding	<0.001/1.32 (1.14, 1.53)	<0.001
No exercise/confined/tethered	<0.001/1.32 (1.28, 1.36)	<0.001
Insufficient food and/or water	<0.001/1.08 (1.05, 1.12)	<0.001
Baiting/poisoning	<0.001/0.36 (0.30, 0.43)	<0.001
Tail docking or other surgical procedure	– [c]	– [c]
No treatment	<0.001/0.59 (0.57, 0.62)	<0.001
Dog fighting or other prohibited offence	0.134/0.79 (0.58, 1.08)	– [c]
No shelter	<0.001/1.19 (1.14, 1.24)	<0.001
Poor living condition	– [c]	0.046
Cruelty [d]	0.004/0.95 (0.91, 0.98)	<0.001
Abandonment	<0.001/1.25 (1.21, 1.29)	<0.001
Knowingly allowing an animal to kill/injure another	0.111/0.86 (0.71, 1.04)	<0.001

[a] Odds ratio refers to N-RB relative to RB; [b] Odds ratio and the 95% confidence interval (CI) of each breed group for every complaint code are presented in Table 4; [c] Breed factor (N-RB/RB or breed group) was not selected in the logistic regression model; [d] A person was reported to have abused an animal.

3.3. *Predispositions of Breed Groups to Welfare Complaints*

When we compared the numbers of different breed groups in our data with the councils' data (Table 2), the following breed groups were over-represented in our data, listed in declining order of overrepresentation coefficient ($p < 0.001$): utility, working dogs and terriers. The following breeds in our database listed in increasing order of overrepresentation coefficient were underrepresented ($p < 0.001$): toys, non-sporting and gundogs.

In the regression model (Table 3), twelve (n = 12/18) complaint codes were predicted by breed group. Detailed results related to breed groups are summarized in Table 4. Toy dogs were more likely to be the subject of: no treatment, hot animal in car and overcrowding; terriers to: abandonment, cruelty, and knowingly allow an animal to kill/injure another; utility dogs to: insufficient food and/or water, no shelter, no exercise/confined/tethered and poor living condition; non-sporting dogs to: no treatment, hot animal in car and poor dog condition; hounds to: knowingly allow an animal to kill/injure another, cruelty and poor dog condition; and working dogs to: no exercise/confined/tethered, no shelter and insufficient food and/or water.

Table 4. Odds ratio and the 95% confidence interval (CI) of each breed group for every complaint code.

	No Treatment	Abandoned	Cruelty [a]	Insufficient Food and/or Water	No Shelter	No Exercise/ Confined/ Tethered	Hot Animal in Car	Baiting and Poisoning	Poor Dog Condition	Overcrowding	Knowingly Allowing an Animal to Kill/Injure Another	Poor Living Condition
Hounds/Gundogs	1.00 (0.89, 1.12)	0.92 (0.82, 1.04)	**1.17 (1.03, 1.33)**	0.96 (0.87, 1.07)	**0.74 (0.64, 0.85)**	1.10 (0.99, 1.23)	**0.82 (0.68, 0.99)**	0.77 (0.54, 1.12)	**1.47 (1.32, 1.62)**	1.56 (0.94, 2.57)	**2.44 (1.38, 4.29)**	1.10 (0.97, 1.24)
Non-Sporting/Gundogs	**1.31 (1.19, 1.45)**	0.94 (0.85, 1.05)	**0.83 (0.74, 0.94)**	**0.88 (0.80, 0.96)**	**0.67 (0.59, 0.76)**	**0.75 (0.68, 0.83)**	**1.62 (1.41, 1.87)**	**0.67 (0.48, 0.94)**	**1.65 (1.51, 1.80)**	1.12 (0.69, 1.80)	0.55 (0.27, 1.13)	1.05 (0.94, 1.17)
Terrier/Gundogs	**1.31 (1.19, 1.45)**	1.08 (0.99, 1.18)	**1.28 (1.16, 1.41)**	1.01 (0.94, 1.09)	**0.88 (0.80, 0.97)**	1.06 (0.98, 1.15)	**1.22 (1.08, 1.38)**	**0.49 (0.37, 0.64)**	1.01 (0.94, 1.09)	0.70 (0.46, 1.06)	**1.76 (1.08, 2.85)**	1.02 (0.94, 1.12)
Toys/Gundogs	**1.50 (1.36, 1.66)**	0.92 (0.83, 1.02)	0.97 (0.86, 1.08)	**0.63 (0.58, 0.70)**	**0.41 (0.36, 0.48)**	**0.51 (0.46, 0.57)**	**2.67 (2.34, 3.05)**	0.85 (0.62, 1.16)	1.07 (0.97, 1.17)	**2.08 (1.35, 3.20)**	0.71 (0.36, 1.38)	0.98 (0.88, 1.09)
Utility/Gundogs	0.98 (0.89, 1.07)	1.03 (0.94, 1.13)	1.01 (0.91, 1.12)	**1.12 (1.03, 1.22)**	**1.13 (1.02, 1.26)**	**1.38 (1.26, 1.50)**	**0.49 (0.42, 0.57)**	**0.50 (0.36, 0.68)**	**1.45 (1.34, 1.57)**	0.76 (0.48, 1.20)	1.21 (0.71, 2.06)	**1.10 (1.00, 1.21)**
Working Dogs/Gundogs	0.93 (0.85, 1.01)	0.95 (0.87, 1.03)	**1.15 (1.04, 1.26)**	1.01 (0.93, 1.09)	1.09 (0.99, 1.20)	**1.32 (1.22, 1.43)**	0.97 (0.85, 1.10)	**0.65 (0.50, 0.85)**	**1.13 (1.04, 1.22)**	1.22 (0.82, 1.83)	1.19 (0.71, 1.97)	1.09 (0.99, 1.19)
Non-Sporting/Hounds	**1.31 (1.18, 1.46)**	1.02 (0.91, 1.15)	**0.71 (0.62, 0.81)**	0.91 (0.82, 1.01)	0.91 (0.78, 1.05)	**0.68 (0.61, 0.76)**	**1.97 (1.66, 2.34)**	0.87 (0.59, 1.28)	**1.13 (1.03, 1.24)**	0.72 (0.45, 1.15)	**0.22 (0.12, 0.44)**	0.96 (0.85, 1.07)
Terrier/Hounds	1.06 (0.96, 1.17)	**1.17 (1.06, 1.30)**	1.09 (0.98, 1.21)	1.05 (0.96, 1.14)	**1.19 (1.05, 1.35)**	0.96 (0.88, 1.05)	**1.49 (1.27, 1.73)**	**0.63 (0.45, 0.88)**	**0.69 (0.64, 0.75)**	**0.45 (0.30, 0.68)**	0.72 (0.49, 1.06)	0.93 (0.84, 1.03)
Toys/Hounds	**1.51 (1.35, 1.68)**	1.00 (0.89, 1.12)	**0.82 (0.73, 0.93)**	**0.66 (0.59, 0.73)**	**0.56 (0.48, 0.66)**	**0.47 (0.42, 0.52)**	**3.25 (2.76, 3.82)**	1.10 (0.76, 1.59)	**0.73 (0.66, 0.80)**	1.33 (0.87, 2.03)	**0.29 (0.16, 0.53)**	0.89 (0.80, 1.00)
Utility/Hounds	0.98 (0.88, 1.09)	**1.12 (1.00, 1.24)**	**0.86 (0.77, 0.96)**	**1.16 (1.06, 1.28)**	**1.54 (1.35, 1.75)**	**1.25 (1.14, 1.37)**	**0.60 (0.49, 0.72)**	**0.64 (0.44, 0.93)**	0.99 (0.91, 1.08)	**0.49 (0.31, 0.77)**	**0.50 (0.32, 0.78)**	1.00 (0.90, 1.12)
Working Dogs/Hounds	0.93 (0.84, 1.02)	1.03 (0.93, 1.13)	0.98 (0.88, 1.09)	1.04 (0.95, 1.14)	**1.48 (1.31, 1.68)**	**1.20 (1.09, 1.31)**	**1.17 (1.00, 1.38)**	0.84 (0.60, 1.17)	**0.77 (0.71, 0.83)**	0.78 (0.53, 1.16)	**0.49 (0.32, 0.74)**	0.99 (0.90, 1.10)
Terrier/Non-Sporting	**0.81 (0.75, 0.87)**	**1.15 (1.06, 1.25)**	**1.53 (1.40, 1.69)**	**1.15 (1.07, 1.24)**	**1.32 (1.18, 1.46)**	**1.41 (1.30, 1.53)**	**0.75 (0.68, 0.84)**	**0.73 (0.54, 0.98)**	**0.61 (0.57, 0.66)**	**0.62 (0.42, 0.92)**	**3.21 (1.77, 5.81)**	0.97 (0.90, 1.06)
Toys /Non-Sporting	**1.15 (1.05, 1.25)**	0.98 (0.88, 1.08)	**1.16 (1.03, 1.30)**	**0.72 (0.66, 0.80)**	**0.62 (0.53, 0.72)**	**0.68 (0.61, 0.76)**	**1.65 (1.47, 1.85)**	1.27 (0.91, 1.77)	**0.65 (0.59, 0.70)**	**1.86 (1.25, 2.76)**	1.30 (0.61, 2.74)	0.94 (0.84, 1.04)
Utility/Non-Sporting	**0.75 (0.69, 0.81)**	**1.09 (1.00, 1.19)**	**1.21 (1.09, 1.34)**	**1.28 (1.18, 1.39)**	**1.70 (1.52, 1.90)**	**1.83 (1.68, 1.99)**	**0.30 (0.26, 0.35)**	0.74 (0.53, 1.03)	**0.88 (0.82, 0.94)**	0.68 (0.44, 1.04)	**2.20 (1.17, 4.16)**	1.05 (0.96, 1.15)
Working Dogs/Non-Sporting	**0.71 (0.65, 0.76)**	1.00 (0.92, 1.09)	**1.38 (1.25, 1.52)**	**1.15 (1.07, 1.24)**	**1.63 (1.47, 1.81)**	**1.76 (1.62, 1.90)**	**0.60 (0.53, 0.66)**	0.97 (0.72, 1.30)	**0.68 (0.64, 0.73)**	1.09 (0.76, 1.58)	**2.17 (1.17, 3.99)**	1.04 (0.95, 1.13)
Toys/Terrier	**1.42 (1.32, 1.53)**	**0.85 (0.78, 0.92)**	**0.75 (0.69, 0.83)**	**0.63 (0.58, 0.68)**	**0.47 (0.41, 0.53)**	**0.48 (0.44, 0.53)**	**2.19 (2.00, 2.40)**	**1.75 (1.33, 2.30)**	1.05 (0.98, 1.13)	**2.98 (2.15, 4.14)**	**0.40 (0.24, 0.68)**	0.96 (0.88, 1.04)
Utility/Terrier	**0.92 (0.87, 0.99)**	0.95 (0.89, 1.01)	**0.79 (0.73, 0.85)**	**1.11 (1.05, 1.18)**	**1.29 (1.20, 1.39)**	**1.29 (1.22, 1.37)**	**0.40 (0.35, 0.46)**	1.02 (0.77, 1.34)	**1.43 (1.35, 1.52)**	1.09 (0.76, 1.57)	**0.69 (0.49, 0.96)**	**1.08 (1.01, 1.15)**
Working Dogs/Terrier	**0.87 (0.83, 0.93)**	**0.87 (0.82, 0.93)**	**0.90 (0.84, 0.95)**	1.00 (0.95, 1.05)	**1.24 (1.16, 1.33)**	**1.24 (1.18, 1.31)**	**0.79 (0.73, 0.86)**	**1.34 (1.07, 1.67)**	**1.11 (1.06, 1.17)**	**1.76 (1.32, 2.34)**	**0.68 (0.50, 0.90)**	**1.06 (1.00, 1.13)**
Utility/Toys	**0.65 (0.60, 0.71)**	**1.12 (1.02, 1.22)**	1.04 (0.94, 1.15)	**1.77 (1.62, 1.92)**	**2.75 (2.42, 3.13)**	**2.68 (2.44, 2.94)**	**0.18 (0.16, 0.21)**	**0.58 (0.43, 0.80)**	**1.36 (1.26, 1.47)**	**0.37 (0.25, 0.53)**	1.70 (0.96, 3.00)	**1.12 (1.03, 1.23)**
Working Dogs/Toys	**0.62 (0.57, 0.66)**	1.03 (0.95, 1.12)	**1.19 (1.09, 1.30)**	**1.59 (1.46, 1.72)**	**2.65 (2.34, 3.00)**	**2.57 (2.35, 2.81)**	**0.36 (0.33, 0.40)**	**0.76 (0.59, 1.00)**	1.05 (0.98, 1.13)	**0.59 (0.44, 0.80)**	1.67 (0.97, 2.87)	**1.11 (1.02, 1.21)**
Working Dogs/Utility	0.95 (0.88, 1.01)	**0.92 (0.86, 0.98)**	**1.14 (1.06, 1.23)**	**0.90 (0.85, 0.95)**	0.96 (0.89, 1.04)	0.96 (0.90, 1.02)	**1.97 (1.73, 2.24)**	**1.31 (1.00, 1.71)**	**0.78 (0.73, 0.82)**	**1.61 (1.14, 2.27)**	0.98 (0.68, 1.43)	0.99 (0.92, 1.06)

The first number is the odds ratio, and the two numbers in the brackets are the CI. Bold: the odds ratio (CI) was marked as statistically significant ($p < 0.05$) when the CI did not cover 1.00. [a] A person was reported to have abused an animal.

4. Discussion

4.1. Breed Identification

The agreement of breed recognition between the general public and RSPCA inspectors was very low, which is in line with a previous study demonstrating inconsistent breed identification [22]. In light of the inconsistency, instead of assigning a breed for each dog, a potential alternative is to group them by function or more general characteristics into a recognised breed group [36]. Various kennel clubs have adopted similar concepts in their breed group criteria. For instance, in the ANKC, dogs that are originally bred to work with livestock are classified as working dogs, and those developed to assist hunters in retrieving game are classified as gundogs [7]. In this study, breed information provided by the general public and RSPCA inspectors was used to categorise dogs into ANKC breed groups. Consequently, the agreement increased from 23.2% to 77.8%, with substantial reliability (Cohen's Kappa = 0.69) between the public and trained inspectors. This result suggests that both the general public and inspectors are able to recognize some important and obvious characteristics of the dogs. Hence breed groups, rather than breeds, may be a better and more practical way of classifying breed for shelter research.

4.2. RB versus N-RB

This study examined the relationship between breed and canine welfare complaints. Specifically, we examined RB/N-RB and different breed groups with respect to their predisposition to be reported for welfare problems. We found there was a greater proportion of RB in the councils' data compared with RSPCA's, which might reflect a low rate of registration of N-RB, mainly crossbred dogs, in the two Queensland regions or that RB dogs were less likely to be involved in poor welfare or cruelty. These findings are supported by a previous study showing that crossbred dogs are at a higher risk of non-accidental injuries as a result of physical abuses, such as beating, throwing and burning, than pure bred dogs [37]. A further analysis of different complaint codes revealed that RB were predisposed to complaints related to "gaming", where owners allowed these dogs to engage in racing, fighting and blood sports. The number of dog fight cases was relatively small [21], so we should interpret them cautiously, although it is likely that dog fighting occurs more frequently than is reported. Credible and actionable evidence of such events is rarely received. One surprising finding was RB dogs were reported more often than N-RB for not receiving adequate veterinary care. Previous research into pet ownership and attitudes to pet care found that owners of shelter-acquired pets, usually mixed breed [38], took their animals for veterinary care more often and were equally willing to spend over $1000 on medical treatment for their pets [39] than pets acquired by other means. Finally, our data suggest that many N-RB dogs are not registered, which may make them more likely to be surrendered to a shelter or abandoned when medical care is required. Previous studies have reported higher rates of surrender of N-RB dogs [38]. N-RB dogs were more likely to be involved in complaints related to husbandry practices and abandonment.

4.3. Breed Groups

RB dogs were divided into breed groups and strong correlations between the groups, characteristics of the breeds and reasons for being reported were observed. For instance, toy dogs are small and possibly travel with owners more often, and thus, as found in our study, were more likely to be left alone in a car in hot weather. Previous research reports that smaller breeds of dogs are popular in Australia [40], which might help explain the number of complaints about toy dogs being left unattended in hot vehicles, as reported previously [21]. However, increased awareness of the dangers for dogs in hot cars through regular campaigning on this issue may also explain the high number of reports received [41]. Many terriers, especially those with some pit bull type characteristics (e.g., Staffordshire bull terrier, American Staffordshire terrier, and pit bull terrier), are considered aggressive and dangerous [15], therefore would be predisposed to being reported for abandonment, dog fights and cruelty [17,19,37,42]. Finally, utility and working dogs are mostly bred for guarding, rescuing or herding functions [7], and are generally energetic and require exercise [43,44]. These breeds were reported for not receiving adequate exercise. In contrast, toy breeds were the least likely to be reported for insufficient exercise, which is in agreement with previous research that found that smaller dogs were likely to have their exercise needs met even though they were walked less frequently [45].

4.4. Practical Application

This study provides fundamental information about the relationship between breed groups and various types of welfare complaints in dogs. The information can be used to develop education campaigns to increase awareness of what is involved in adequately and appropriately caring for dogs. Specific breeds have specific needs with respect to, for example, exercise requirements and cognitive enrichment to ensure their welfare is good. Information could be made available to prospective new owners of specific breeds to improve their understanding of the breed and its care requirements. Such information could also inform decision-making around breed choice as requirements of the breed could be matched with the ability of the owner to provide these needs.

4.5. Limitations and Need for Future Research

This was the first study providing fundamental information of the relationship between dog breed and welfare complaints made to a welfare organization with the responsibility of administering the Animal Welfare Act. Future research could focus on common breeds and explore the welfare issues in more detail.

There were several limitations of this present study. First, N-RB dogs contributed 32.7% of our dataset but only 1.7% of the total population in the councils' data. This major difference might indicate that N-RB dogs were indeed more susceptible to animal welfare concerns, but might also be affected by: (1) the difficulty of breed recognition [22], (2) the potentially different criteria of breed classification in our data and the councils' data and (3) the possibly lower registration rate of N-RB. Second, complaint codes were made based on public reports, which are likely to be inaccurate, at least in some cases. Third, we compared our data with reference data from the Gold Coast and Ipswich City Councils, urban areas in South East Queensland. The RSPCA cases were collected from a broad geographical area including both urban and rural areas, which may cause some regional bias. There were only 0.76% (n = 814/107,597) of cases from the Gold Coast and Ipswich regions. If we only compare the cases in these two regions, our results may be skewed by some less-common breeds that are either reported in our data or the councils' registration. Our data cover the Queensland regions along the East Australian seashore, which are nearly identical to the previous research [46]. Therefore, we decided to use similar methods by comparing our entire data with the councils' data. Given these limitations, the data reported here included canine welfare complaints only in Queensland, and national or global generalization should be made with caution.

5. Conclusions

Dog identification classified on the basis of breed groups rather than specific breed had a higher agreement between the public and shelter staff, and thus may serve as a better method of describing dogs involved in welfare reports. N-RB dogs, mainly crossbred dogs, were significantly more likely to be reported for alleged animal welfare concerns, especially poor living conditions and abandonment than RB. In addition, the characteristics of specific breeds, such as size, physical traits and exercise demands, were correlated to the reported complaints. Our results can help to improve public education and awareness raising. Finally, future studies are encouraged to explore in more detail the relationships between breed and welfare issues in dogs.

Author Contributions: Conceptualization, C.J.C.P., H.Y.S. and M.B.A.P.; methodology, H.Y.S., C.J.C.P. and M.B.A.P.; software, H.Y.S. and C.J.C.P.; validation, H.Y.S., C.J.C.P. and M.B.A.P.; formal analysis, H.Y.S.; investigation, H.Y.S.; resources, M.B.A.P. and C.J.C.P.; data curation, C.J.C.P., H.Y.S. and M.B.A.P.; writing—original draft preparation,

H.Y.S.; writing—review and editing, C.J.C.P., M.B.A.P. and H.Y.S.; visualisation, H.Y.S.; supervision, C.J.C.P. and M.B.A.P.; project administration, C.J.C.P. and M.B.A.P.

Funding: This research received no external funding.

Acknowledgments: We thank RSPCA, Qld, for providing the database of canine welfare complaints, RPSCA inspectors for the consultation, and Gold Coast and Ipswich councils for providing the breed registration data.

Conflicts of Interest: Mandy Paterson is employed as the principal scientist by RSPCA, Qld, but none of the authors received any interest or financial support from people or organisations that inappropriately influenced this study.

Appendix A

Table A1. Description of each complaint code alleging a welfare issue.

Complaint Code	Description
Abandonment	An animal was abandoned/left by the owner either at their abode or somewhere else such as in the bush.
Baiting/poisoning	An animal was poisoned or planned to be poisoned.
Cause captive animal to be injured/killed by dog	A person let a captive animal be injured/killed by a dog.
Cruelty	A person was reported to have abused an animal.
Dog fighting or other prohibited offence	A person was reported as allowing dogs to fight or conducting other specifically prohibited acts.
Emergency relief	Emergency relief is required for an animal left unattended because its owner experienced an emergency (e.g., flood or being hit by a car).
Hot animal in car	An animal was left unattended in a car during hot weather.
Insufficient food and/or water	An animal has insufficient food and/or water.
Keep or use animal for blooding/coursing a dog	A person used a live bait for blooding/coursing a dog.
Knowingly allow an animal to kill/injure another	A person allows one animal to kill/injuring another one, and does nothing to stop them.
No exercise/confined/tethered	An animal is confined or tethered and not given a suitable amount of exercise.
No shelter	An animal is not provided with suitable shelter provisions.
No treatment	An animal did not receive appropriate medical treatment when needed.
Overcrowding	The number of animals was too high for the living space provided.
Poor dog condition	The general condition of an animal is poor. (e.g., messy/matted coat, pussy eyes, etc.)
Poor living condition	The living environment of the animal is poor.
Prohibition order breached	An owner violated a prohibition order. [a]
Tail docking or other surgical procedure	Tail docking or other surgical procedure (e.g., declaw removal, etc.) was conducted on an animal.

[a] Prohibition order: A prohibition order is given by the court when a person convicted of an animal welfare offense must not possess any or specific animal for a prescribed period of time [2].

Table A2. Breed list.

Comment from the Public	Breed List	Breed Group
Affenpinscher	Affenpinscher	Toys
Afghan hound	Afghan hound	Hounds
Airedale terrier	Airedale terrier	Terrier
Akita	Akita	Utility
Alaskan husky	Siberian husky	Utility
Alaskan malamute	Alaskan malamute	Utility
American bulldog	American bulldog	Non-sporting
American foxhound	Foxhound	Hounds
American pit bull terrier	Pit bull terrier	Terrier
American Staffordshire terrier	Staffordshire terrier	Terrier
American water spaniel	American water spaniel	Gundogs
Anatolian shepherd dog	Anatolian shepherd dog	Utility

Table A2. *Cont.*

Comment from the Public	Breed List	Breed Group
Australian bandog	Cross breed	N-RB
Australian bulldog	Australian bulldog	Non-sporting
Australian bulldog cross		
Australian cattle dog	Australian cattle dog	Working dogs
Australian koolie	Coolie/koolie	N-RB
Australian sheepdog	Australian sheepdog	Working dogs
Australian shepherd	Australian shepherd	Working dogs
Australian silky terrier	Australian silky terrier	Toys
Australian stumpy tail cattle dog	Australian stumpy tail cattle dog	Working dogs
Australian terrier	Australian terrier	Terrier
Bandogge mastiff	Cross breed	N-RB
Basenji	Basenji	Hounds
Basset fauve de Bretagne	Basset fauve de Bretagne	Hounds
Basset hound	Basset hound	Hounds
Beagle	Beagle	Hounds
Bearded collie	Bearded collie	Working dogs
Bedlington terrier	Bedlington terrier	Terrier
Belgian shepherd	Belgian shepherd	Working dogs
Belgian shepherd—Groenendael		Working dogs
Belgian shepherd—Laekenois		Working dogs
Belgian shepherd—Malinois		Working dogs
Belgian shepherd—Tervueren		Working dogs
Bernese mountain dog	Bernese mountain dog	Utility
Bichon frise	Bichon frise	Toys
Bloodhound	Bloodhound	Hounds
Bluetick coohound	Bluetick coohound	Hounds
Border collie	Border collie	Working dogs
Border collie × Labrador		
Border collie, miniature		
Border terrier	Border terrier	Terrier
Borzoi	Borzoi	Hounds
Boston terrier	Boston terrier	Non-sporting
Bouvier des flandres	Bouvier des Flandres	Working dogs
Boxer	Boxer	Utility
Boxer cross		
Boxer × bullmastiff		
Boxer × American Staffordshire terrier		
Bracco Italiano	Bracco Italiano	Gundogs
Briard	Briard	Working dogs
British bulldog	British bulldog	Non-sporting
Brittany	Brittany	Gundogs
Bull Arab	Bull Arab	N-RB
Bull Arab × greyhound		
Bull terrier	Bull terrier	Terrier
Bull terrier cross		
Bull Terrier, miniature		
Bulldog	British bulldog	Non-sporting
Bulldog cross		
Bullmastiff	Bullmastiff	Utility
Bullmastiff cross		
Bullmastiff × wolfhound × Great Dane		
Cane corso (Italian mastiff)	Cane corso	Utility
Canaan dog	Canaan dog	Non-sporting
Cairn terrier	Cairn terrier	Terrier
Cattle dog	Australian cattle dog	Working dogs
Cattle dog cross		

Table A2. *Cont.*

Comment from the Public	Breed List	Breed Group
Cavalier King Charles spaniel	Cavalier King Charles spaniel	Toys
Central Asian shepherd dog	Central Asian shepherd dog	Utility
Cesky terrier	Cesky terrier	Terrier
Chesapeake Bay retriever	Chesapeake Bay retriever	Gundogs
Chihuahua Chihuahua cross Chihuahua × Jack Russell Long hair chihuahua	Chihuahua	Toys
Chinese crested dog Chinese crested dog—powder puff	Chinese crested dog	Toys
Chow chow	Chow chow	Non-sporting
Clumber spaniel	Clumber spaniel	Gundogs
Cocker spaniel Cocker spaniel, American Cocker spaniel, English	Cocker spaniel	Gundogs
Collie Collie rough Collie smooth	Collie	Working dogs
Corgi Corgi, Cardigan Welsh Corgi, Pembroke Welsh Corgi × fox terrier	Corgi	Working dogs
Coton de Tulear	Coton de Tulear	Toys
Cross breed	Cross breed	N-RB
Curly coated retriever	Curly coated retriever	Gundogs
Dachshund Dachshund, long-haired Dachshund, miniature	Dachshund	Hounds
Dalmatian Dalmatian cross	Dalmatian	Non-sporting
Dandie dinmont terrier	Dandie dinmont terrier	Terrier
Deerhound	Deerhound	Hounds
Dingo Dingo cross	Cross breed	N-RB
Dobermann	Dobermann	Utility
Dogue de Bordeaux	Dogue de Bordeaux	Utility
Dunker	Dunker	Hounds
Dutch shepherd	Dutch shepherd	Working dogs
English foxhound	Foxhound	Hounds
English pointer	English pointer	Gundogs
English mastiff	English mastiff	Utility
English setter	English setter	Gundogs
English springer spaniel	Springer spaniel	Gundogs
English toy terrier	English toy terrier	Toys
Field spaniel	Field spaniel	Gundogs
Finnish lapphund	Finnish lapphund	Working dogs
Flat coated retriever	Flat coated retriever	Gundogs
Formosan mountain dog (Taiwan dog)	Formosan mountain dog	Utility
Fox terrier Fox terrier, smooth	Fox terrier	Terrier
Foxhound	Foxhound	Hounds
French bulldog	French bulldog	Non-sporting
German coolie	Coolie/koolie	N-RB
German hunting terrier	German hunting terrier	Terrier
German pinscher	German pinscher	Utility
German shepherd German shepherd cross	German shepherd	Working dogs
German shorthaired pointer	German shorthaired/wirehaired pointer	Gundogs
German spitz	Spitz	Non-sporting

Table A2. *Cont.*

Comment from the Public	Breed List	Breed Group
German wirehaired pointer	German shorthaired/wirehaired pointer	Gundogs
Glen of Imaal terrier	Glen of Imaal terrier	Terrier
Golden retriever	Golden retriever	Gundogs
Gordon setter	Gordon setter	Gundogs
Great Dane	Great Dane	Non-sporting
Great dane × bull Arab		
Great dane × bullmastiff		
Great Pyrenees	Great Pyrenees	Working dogs
Greater Swiss mountain dog	Greater Swiss mountain dog	Working dogs
Greyhound	Greyhound	Hounds
Griffon Bruxellois	Griffon Bruxellois	Toys
Harrier	Harrier	Hounds
Havanese	Havanese	Toys
Hungarian vizsla	Hungarian vizsla	Gundogs
Husky	Siberian husky	Utility
Husky cross		
Ibizan hound	Ibizan hound	Hounds
Irish red and white setter	Irish setter	Gundogs
Irish setter		Gundogs
Irish terrier	Irish terrier	Terrier
Irish water spaniel	Irish water spaniel	Gundogs
Irish wolfhound	Irish wolfhound	Hounds
Italian greyhound	Italian greyhound	Toys
Italian spinone	Italian spinone	Gundogs
Jack Russell terrier	Jack Russell terrier	Terrier
Japanese chin	Japanese chin	Toys
Japanese spitz	Spitz	Non-sporting
Kangal shepherd dog	Kangal shepherd dog	Utility
Keeshond	Keeshond	Non-sporting
Kelpie	Kelpie	Working dogs
Kelpie cross		
Kelpie × staffordshire terrier		
Kelpie × border collie		
Kelpie × cattle dog		
Kelpie × labrador		
Kelpie × dingo		
Kerry blue terrier	Kerry blue terrier	Terrier
King Charles spaniel	King Charles spaniel	Toys
Kuvasz	Kuvasz	Working dogs
Labrador retriever	Labrador retriever	Gundogs
Labrador retriever cross		
Labradoodle		
Lagotto Romagnolo	Lagotto Romagnolo	Gundogs
Lakeland terrier	Lakeland terrier	Terrier
Large Munsterlander	Large Munsterlander	Gundogs
Leonberger	Leonberger	Utility
Large terrier cross	Terrier	Terrier
Lancashire heeler	Lancashire heeler	Working dogs
Lhasa apso	Lhasa apso	Non-sporting
Louisiana Catahoula leopard dog	Louisiana Catahoula leopard dog	Working dogs
Löwchen	Löwchen	Toys
Lurcher	Cross breed	N-RB
Maltese	Maltese	Toys
Maltese cross		
Manchester terrier	Manchester terrier	Terrier
Maremma sheepdog	Maremma sheepdog	Working dogs
Mastiff	Mastiff	Utility
Mastiff cross		
Mastiff × bull Arab		

Table A2. *Cont.*

Comment from the Public	Breed List	Breed Group
Medium terrier	Terrier	Terrier
Medium terrier cross		
Miniature fox terrier	Fox Terrier	Terrier
Miniature pinscher	Miniature pinscher	Toys
Neapolitan mastiff	Neapolitan mastiff	Utility
New Zealand huntaway	New Zealand huntaway	Working dogs
Newfoundland	Newfoundland	Utility
Norfolk terrier	Norfolk terrier	Terrier
North Queensland bullhound	Cross breed	N-RB
Norwegian elkhound	Norwegian elkhound	Hounds
Norwich terrier	Norwich terrier	Terrier
Nova Scotia duck tolling retriever	Nova Scotia duck tolling retriever	Gundogs
Old English sheepdog	Old English sheepdog	Working dogs
Papillon	Papillon	Toys
Parson Russell terrier	Parson Russell terrier	Terrier
Pekingese	Pekingese	Toys
Peruvian hairless dog	Peruvian hairless dog	Hounds
Petit basset griffon vendeen	Petit basset griffon vendeen	Hounds
Pharaoh hound	Pharaoh hound	Hounds
Pit bull terrier	Pit bull terrier	Terrier
Pig dog	Cross breed	Terrier
Pointer	Pointer	Gundogs
Polish lowland sheepdog	Polish lowland sheepdog	Working dogs
Pomeranian	Pomeranian	Toys
Poodle	Poodle	Non-sporting
Poodle toy		
Poodle miniature		
Poodle standard		
Poodle × shih tzu		
Portugese podengo	Portugese podengo	Hounds
Portuguese water dog	Portuguese water dog	Utility
Pug	Pug	Toys
Puli	Puli	Working dogs
Prague ratter	Cross breed	N-RB
Pyrenean mastiff	Pyrenean mastiff	Utility
Pyrenean mountain dog	Pyrenean mountain dog	Utility
Rhodesian ridgeback	Rhodesian ridgeback	Hounds
Rottweiler	Rottweiler	Utility
Rottweiler × mastiff		
Russian black terrier	Russian black terrier	Utility
Saint bernard	Saint bernard	Utility
Saluki	Saluki	Hounds
Samoyed	Samoyed	Utility
Sarplaninac	Sarplaninac	Utility
Schipperke	Schipperke	Non-sporting
Schnauzer	Schnauzer	Utility
Schnauzer, miniature		
Schnauzer, standard		
Schnauzer, giant		
Scottish terrier	Scottish terrier	Terrier
Sealyham terrier	Sealyham terrier	Terrier
Shar pei	Shar pei	Non-sporting
Shar Pei cross		
Shetland sheepdog	Shetland sheepdog	Working dogs
Shiba inu	Shiba inu	Utility
Shih tzu	Shih tzu	Non-sporting
Shih tzu × maltese		
Siberian husky	Siberian husky	Utility
Skye terrier	Skye terrier	Terrier

Table A2. *Cont.*

Comment from the Public	Breed List	Breed Group
Sloughi	Sloughi	Hounds
Small terrier cross	Terrier	Terrier
Smithfield cattle dog	Cross breed	N-RB
Soft coated wheaten terrier	Soft coated wheaten terrier	Terrier
Spaniel	Spaniel	Gundogs
Spanish water dog	Spanish water dog	Gundogs
Spitz	Spitz	Non-sporting
Spoodle	Cocker spaniel	Gundogs
Staffordshire bull terrier Staffordshire bull terrier × labrador	American Staffordshire bull terrier	Terrier
Staghound	Staghound	N-RB
Swedish vallhund	Swedish vallhund	Working dogs
Tenterfield terrier	Tenterfield terrier	Terrier
Terrier	Terrier	Terrier
Thai ridgeback	Thai ridgeback	Hounds
Tibetan mastiff	Tibetan mastiff	Utility
Tibetan spaniel	Tibetan spaniel	Toys
Tibetan terrier	Tibetan terrier	Non-sporting
Timber shepherd	Cross breed	N-RB
Weimaraner	Weimaraner	Gundogs
Welsh springer spaniel	Springer spaniel	Gundogs
Welsh terrier	Welsh terrier	Terrier
West highland white terrier	West highland white terrier	Terrier
Whippet	Whippet	Hounds
White Swiss shepherd dog	White Swiss shepherd dog	Working dogs
Wirehaired fox terrier	Fox terrier	Terrier
Xoloitzcuintle	Xoloitzcuintle	Non-sporting
Yorkshire terrier	Yorkshire terrier	Toys

ANKC: Australian National Kennel Council (http://ankc.org.au/); AMKC: American Kennel Club (https://www.akc.org/dog-breeds/); UKC: United Kennel Club (https://www.ukcdogs.com/breed-standards); NZKC: New Zealand Kennel Club (https://www.dogsnz.org.nz/home/home).

References

1. Tallichet, S.E.; Hensley, C. Rural and urban differences in the commission of animal cruelty. *Int. J. Offender Ther.* **2005**, *49*, 711–726. [CrossRef] [PubMed]
2. Queensland Government–Department of Agriculture and Fisheries. Animal Care and Protection Act 2001. Available online: https://www.legislation.qld.gov.au/view/pdf/inforce/current/act-2001-064 (accessed on 22 April 2019).
3. Mogbo, T.C.; Oduah, F.N.; Okeke, J.J.; Ufele, A.N.; Nwankwo, O.D. Animal cruelty: A review. *J. Nat. Sci. Res.* **2013**, *3*, 94–98.s.
4. Newberry, M. Associations between different motivations for animal cruelty, methods of animal cruelty and facets of impulsivity. *Psychol. Crime Law* **2018**, *24*, 500–526. [CrossRef]
5. Hare, B. Domestication experiments reveal developmental link between friendliness and cognition. *J. Bioecon.* **2018**, *20*, 159–163. [CrossRef]
6. Savolainen, P.; Zhang, Y.-P.; Luo, J.; Lundeberg, J.; Leitner, T. Genetic evidence for an east asian origin of domestic dogs. *Science* **2002**, *298*, 1610–1613. [CrossRef] [PubMed]
7. Australian National Kennel Council LTD. Breeds. Available online: http://ankc.org.au/Home (accessed on 19 November 2018).
8. American Kennel Club. Greyhound. Available online: https://www.akc.org/dog-breeds/greyhound/ (accessed on 19 November 2018).
9. Arvelius, P.; Malm, S.; Svartberg, K.; Strandberg, E. Measuring herding behavior in Border collie—effect of protocol structure on usefulness for selection. *J. Vet. Behav.* **2013**, *8*, 9–18. [CrossRef]
10. United Kennel Club. Border Collie. Available online: https://www.ukcdogs.com/border-collie (accessed on 13 March 2019).

11. McEwan, A.; Skandakumar, K. The welfare of greyhounds in Australian racing: Has the industry run its course? *Aust. Anim. Prot. Law J.* **2011**, *6*, 52–74.
12. Stafford, K. The welfare of the athletes; greyhounds and sled dogs. In *The welfare of dogs*; Phillips, C., Ed.; Springer: Dordrecht, The Netherlands, 2006; pp. 143–160, ISBN 978-140-204-361-1.
13. Atkinson, M.; Young, K. Reservior dogs: Greyhound racing, mimesis and sports-related violence. *Int. Rev. Sociol. Sport* **2005**, *40*, 335–356. [CrossRef]
14. Zilocchi, M.; Tagliavini, Z.; Cianni, E.; Gazzano, A. Effects of physical activity on dog behavior. *Dog Behav.* **2016**, *2*, 9–14. [CrossRef]
15. Collier, S. Breed-specific legislation and the pit bull terrier: Are the laws justified? *J. Vet. Behav.* **2006**, *1*, 17–22. [CrossRef]
16. Australian Government-Federal Register of Legislation. Animals (Importation of Certain Dog Breeds) Regulations 2004 (NI). Available online: https://www.legislation.gov.au/Details/C2018Q00020 (accessed on 31 May 2019).
17. Tarver, E.C. The dangerous individual('s) dog: Race, criminality and the 'Pit bull'. *Cul. Theor. Criti.* **2014**, *55*, 273–285. [CrossRef]
18. Devin, B. Breed specific legislation: Unfair prejudice & (and) ineffective policy. *Animal l.* **2004**, *10*, 313.
19. Kalof, L.; Taylor, C. The discourse of dog fighting. *Humanit. Soc.* **2007**, *31*, 319–333. [CrossRef]
20. Johnson, O. Dog fighting and the growing social epidemic of animal cruelty. *J. Law Enf.* **2014**, *3*, 1–9.
21. Shih, H.Y.; Paterson, M.B.A.; Phillips, C.J.C. A Retrospective Analysis of Complaints to RSPCA Queensland, Australia, about Dog Welfare. *Animals* **2019**, *9*, 282. [CrossRef] [PubMed]
22. Voith, V.L.; Trevejo, R.; Dowling-Guyer, S.; Chadik, C.; Marder, A.; Johnson, V.; Irizarry, K. Comparison of visual and DNA breed identification of dogs and inter-observer reliability. *Am. J. Sociol. Res.* **2013**, *3*, 17–29. [CrossRef]
23. Mehrkam, L.R.; Clive, D.L.W. Behavioral differences among breeds of domestic dogs (Canislupus familiaris): Current status of the science. *Appl. Anim. Behav. Sci.* **2014**, *155*, 12–27. [CrossRef]
24. Gunter, L.M.; Barber, R.T.; Wynne, C.D.L. What's in a name? Effect of breed perceptions & labeling on attractiveness, adoptions & length of stay for pit-bull-type dogs. *PLoS ONE* **2015**, *11*, e0146857. [CrossRef]
25. Clarke, T.; Mills, D.; Cooper, J. "Type" as central to perceptions of breed differences in behavior of domestic dog. *Soc. Anim.* **2016**, *24*, 467–485. [CrossRef]
26. Olson, K.R.; Levy, J.K.; Norby, B.; Crandall, M.M.; Broadhurst, J.E.; Jacks, S.; Barton, R.C.; Zimmerman, M.S. Inconsistent identification of pit bull-type dogs by shelter staff. *Vet. J.* **2015**, *206*, 197–202. [CrossRef]
27. Shih, H.Y.; Paterson, M.B.A.; Phillips, C.J.C. Socioeconomic influences on reports of canine welfare concerns to the RSPCA in Queensland, Australia. *Animals* **2019** (under review).
28. New Zealand Kennel Club. Recognised Pedigree Breeds. Available online: https://www.dogsnz.org.nz/home/home (accessed on 1 August 2018).
29. American Kennel Club. Dog Breeds. Available online: https://www.akc.org/dog-breeds/ (accessed on 1 Augst 2018).
30. United Kennel Club. Breed Standards. Available online: https://www.ukcdogs.com/breed-standards (accessed on 1 August 2018).
31. Landis, J.R.; Koch, G.G. The measurement of observer agreement for categorical data. *Int. Biom. Soc.* **1997**, *33*, 159–174. [CrossRef]
32. City of Gold Coast. Dog Registration. Available online: https://cogc.cloud.infor.com/ePathway/epthprod/Web/Animals/AnimalClasses.aspx?js=1900370700 (accessed on 17 June 2019).
33. City of Ipswich. New Dog Registration. Available online: https://epathway.ipswich.qld.gov.au/ePathway/Production/Web/Animals/AnimalClasses.aspx?js=1151207821 (accessed on 17 June 2019).
34. Minitab®, Salford Predictive Modeler®. Introduction to Logistic Regression Modeling. Available online: https://minitab.com/uploadedFiles/Content/Products/SPM/IntroLOGIT.pdf (accessed on 17 June 2019).
35. Minitab® 18. Perform Stepwise Regression for Fit Regression Model. Available online: https://support.minitab.com/en-us/minitab/18/help-and-how-to/modeling-statistics/regression/how-to/fit-regression-model/perform-the-analysis/perform-stepwise-regression/ (accessed on 17 June 2019).
36. Simpson, R.J.; Simpson, K.J.; VanKavage, L. Rethinking dog breed identification in veterinary practice. *J. Am. Vet. Med. Assoc.* **2012**, *241*, 1163–1166. [CrossRef] [PubMed]

37. Murno, H.M.C.; Thrusfield, M.V. Battered pets': Non-accidental physical injuries found in dogs and cats. *J. Small Anim. Pract.* **2001**, *42*, 279–290. [CrossRef]
38. New, J.C.J.; Salman, M.D.; King, M.; Scarlett, J.M.; Kass, P.H.; Hutchison, J.M. Characteristics of shelter-relinquished animals and their owners compared with animals and their owners in U.S. pet-owning households. *J. Appl. Anim. Welf. Sci.* **2000**, *3*, 179–201. [CrossRef]
39. Freiwald, A.; Litster, A.; Weng, H.Y. Survey to investigate pet ownership and attitudes to pet care in metropolitan Chicago dog and/or cat owners. *Prev. Vet. Med.* **2014**, *115*, 198–204. [CrossRef] [PubMed]
40. Teng, K.T.; McGreevy, P.D.; Toribio, J.-A.L.M.L.; Dhand, N.K. Trends in popularity of some morphological traits of purebred dogs in Australia. *Canine Genet. Epidemiol.* **2016**, *3*, 1–9. [CrossRef] [PubMed]
41. RSPCA Queensland. Keep Pets out of Hot Cars. Available online: https://www.rspcaqld.org.au/blog/pet-care/dogs-in-hot-cars (accessed on 19 June 2019).
42. Burley, S. My dog's the champ: An analysis of young people in urban settings and fighting dog breeds. *Anthropol. Matters J.* **2008**, *10*, 1–18.
43. Lim, C.; Rhodes, R.E. Sizing up physical activity: The relationships between dog characteristics, dog owners' motivations, and dog walking. *Psychol. Sport Exerc.* **2016**, *24*, 65–71. [CrossRef]
44. Westgarth, C.; Christian, H.E.; Christley, R.M. Factors associated with daily walking of dogs. *BMC Vet. Res.* **2015**, *11*, 116–128. [CrossRef]
45. Pickup, E.; German, A.J.; Blackwell, E.; Evans, M.; Westgarth, C. Variation in activity levels amongst dogs of different breeds: results of a large online survey of dog owners from the UK. *J. Nutr. Sci.* **2017**, *6*, 1–7. [CrossRef] [PubMed]
46. Schot, A.A.W.-v.d.; Day, C.; Morton, J.M.; Rand, J.; Phillips, C.J.C. Risk factors for behavior problems in cats presented to an Australian companion animal behavior clinic. *J. Vet. Behav.* **2016**, *14*, 34–40. [CrossRef]

animals

MDPI

Article

Determination of Prolactin in Canine Saliva: Is It Possible to Use a Commercial ELISA Kit?

Jara Gutiérrez *, Angelo Gazzano, Beatrice Torracca, Valentina Meucci and Chiara Mariti

Dipartimento di Scienze Veterinarie, Università di Pisa, 56124 Pisa, Italy

* Correspondence: jara.gutierrez@vet.unipi.it

Received: 13 May 2019; Accepted: 2 July 2019; Published: 4 July 2019

Simple Summary: Prolactin is considered a remarkable index of stress response, both acute and chronic, in several species. Some studies have investigated the possibility of measuring prolactin in saliva in human beings and Rhesus macaque. The possibility of measuring it in dog saliva would provide a non-invasive, helpful tool for the assessment of dog welfare. The aims of this research article are to study (1) the possibility of quantifying canine prolactin in saliva using a prolactin canine ELISA (enzyme-linked immunosorbent assay) kit validated for measuring prolactin in canine blood and (2) the potential presence of a correlation between prolactin levels in saliva and plasma.

Abstract: Prolactin has been reported to be a remarkable index of stress response, both acute and chronic, in several species. The use of biological matrixes other than blood is receiving increasing interest in the study of hormones, due to the lower invasiveness in collection. This research aimed to investigate the possibility of using a commercial ELISA (enzyme-linked immunosorbent assay) kit for measuring canine prolactin in blood for the quantification of canine prolactin in saliva. Study 1 consisted of a validation protocol, using saliva samples collected from lactating and non-lactating dogs. Study 2 was conducted to investigate a possible correlation between prolactin concentration in saliva and plasma in sheltered dogs by using the same kit. Prolactin values were reliably read only when they came from blood samples, not from saliva, but tended to be low in most of the cases. Study 1 showed that saliva had a matrix effect. In study 2, saliva prolactin levels were low and in 42.9% of cases, not readable. No correlation between prolactin values in plasma and saliva was found ($\rho = 0.482$; $p = 0.274$). These findings suggested that the determination of prolactin in dog saliva through an ELISA kit created for measuring prolactin in dog blood was unreliable.

Keywords: blood; dogs; prolactin; saliva; stress

1. Introduction

The best-known role of prolactin in the dog is the stimulation of the growth of the mammary gland and the lactation processes. Nevertheless, prolactin has over 300 separate biological activities and it plays multiple homeostatic roles and physiological functions in the organism, such as the electrolyte balance, luteal function, regulation of the immune system, osmoregulation, angiogenesis, maintenance of the inter-oestrous interval, etc. [1].

In addition, prolactin is considered an index of acute stress in some species. For instance, prolactin concentrations increased after various stressful stimuli in humans [2] and rats [3–6], including male rats [7]. In lactating rats, hyperprolactinemia seemed to be a significant factor for the decrease of plasma oxytocin response to acute stress [8]. Furthermore, prolactin seems to reduce anxiety-related behavior in both female and male rats, maybe because prolactin acts as an endogenous anxiolytic, both in males and non-pregnant female rats [9]. The decrease in stress-induced secretion of ACTH from the adenohypophysis and reduced corticosterone secretion from the adrenal gland in lactating rats is also manifested in mid-gestation, from day 15 until day 21 of pregnancy [10].

In human beings, prolactin-releasing stimuli include suckling, perception of visual, acoustic, and olfactory stimuli and stress [11]. For instance, it has been suggested that human serum prolactin concentrations may be elevated by psychological stressors, as well as by psychosocial stress (Trier Social Stress Test) [12]. In addition, in human beings, it is known that prolactin is in response to severe experimental stress induced by hypoglycaemia [2], surgery [2], parachute jumping in military recruits [13], and compulsory swimming in non-swimmers [14]. A significant correlation was found between day-to-day changes in anxiety measured by questionnaires as well as by stress hormones, cortisol, and prolactin [15].

Prolactin also increases in response to psychological stressors like restraint and transport in rats [16], heat stressin both rats and domestic ruminants [16,17], and stressful situations in donkeys [18], dromedaries [19], cattle [20], and sheep [21].

In dogs, both cortisol and prolactin decrease immediately after parturition [22]. Moreover, prolactin increases just after delivery due to the pups' suckling stimulation [22]. In lactating bitches, high levels of prolactin and increased expression of prolactin receptors in the paraventricular nucleus may produce a decrease in the stress response during lactation [23,24]. In addition, anxious dogs displaying signs such as stereotypes, displacement activities, various autonomic disorders, and fear aggression, have an increase in prolactin blood levels [25]. Assistance dogs seem to have higher mean prolactin blood levels than pet dogs, suggesting a possible role of canine blood prolactin as an index of stress-related responses in dogs [26].

When quantifying certain physiological parameters, blood has often been used as the best body fluid to evaluate different biomarkers. However, in recent years, the replacement of blood by saliva samples has achieved growing interest due to its reduced invasiveness, cost, and risk of infection compared to blood collection [27]. Nevertheless, the concentration of these specific biomarkers may often differ between blood and saliva [28]. When multiple human biomarkers were compared in plasma and saliva samples, consistent correlations were found between both types of saliva sampling (passive drool and filter paper), but little correlation was found between plasma and saliva [28]. Korot'ko and Gotovtseva [29] found that human prolactin levels were lower in saliva than serum. In other cases, some methods were reported to be unable to quantify a certain biomarker in the saliva matrix, such as it being reported for salivary oxytocin when measured by immunoassay [30].

As prolactin is usually measured in blood to reduce the stress of blood sampling, it would be useful to find different biological matrixes in which prolactin could be reliably measured. The possibility of quantifying prolactin in human saliva has been evaluated using four commercially available methods different from ELISA (enzyme-linked immunosorbent assay), but none of which could detect it [31]. Salivary prolactin has instead been successfully measured in Rhesus macaques (*Macaca mulatta*) by radioimmunoassay, but a positive correlation between serum and salivary prolactin was not found, perhaps because the pulsatile variation of prolactin in the blood may be an impediment to detecting such a correlation [32]. This pulsatile secretion for prolactin in blood has also been found for dogs [33].

The aim of this study was to evaluate whether a commercially available ELISA kit for measuring canine prolactin in blood is also suitable for measuring canine prolactin in saliva. To do that, the research was divided into two parts: Study 1) consisted of a validation protocol using saliva samples collected from lactating and non-lactating dogs, and study 2 investigated the possible correlation between prolactin concentration in saliva and plasma in sheltered dogs.

2. Materials and Methods

The procedure was communicated to the Ethics Committee of the University of Pisa, Italy (OPBA, Organismo Preposto per il Benessere Animale) and received a favorable opinion with Decision N.09/2018.

2.1. ELISA Kit

The prolactin ELISA kit used in this study is an enzyme immunoassay for the detection of canine prolactin in serum developed by Demeditec Diagnostic GmbH (Kiel, Germany). For this research, 2 kits were used and the assay was performed according to the manufacturer's instructions.

The microplate was coated with a monoclonal antibody specific for canine prolactin. Calibrators and samples were placed in front of the 96-well plate. A total of 25 µl calibrators and samples were added in triplicate in successive wells and incubated for 2 hours at room temperature. Endogenous canine prolactin in the sample binds to the antibodies fixed on the inner surface of the wells. Non-reactive sample components were removed by a washing step.

Afterwards, a second polyclonal horseradish peroxidase-labeled antibody, directed against another epitope of the prolactin molecule, was added. During an hour of incubation, a sandwich complex consisting of the 2 antibodies and the canine prolactin was formed. After incubation, the plate was washed with the provided wash buffer and a substrate solution (3, 3′, 5, 5′-tetramethylbenzidine) was added, followed by another 30 min of incubation time. Finally, stop solution (hydrochloric acid) was added and the absorption was measured at 450 nm within 30 min with a Multiskan™ FC Microplate Photometer (ThermoFisher Scientific, Waltham, MA, USA). Concentrations of prolactin were estimated from a calibration curve obtained by plotting the optical density versus the concentration for each one of the calibrators (80.0, 40.0, 20.0, 10.0, 5.0, and 2.5 ng/ml).

2.2. Saliva and Plasma Samples Collection

For study 1, saliva samples from 21 healthy adult dogs (7 males and 14 females in anoestrus, 1–11 years old) were collected and pooled to establish a sample with regular values (non-lactating saliva, NLS). Saliva samples from 5 bitches in a lactation period (1–7 years old) were collected and pooled to establish a sample with assumed high levels of prolactin (lactating saliva, LS). Saliva was always collected in the morning by the same person, except for some lactating females, whose samples were collected by the person in charge. Saliva was collected using flat cottons (Salivette®Cotton swab, neutral 51.1534, Sarstedt) and gloves, in order to avoid variability in the results and contamination. Samples remained in a cold chain (0 to +4 °C) until they reached the laboratory, where they were centrifuged two consecutive times (7000 rpm, 10–15 min). Saliva was obtained after centrifugation and stored at −20°C until analysis.

For study 2, saliva and plasma samples were collected from 10 healthy adult mixed breed dogs (1 female and 9 males, 1–11 years old) single-housed in a municipal shelter, between 11:00 and 12:00 a.m. and always followed the same order: First blood, and then saliva, with an interval no longer than 2 minutes between them. With the exception of 3 dogs, from which we did not get enough saliva, both saliva and blood were analyzed for each dog. The extraction of saliva was carried out following the same protocol described for study 1.

2.3. Validation Parameters

The possible application of an ELISA kit for canine blood prolactin to canine saliva samples (study 1) was determined by evaluating the linearity, limit of quantification, matrix effect, and spiking recovery.

The lyophilized master calibrator (80 ng of lyophilized in serum/buffer matrix containing highly purified canine prolactin) was reconstituted with 1 ml of distilled water. To evaluate linearity, this prolactin standard solution was diluted with the provided sample diluent to obtain the solutions for the calibration curves (40.0, 20.0, 10.0, 5.0, and 2.5 ng/ml).

Previous literature shows that in dogs, blood prolactin levels tend to be low. In addition, unpublished data obtained by the authors of the current study showed that salivary prolactin levels are commonly low. Consequently, for this validation method, linearity could not be reliably assessed with serial dilutions of normal samples. In order to reach higher values that could improve the reading of the kit, it was decided to dilute standard solutions to obtain different final-added prolactin concentrations in artificial saliva (AS_{20}, AS_{10}, AS_5, $AS_{2.5}$; Pickering Laboratories©, Space Park Way, Mountain View, CA, USA),and in non-lactating saliva (NLS_{20}, NLS_{10}, NLS_5). Parallelism between the curves obtained with standard solutions and the ones obtained with the spiked saliva pool was also assessed.

Lactating saliva pool (LS), with assumed high prolactin concentration, was diluted (2:3 and 1:2 for kit 1; 2:3 for kit 2) using artificial saliva as a diluent to evaluate linearity. The goal of analyzing lactating dog saliva as such (LS) and in two different dilutions (LS_{d1}, LS_{d2}) was to find out the minimum concentration of salivary prolactin that could be read by the kit.

To assess the possible matrix effect and recovery for each kit, 3 repetitions of a lyophilized control from the same batch (Demeditec Diagnostics GmbH®, Kiel, Germany) were reconstituted (1 ml) with water (control reconstituted in water = CW; as specified by the manufacturer instructions), AS (control reconstituted in artificial saliva = CAS),and NLS (control reconstituted in saliva of non-lactating dogs = CNLS). For the first kit, 3 repetitions for each reconstitution were made using a control batch corresponding to 15.14 ng/ml, and for the second kit, a control batch of 8.16 ng/ml was used (Appendix A).

The lower limit of sensitivity was determined as the mean concentration obtained interpolating the optical density plus 2 SDs for all replicates of the 0 standard.

2.4. Statistical Analysis

The possible presence of correlation between the prolactin values in blood and in saliva for study 2 has been analyzed through the Spearman's Rho test ($p < 0.05$).

3. Results

Linearity was acceptable for all the regression curves for prolactin concentrations in sample diluent, in AS and in NLS, for both kits (R^2 values: Sample diluents kit 1 = 0.995, AS kit 1 = 0.99, NLS kit 1= 0.994; sample diluent kit 2= 0.996, AS kit 2 = 0.992, NLS kit 2 = 0.993). Slopes between these regression curves were parallel for each kit, respectively (sample diluent kit 1 = 0.033, AS kit 1 = 0.034, NLS kit 1 = 0.037; sample diluent kit 2 = 0.032, AS kit 2 = 0.035, NLS kit 2 = 0.037) and R^2 coefficients for the same curves were acceptable for both kits, meaning that kits ran similarly.

Levels of prolactin measured in non-lactating (NLS) and lactating (LS) saliva pools were similar and, in both cases, low (mean ± standard deviation: NLS = 0.34 ± 0.30 ng/ml; LS = 1.03 ± 1.22 ng/ml for kit 1; NLS = 1.07 ± 0.20 ng/ml; LS = 1.17 ± 0.07 ng/ml for kit 2). Due to these low levels, it was not possible to conduct a linearity analysis based on physiological sample dilutions.

Values for the non-lactating saliva pool with the addition of the standards to obtain different final added concentrations (kit 1: NLS_{20} = 23.4 ng/ml, NLS_{10} = 11.4 ng/ml, NLS_5 = 6.4 ng/ml; kit 2: NLS_{20} = 24.1 ng/ml, NLS_{10} = 12.3 ng/ml, NLS_5 = 6.4 ng/ml) were higher than the normal added prolactin concentrations (20, 10, and 5 ng/ml).

Prolactin concentration values obtained in CW, CAS, and CNLS (Table 1) were within the target ranges, except for CNLS in kit 2 which was higher (12.1 ng/ml). CAS obtained virtually the same value than CW in both kits. Prolactin concentration values in CNLS were slightly higher than CW, in both kits (kit 1: CNLS–CW = 18.7–16.5 = 2.2 ng/ml; kit 2: CNLS – CW = 12.1–9.7 = 2.4 ng/ml). Values of CNLS were slightly higher than CAS (Table 1).

Table 1. Expected and observed prolactin values when control was dissolved in water (CW), artificial saliva (CAS), and non-lactating saliva (CNLS). Mean and standard deviation for the observed values and recovery percentage.

Sample		Expected: Mean (ng/ml)	Expected: Target Range (ng/ml)	Observed (ng/ml)	CV	Recovery
Control dissolved in water (CW)	Kit 1:	15.14	9.1–21.2	16.5 (s.d. 0.45)	2.8%	109% (s.d. 3)
	Kit 2:	8.16	4.9–11.4	9.7 (s.d. 0.21)	2.2%	119% (s.d. 3)
Control dissolved in artificial saliva (CAS)	Kit 1:	15.14	9.1–21.2	16.3 (s.d. 0.84)	5.2%	107% (s.d. 6)
	Kit 2:	8.16	4.9–11.4	9.9 (s.d. 0.59)	5.9%	122% (s.d. 7)
Control dissolved in non-lactating saliva (CNLS)	Kit 1:	15.14	9.1–21.2	18.7 (s.d. 1.09)	5.8%	124% (s.d. 7)
	Kit 2:	8.16	4.9–11.4	12.1 (s.d. 0.61)	5.0%	148% (s.d. 7)

The lower limit of sensitivity was calculated as the mean concentration obtained interpolating the optical density plus 2 SDs for all replicates of the 0 standard was 1.10 ng/ml.

A reliable inter-assay repeatability was recorded for values whose concentrations were higher than 5 ng/ml (e.g., AS_{20}, AS_{10}, AS_5, F_{10}, F_5), with coefficients of variation lower than 12% (mean %CV ± standard deviation = 8.13% ± 0.03), whereas for lower concentrations (e.g., $AS_{2.5}$, LC, LC_{Dil1}, F_{20}, F), high coefficients of variation were recorded (mean %CV ± standard deviation = 71.54% ± 0.32).

Prolactin concentration values in saliva and plasma samples obtained in study 2 are shown in Table 2. Prolactin values from plasma samples were below the limit of the kit's detection (0.4 ng/ml) in 20.0% of cases (mean ± standard deviation = 4.69 ± 5.37 ng/ml), and were especially high for dog 1 (17.8 ng/ml). Saliva prolactin levels were low too (mean ± standard deviation = 1.94 ± 1.96 ng/ml) and, in 42.9%of cases, not detectable. In fact, prolactin concentration in saliva samples without any additions resulted in very low values in both study 1 and 2.

Table 2. Prolactin concentrations in saliva and plasma samples (ng/ml) using a canine prolactin ELISA (enzyme-linked immunosorbent assay) kit. (BLD = below the limit of detection).

	Plasma		Saliva	
	Concentration (ng/ml)	Coefficient of Variation	Concentration (ng/ml)	Coefficient of Variation
Dog 1	17.8	1.3%	BLD	-
Dog 2	1.5	1.1%	BLD	-
Dog 3	3.6	1.9%	BLD	-
Dog 4	BLD	-	4.6	8.1%
Dog 5	2.6	2.6%	0.0	3.0%
Dog 6	2.1	1.6%	2.0	1.6%
Dog 7	4.5	1.8%	1.1	2.2%
Dog 8	BLD	-	-	-
Dog 9	2.8	1.5%	-	-
Dog 10	2.5	9.3%	-	-

Values from Table 2 clearly show that there is no correspondence between prolactin concentrations in plasma and in saliva: On one hand, dogs 1, 3, and 7 had the highest values of plasma prolactin, but low saliva concentrations, and on the other hand, dog 4 had a relatively high value of saliva prolactin, but prolactin in plasma was not even detectable. This was confirmed by a lack of correlation between plasma and saliva values ($\rho = 0.482$; $p = 0.274$).

4. Discussion

Both kit 1 and 2 were able to read the control within the optimal target ranges. Artificial saliva (AS) did not seem to interfere with the reading of the kit, as the reading of control joined to artificial saliva (CAS) was almost the same for the reading for control added to water (CW). In other terms, artificial saliva did not seem to have a matrix effect. However, saliva is a complex matrix and for this reason further measurements were done using real saliva, which will be discussed later.

In both studies, prolactin concentration in saliva samples without any additions resulted in very low values, often below the limit of detection of the kit, meaning that the kit cannot feasibly read prolactin concentrations in natural saliva samples.

A difference emerged between the values of CW and the values of control joined to non-lactating saliva pool (CNLS). It could be hypothesized that this difference was due to the fact that the kit was reading the prolactin present in the non-lactating saliva pool. However, this does not seem to be a justifiable explanation, since the concentration of prolactin in the non-lactating saliva pool alone (NLS) was lower than the difference between CNLS and CW. Since NLS was used as an example of natural saliva, it can be deduced that such a difference was due to a matrix effect of the natural canine saliva itself, and consequently the kit could not read saliva samples properly. Sometimes, a matrix effect can disappear by diluting the sample serially until a linearity of the results is obtained. However, serial dilutions of the lactating saliva pool (LS) did not exhibit a linear dilution, indicating that a matrix component was interfering with an accurate detection, causing a loss in reading sensitivity.

The variation between the values for non-lactating saliva pool with addition of the standards to obtain different final added concentrations (NLS_{20}, NLS_{10}, NLS_5) and those of the corresponding standards (20, 10, and 5 ng/ml) are not due to the presence of natural prolactin, since calculated NLS values were different when using the different additions. For this reason, natural prolactin could not be reliably determined with the addition of the standards and the observed difference was likely not reflecting the real value of prolactin, due to the presence of a matrix effect of the saliva. In other terms, although the addition of the prolactin standard allowed higher values to be obtained and, therefore, a reading of prolactin results feasible, a large matrix effect was observed, preventing the assertion that the value obtained was reliable.

Furthermore, we expected prolactin concentrations in non-lactating saliva to be lower than prolactin concentration of saliva samples in lactating dogs, since prolactin levels in blood have been reported to have an increase in lactating dogs [33–35]. Indeed, saliva from lactating dogs had values that were only slightly higher than those of non-lactating dogs and with a high variation. In addition, the reliability of NLS and LS readings was not adequate because their coefficients of variation were quite high.

In most of the cases, the kit was not able to detect prolactin in saliva samples as obtained values were very low and some of them did not reach the limit of the kit's detection. In study 2, when prolactin in saliva was detected by the kit, most of these values were higher than the ones from the lactating saliva pool, as mirrored by mean values. A possible explanation might be that shelter dogs are often exposed to stressful conditions, leading to higher prolactin circulating levels [2,14,15]. However, the lack of correlation with plasma concentrations did not allow us to draw reliable conclusions.

In study 2, plasma prolactin concentrations, when readable, agreed with values reported by other authors for canine prolactin in serum, ranging 1–6.3 ng/ml [36–39]. One exception in the current study is represented by dog 1, with high plasmatic prolactin values, probably due to its behavioral problems since it showed a phobia of numerous stimuli, and possibly a state of anxiety (not assessed as it was out if the scope of this study). The relationship between phobia, anxiety, and prolactin has been investigated by Pageat, 2007 [25], however further research is needed for a better understanding of their possible links.

A correlation between plasmatic and salivary prolactin levels from paired single samples was not found. This fits with the results of Lindell et al. [32], who did not find a positive correlation between values of prolactin in the serum and saliva of Rhesus macaques, discussing the possibility that the pulsatile variation of prolactin in serum impedes the detection of a significant relationship with the pooled saliva source, since single blood samples were used, eventually suggesting that a repeated blood sampling might lead to a significant correlation [32].

In this study, blood and saliva were only collected from non-lactating stressed dogs at the shelter, while for the two pools used for validation (lactating dogs and non-lactating dogs), individual samples were not collected. Although having the results of blood prolactin levels from the same samples of saliva used for the validation assay would have been optimal, blood collection was avoided in lactating bitches for ethical reasons. We also avoided collecting multiple saliva samples from lactating bitches, in order to minimize interference with their nursing and lactation. Due to the low values obtained in study 1 for both non-lactating and lactating dogs, for study 2, it was decided to use individual samples of blood and saliva from sheltered dogs, which possibly had higher prolactin values due to stress.

Moreover, the use of Salivette for collecting saliva samples may have caused a bias in the results. The different available saliva collection methods were reported to possibly interfere in the biomarker of interest present in the saliva. For this reason, in certain cases, a correspondence between methods of collection was found. For example, Salivette was reported to be a reliable and predictable method of total and quantified free serum cortisol levels [40]. However, in other cases this correspondence was not found. For instance, nephelometrically determined IgA concentrations were significantly lower in saliva when collected by the Salivette than by a suction or spitting method [41]. When quantifying cortisol and dehydroepiandrosterone (DHEA), results using two different saliva collection methods ('passive drool' method and a citric acid-treated salivette) correlated highly with plasmatic levels and with each other, whereas results using Salivette did not correlate significantly with plasma values [42]. Future research using other methods of saliva collection may be useful for knowing if the method of collection interferes with the measurement of prolactin in saliva.

Taken together, all these results suggest that prolactin cannot be reliably measured in saliva by the ELISA kit used in the study, which is validated for measuring prolactin in dog blood, meaning that canine saliva was not a suitable matrix for the kit. The quantification of prolactin though ELISA kits out of the intended use needs caution. In fact, a previous study found that prolactin in horses cannot be reliably measured using a canine prolactin ELISA kit, due to a matrix effect [43].

5. Conclusions

In summary, for saliva the study found: A matrix effect, very low prolactin concentrations (often under the limit of detection), and a lack of correlation with prolactin plasmatic levels. Validation of the kit showed that prolactin in saliva could be read under certain conditions (standard addition) but without reliability (matrix effect). In the absence of standard addition, prolactin values were too low to be read.

These results suggest that saliva was not a suitable matrix to measure prolactin levels using an ELISA kit created for measuring canine blood prolactin concentrations.

Author Contributions: Conceptualization, A.G. and C.M.; Data curation, J.G.; Investigation, J.G. and C.M.; Methodology, B.T., V.M. and C.M.; Supervision, C.M.; Writing—original draft, J.G. and B.T.; Writing—review & editing, J.G., A.G., and C.M.

Funding: This research received no external funding.

Acknowledgments: We appreciate the collaboration of the dog shelter Canile Soffio di Vento(Pisa, Italy) and to the dog breeding centers *Allevamento della Maschera di Tutankhamon* (Palaia, Pisa, Italy) and *Allevamento Labrador Sindia's Labradors* (Pistoia, Italy), as well as the group of dog owners who collaborated altruistically for this study.

Conflicts of Interest: The authors declare no conflict of interest.

Appendix A

Table A1. Outline distribution adopted for the validation procedure. Details of the wells for kit 1 and kit 2. The only difference is that the second dilution for lactating female saliva (LS_{dil2}) was replaced in kit 2 by saliva from a different pool of domestic dogs (saliva as such -P-, and two dilutions of it -P_{20} and P_{10}-). Caption: A-G (reconstitute lyophilized canine prolactin master calibrators), CW (control dissolved in water), CAS (control dissolved in artificial saliva), CNLS (control dissolved in non-lactating dogs saliva), AS_{20}-$AS_{2.5}$ (artificial saliva with addition of the standards), LS (lactating dogs saliva pool), LS_{d1}-LS_{d2} (dilutions of lactating dogs saliva pool), NLS (non-lactating dogs saliva pool), N_{20}-N_{10}-N_5 (non-lactating dogs saliva pool with addition of the standards, respectively 20, 10 and 5 ng/ml), NLS+CNLS (non-lactating dogs saliva pool + control dissolved in saliva, 1:1).

Kit 1	1	2	3	4	5	6	7	8	9	10	11	12
A	A_0	E_{20}	CW	CNLS	AS_{20}	AS_{10}	$AS_{2.5}$	LS	LS_{d2}	N	NLS_{20}	NLS_5
B	A_0	E_{20}	CW	CNLS	AS_{20}	AS_{10}	$AS_{2.5}$	LS	LS_{d2}	NLS	NLS_{20}	NLS_5
C	$B_{2.5}$	F_{40}	CAS	AS	AS_{20}	AS_5	$AS_{2.5}$	LS_{d1}	LS_{d2}	NLS	NLS_{10}	NLS_5
D	$B_{2.5}$	F_{40}	CAS	AS	AS_{20}	$AS_{2.5}$	$AS_{2.5}$	LS_{d1}	LS_{d2}	NLS	NLS_{10}	NLS_5
E	C_5	G_{80}	CAS	AS	AS_{10}	LS	LS	LS_{d1}	LS_{d2}	NLS_{20}	NLS_{10}	NLS_5
F	C_5	G_{80}	CAS	AS	AS_{10}	LS	LS	LS_{d1}	LS_{d2}	NLS_{20}	NLS_{10}	NLS_5
G	D_{10}	CW	CNLS	AS_{20}	AS_{10}	LS	LS	LS_{d1}	NLS	NLS_{20}	NLS_{10}	NLS + CNLS
H	D_{10}	CW	CNLS	AS_{20}	AS_{10}	LS	LS	LS_{d1}	NLS	NLS_{20}	NLS_{10}	NLS + CNLS
Kit 2	**1**	**2**	**3**	**4**	**5**	**6**	**7**	**8**	**9**	**10**	**11**	**12**
A	A_0	E_{20}	CW	CNLS	AS_{20}	AS_{10}	$AS_{2.5}$	LC	P	NLS	NLS_{20}	NLS_5
B	A_0	E_{20}	CW	CNLS	AS_{20}	AS_{10}	$AS_{2.5}$	LC	P	NLS	NLS_{20}	NLS_5
C	$B_{2.5}$	F_{40}	CAS	AS	AS_{20}	AS_5	$AS_{2.5}$	LS_{d1}	P_{20}	NLS	NLS_{10}	NLS_5
D	$B_{2.5}$	F_{40}	CAS	AS	AS_{20}	$AS_{2.5}$	$AS_{2.5}$	LS_{d1}	P_{20}	NLS	NLS_{10}	NLS_5
E	C_5	G_{80}	CAS	AS	AS_{10}	LS	LS	LS_{d1}	P_{20}	NLS_{10}	NLS_{10}	NLS_5
F	C_5	G_{80}	CAS	AS	AS_{10}	LS	LS	LS_{d1}	P_{10}	NLS_{20}	NLS_{10}	NLS_5
G	D_{10}	CW	CNLS	AS_{20}	AS_{10}	LS	LS	LS_{d1}	NLS	NLS_{20}	NLS_{10}	NLS + CNLS
H	D_{10}	CW	CNLS	AS_{20}	AS_{10}	LS	LS	LS_{d1}	NLS	NLS_{20}	NLS_{10}	NLS + CNLS

References

1. Bole-Feysot, C.; Goffin, V.; Edery, M.; Binart, N.; Kelly, P.A. Prolactin (PRL) and Its Receptor: Actions, Signal Transduction Pathways and Phenotypes Observed in PRL Receptor Knockout Mice. *Endocr. Rev.* **1998**, *19*, 225–268. [CrossRef] [PubMed]
2. Noel, G.L.; Suh, H.K.; Stone, J.G.; Frantz, A.G. Human Prolactin and Growth Hormone Release during Surgery and other Conditions of Stress. *J. Clin. Endocrinol. Metab.* **1972**, *35*, 840–851. [CrossRef] [PubMed]
3. Colborn, D.R.; Thompson, D.L.; Roth, T.L.; Capehart, J.S.; White, K.L. Responses of cortisol and prolactin to sexual excitement and stress in stallions and geldings. *J. Anim. Sci.* **1991**, *69*, 2556–2562. [CrossRef] [PubMed]
4. Neill, J.D. Effect of "Stress" on Serum Prolactin and Luteinizing Hormone Levels During the Estrous Cycle of the Rat. *Endocrinology* **1970**, *87*, 1192–1197. [CrossRef] [PubMed]
5. Euker, J.S.; Meites, J.; Riegle, G.D. Effects of Acute Stress on Serum LH and Prolactin in Intact, Castrate and Dexamethasone-Treated Male Rats. *Endocrinology* **1975**, *96*, 85–92. [CrossRef] [PubMed]
6. Döhler, K.D.; Gärtner, K.; Von zurMühlen, A.; Döhler, U. Activation of anterior pituitary, thyroid and adrenal gland in rats after disturbance stress. *Acta Endocrinol.* **1977**, *86*, 489–497. [CrossRef]
7. Pintér-Kübler, B.; Ferenczi, S.; Nuñez, C.; Zelei, E.; Polyák, Á.; Milanés, M.V.; Kovács, K.J. Differential Changes in Expression of Stress- and Metabolic-Related Neuropeptides in the Rat Hypothalamus during Morphine Dependence and Withdrawal. *PLoS ONE* **2013**, *8*, e67027. [CrossRef] [PubMed]
8. Carter, D.A.; Lightman, S.L. Oxytocin Responses to Stress in Lactating and Hyperprolactinaemic Rats. *Neuroendocrinology* **1987**, *46*, 532–537. [CrossRef] [PubMed]
9. Torner, L.; Toschi, N.; Pohlinger, A.; Landgraf, R.; Neumann, I.D. Anxiolytic and Anti-Stress Effects of Brain Prolactin: Improved Efficacy of Antisense Targeting of the Prolactin Receptor by Molecular Modeling. *J. Neurosci.* **2001**, *21*, 3207–3214. [CrossRef]

10. Neumann, I.D.; Johnstone, H.A.; Hatzinger, M.; Liebsch, G.; Shipston, M.; Russell, J.A.; Landgraf, R.; Douglas, A.J. Attenuated neuroendocrine responses to emotional and physical stressors in pregnant rats involve adenohypophysial changes. *J. Physiol.* **1998**, *508*, 289–300. [CrossRef]
11. Vega, S.R.; Hollmann, W.; Strüder, H.K. Influences of Exercise and Training on the Circulating Concentration of Prolactin in Humans. *J. Neuroendocr.* **2012**, *24*, 395–402. [CrossRef] [PubMed]
12. Lennartsson, A.-K.; Jonsdottir, I.H. Prolactin in response to acute psychosocial stress in healthy men and women. *Psychoneuroendocrinology* **2011**, *36*, 1530–1539. [CrossRef] [PubMed]
13. Noel, G.L.; Dimond, R.C.; Earll, J.M.; Frantz, A.G. Prolactin, thyrotropin, and growth hormone release during stress associated with parachute jumping. *Aviat. Space Environ. Med.* **1976**, *47*, 534–537. [PubMed]
14. Vaernes, R.; Ursin, H.; Darragh, A.; Lambe, R. Endocrine response patterns and psychological correlates. *J. Psychosom. Res.* **1982**, *26*, 123–131. [CrossRef]
15. Jeffcoate, W.; Lincoln, N.; Selby, C.; Herbert, M. Correlation between anxiety and serum prolactin in humans. *J. Psychosom. Res.* **1986**, *30*, 217–222. [CrossRef]
16. Van Vugt, D.; Bruni, J.; Meites, J. Naloxone inhibition of stress-induced increase in prolactin secretion. *Life Sci.* **1978**, *22*, 85–89. [CrossRef]
17. Alamer, M. The Role of Prolactin in Thermoregulation and Water Balance during Heat Stress in Domestic Ruminants. *Asian J. Anim. Veter Adv.* **2011**, *6*, 1153–1169. [CrossRef]
18. Kataria, N.; Kataria, A.K. Assessment of stress due to hot ambience in donkeys from arid tracts in India. *J. Stress Physiol. Biochem.* **2010**, *6*, 12–17.
19. Kataria, N.; Kataria, A.K. Can prolactin be a measurable marker of stress in dromedaries? *Slov. Vet. Res.* **2010**, *47*, 133–138.
20. Yayou, K.-I.; Ito, S.; Yamamoto, N.; Kitagawa, S.; Okamura, H. Relationships of stress responses with plasma oxytocin and prolactin in heifer calves. *Physiol. Behav.* **2010**, *99*, 362–369. [CrossRef] [PubMed]
21. Kataria, N.; Kataria, A.K. Use of prolactin as an indicator of stress in Marwari sheep from arid tracts in India. *Comp. Clin. Pathol.* **2011**, *20*, 333–336. [CrossRef]
22. Concannon, P.W. Parturition and Lactation in the Bitch: Serum Progesterone, Cortisol and Prolactin. *Boil. Reprod.* **1978**, *19*, 1113–1118. [CrossRef] [PubMed]
23. Lightman, S.L. Alterations in Hypothalamic- Pituitary Responsiveness during Lactation. *Ann. N. Y. Acad. Sci.* **1992**, *652*, 340–346. [CrossRef] [PubMed]
24. Grattan, D.R. Behavioural significance of prolactin signalling in the central nervous system during pregnancy and lactation. *Reproduction* **2002**, *123*, 497–506. [CrossRef] [PubMed]
25. Pageat, P.; Lafont, C.; Falewee, C.; Bonnafous, L.; Gaultier, E.; Silliart, B. An evaluation of serum prolactin in anxious dogs and response to treatment with selegiline or fluoxetine. *Appl. Anim. Behav. Sci.* **2007**, *105*, 342–350. [CrossRef]
26. Mengoli, M.; Mendonça, T.; Oliva, J.L.; Bienboire-Frosini, C.; Chabaud, C.; Codecasa, E.; Pageat, P. Do assistance dogs show work overload? Canine blood prolactin as a clinical parameter to detect chronic stress-related response. In Proceedings of the 11th International Veterinary Behaviour Meeting, Samorin, Slovakia, 14–16 September 2017.
27. Liu, J.; Duan, Y. Saliva: A potential media for disease diagnostics and monitoring. *Oral Oncol.* **2012**, *48*, 569–577. [CrossRef] [PubMed]
28. Williamson, S.; Munro, C.; Pickler, R.; Grap, M.J.; Elswick, R.K. Comparison of Biomarkers in Blood and Saliva in Healthy Adults. *Nurs. Res. Pr.* **2012**, *2012*, 1–4. [CrossRef] [PubMed]
29. Korot'ko, G.F.; Gotovtseva, L.P. Pituitary, adrenal, and sex hormones in saliva. *Hum. Physiol.* **2002**, *28*, 375–377. [CrossRef]
30. Horvat-Gordon, M.; Granger, D.A.; Schwartz, E.B.; Nelson, V.J.; Kivlighan, K.T. Oxytocin is not a valid biomarker when measured in saliva by immunoassay. *Physiol. Behav.* **2005**, *84*, 445–448. [CrossRef]
31. Thijssen, J.H.; Van Goozen, S.H.; Van Engeland, H.; Matute, L.M.; A Blankenstein, M. None of four commercially available assays detects prolactin in human saliva. *Clin. Chem.* **2000**, *46*, 1409–1410.
32. Lindell, S.G.; Suomi, S.J.; Shoaf, S.; Higley, J.; Linnoila, M. Salivary prolactin as a marker for central serotonin turnover. *Boil. Psychiatry* **1999**, *46*, 568–572. [CrossRef]
33. Kooistra, H.; Okkens, A. Secretion of Prolactin and Growth Hormone in Relation to Ovarian Activity in the Dog. *Reprod. Domest. Anim.* **2001**, *36*, 115–119. [CrossRef] [PubMed]

34. Gräf, K.-J. Serum oestrogen, progesterone and prolactin concentrations in cyclic, pregnant and lactating beagle dogs. *Reproduction* **1978**, *52*, 9–14. [CrossRef]
35. Kooistra, H.; Okkens, A. Secretion of growth hormone and prolactin during progression of the luteal phase in healthy dogs: a review. *Mol. Cell. Endocrinol.* **2002**, *197*, 167–172. [CrossRef]
36. Corrada, Y.; Rimoldi, I.; Arreseigor, S.; Marecco, G.; Gobello, C. Prolactin reference range and pulsatility in male dogs. *Theriogenology* **2006**, *66*, 1599–1602. [CrossRef]
37. Urhausen, C.; Seefeldt, A.; Eschricht, F.M.; Koch, A.; Hoppen, H.O.; Beyerbach, M.; Möhrke, C.; Dieleman, S.J.; Günzel-Apel, A.R. Concentrations of prolactin, LH, testosterone, TSH and thyroxine in normospermic dogs of different breeds. *Reprod. Domest. Anim.* **2009**, *44*, 279–282. [CrossRef] [PubMed]
38. Gobello, C.; Colombani, M.; Scaglia, H.; De La Sota, R.L.; Goya, R.G. Heterogeneity of circulating prolactin in the bitch. *Reprod. Nutr. Dev.* **2001**, *41*, 505–511. [CrossRef]
39. Gräf, K.-J.; Friedreich, E.; Matthes, S.; Hasan, S.H.; Gräf, K.-J. Homologous *radioimmunoassay for* canine prolactin and its application in various physiological states. *J. Endocrinol.* **1977**, *75*, 93–103. [CrossRef]
40. Poll, E.-M.; Kreitschmann-Andermahr, I.; Langejuergen, Y.; Stanzel, S.; Gilsbach, J.M.; Gressner, A.; Yagmur, E. Saliva collection method affects predictability of serum cortisol. *Clin. Chim. Acta* **2007**, *382*, 15–19. [CrossRef]
41. Aufricht, C.; Tenner, W.; Salzer, H.R.; E Khoss, A.; Wurst, E.; Herkner, K. Salivary IgA concentration is influenced by the saliva collection method. *Eur. J. Clin. Chem. Clin. Biochem. J. Forum Eur. Clin. Chem. Soc.* **1992**, *30*, 81–83.
42. Gallagher, P.; Leitch, M.M.; Massey, A.E.; McAllister-Williams, R.H.; Young, A.H. Assessing cortisol and dehydroepiandrosterone (DHEA) in saliva: effects of collection method. *J. Psychopharmacol.* **2006**, *20*, 643–649. [CrossRef] [PubMed]
43. Bienboire-Frosini, C.; Chabaud, C.; Cozzi, A.; Pageat, P. Is equine prolactin measurable using a commercially available canine prolactin ELISA kit? In Proceedings of the AWSELVA-ECAWBM-ESVCE Congress, Bristol, UK, 30 September–3 October 2015.

animals

MDPI

Article

The Perceived Value of Behavioural Traits in Australian Livestock Herding Dogs Varies with the Operational Context

Jonathan Early [1,*], Elizabeth Arnott [1], Bethany Wilson [1], Claire Wade [2] and Paul McGreevy [1]

1 Sydney School of Veterinary Science, Faculty of Science, University of Sydney, Sydney, NSW 2006, Australia
2 Sydney School of Life and Environmental Sciences, Faculty of Science, University of Sydney, Sydney, NSW 2006, Australia
* Correspondence: jonathan.early@sydney.edu.au

Received: 25 June 2019; Accepted: 11 July 2019; Published: 16 July 2019

Simple Summary: Information on Australian livestock herding dogs and their handlers and breeders is limited. This study aimed to collate baseline information on how handlers and breeders value various behavioural traits relevant to the work of these dogs. A survey was presented to explore herding dog behaviour in four contexts including work and competition. The behavioural traits were divided into three groups: working manoeuvres, working attributes and general attributes. Data from 811 respondents revealed that several behavioural traits were of high and low value to handlers and breeders across all contexts, while others were unique to only one or two contexts. For example, *cast*, *force*, *gather*, *trainable*, *confident* and *friendly* were of most value; whereas *bite*, *bark* and *back* were of less value. Further analysis revealed that respondents can be considered as coming from two main groups: firstly, handlers with a preference for specialised dogs in the utility context and, secondly, handlers focussed on the yard context, who need dogs that have a broad range of skills and that are easy to work with. This information may assist in matching handlers with suitable dogs. Future research should clarify handlers' understanding of innate and learnt behaviours.

Abstract: This study investigated the value that handlers and breeders assign to various behavioural traits in Australian livestock herding dogs. Data were obtained from 811 handlers and breeders through the 'Australian Farm Dog Survey'. Respondents were asked to consider dogs within four contexts: utility (livestock herding in both paddocks and yards), mustering (livestock herding in paddocks and along livestock routes), yards (in and around sheds, sale-yards and transport vehicles), and trial (specifically a standard 3-sheep trial), and to rate the value of 16 working manoeuvres (movement sequences used in herding), 11 working attributes (skills or attributes used in herding) and five general attributes (personality traits ascribed to an individual dog). The most valued working manoeuvres were *cast*, *force* and *gather*. *Bite*, *bark* and *backing* were considered of little value in certain contexts, notably the trial context. Across all four contexts, the general attributes most valued in dogs were being *trainable*, *motivated*, *confident* and *friendly*, while *control* and *trainability* were the working attribute traits considered to be of most value. *Excitability* was revealed to be a 'Goldilocks' trait in that respondents preferred not too much or too little but a 'just right' amount in their preferred dog. Analysis indicated a handler preference for either specialised dogs for the utility context or dogs who are easy to work with because of a broad range of traits favoured in the yard context. These results reveal both generalities across and the need for specialisation within these four herding contexts. Further investigation may help to reveal how well handlers distinguish between innate and learnt behaviours when selecting and training livestock herding dogs. Identifying which group handlers fit into optimally may assist in selecting suitable dog–human dyads.

Keywords: herding; livestock; working dog; survey; traits; boldness

1. Introduction

The national population of working livestock herding dogs in Australia has been estimated at more than 94,000 individuals [1]. Over an average working career, the estimated economic value of these dogs' work is over A$40,000 per individual [2]. Successful partnerships between dog and handler reflect the quality of the match between the personality and behaviour profile of the dog and the preferences, experience and skills of its handler as well as the perceived financial value of the dog [3].

Livestock herding dogs are routinely used to move livestock in three over-arching contexts that are also used to label specific working skill-sets: utility (both paddock and yard), mustering (paddock and livestock routes) and yard (in and around sheds, sale-yards and transport vehicles). They are selected primarily for performance and health rather than morphological traits [4], an approach that has resulted in the prevalence of a suite of behaviours thought to be stylised elements of the predatory sequence exhibited by *Canis lupus familiaris* [5–7].

Natural instinct, combined with opportunities to regularly practise and be reinforced for herding behaviours, is fundamental to any dog's performance in a herding task. The unique triadic interaction of humans, dogs, livestock and sometimes handlers on horseback has been referred to as a 'mutually adjusted system' [8]. Insufficient or poor quality training may jeopardise dog and livestock welfare and compromise learning outcomes [9–11]. Investigations into handler–dog interactions during livestock herding training have focussed on moderating access to livestock through negative punishment (interrupting access to livestock) or positive reinforcement (allowing continued access to livestock) [5,12]. These operant techniques reveal the reinforcing value of access to livestock in dogs that have been selected to relish this sort of work [13].

Data on the ease or difficulty with which livestock herding dog handlers can condition dogs to perform certain working behaviour traits may reveal areas in which trainer education may be especially beneficial. They may also identify which traits deserve a particular focus in breeding and selection to ensure livestock herding dogs can perform the task for which they are being bred.

Peer-reviewed studies on the behaviour of livestock herding dogs are rare (*see* [5,8,12,14–16]) and, among them, few are easily transferable to the Australian context. Importantly, only two studies [5,8] defined the dogs used in their studies as working dogs or from working dog lines, rather than companion dogs of herding breeds. One Australian study, which was not subject to peer-review, examined the inheritance of the behavioural trait *eye* (commonly defined as a dog's ability to hold sheep together by staring at them) [17]. It reported that, when using a six-point scale to score *eye*, the 28 dogs tested were most likely to be scored as intermediate to or aligned with one of their parent's scores.

Popular literature on livestock herding in Australia suggests that certain behavioural traits such as *eye*, *force*, *boldness*, *anticipation* and *cast* are pivotal to successful herding ability [18,19]. However, the definition, interpretation, perceived value and relevance of these and other traits varies among authors [17,18,20–25]. For example, a recent study of eight Australian herding manuals identified a significant discordance in the frequency of the use of such popular terms [26].

The current study used a questionnaire to identify the ideal position for a livestock herding dog on the shyness–boldness continuum, the value that handlers place in the ideal dog on the expression of five general attributes; 11 working attributes; 16 working manoeuvres and the ease with which dogs can be trained to show each of these in four distinct herding contexts. A central hypothesis for the current survey study was that respondents would, having some knowledge of innate versus learnt behaviours, report innate behaviours as being more difficult to train.

2. Materials and Methods

The 'Australian Farm Dog Survey' was designed to investigate the distribution of farm dogs in Australia, their usage, their management and the views of their owners along with demographic information relating to the breeder/handler (for other publications using this survey *see* [2,3,27]). In particular, respondents were asked about five general attributes, 11 working attributes and

16 working manoeuvres in their dogs. The questions were grouped within four herding contexts: utility, mustering, yard, and competition (3-sheep trial).

Prior to publication of the survey, popular working-dog training manuals were consulted and advice was sought from members of the Working Kelpie Council of Australia to ensure that the terminology in the survey was appropriate for the target audience. A pilot distribution of the survey to 125 participants led to some minor modifications prior to widespread distribution.

The online version of the survey was available over a three-month period from 10 March to 10 June 2013. All promotional materials relating to the survey indicated that a hard copy of the survey with a reply-paid envelope would be provided to participants if they requested one by telephone. Approval for this study was granted from the University of Sydney Human Research Ethics Committee (Approval number 15474).

The target population for the survey was livestock herding dog users across Australia. Participation was encouraged with entry into a prize draw to win commercial working-dog food at the end of the survey period. An introductory message gave participants the option to respond anonymously with an assurance of confidentiality were they to choose to leave their details to enter the prize draw.

A link to the online survey was posted on the websites of the University of Sydney [28] Meat and Livestock Australia [29] and the Working Kelpie Council of Australia (WKCA) [30]. It was advertised through stories in multiple regional newspapers, on three nation-wide television programs and in two national agricultural magazines. The committee of the 2013 Casterton Kelpie Auction (one of Australia's leading livestock herding dog auction events) promoted the survey in a mail-out to past and current vendors and purchasers. The researchers also recruited survey participants, in person, at livestock herding dog trials during the study period.

The online version of the survey was constructed using the survey system QSmart (Torque Management Systems Limited, Auckland, New Zealand). The entire questionnaire had a maximum of 143 items assigned to 10 sections. However, participants needed to answer fewer questions if they responded in the negative to questions about certain activities, such as breeding or trialling of dogs. The logic system of the online survey permitted the redirection of participants to questions of relevance. (To view the complete survey, *see* [31].

In Early et al. [26], we defined working manoeuvres, working attributes (referred then as working skills) and general attributes: working manoeuvres represent a sequence of movements used in herding; working attributes reflect an ability used in herding; general attributes are personality traits ascribed to an individual. Where a trait might fit into both working and general attributes or might lie on the same spectrum (e.g., *boldness* and *cautiousness*), the authors made a decision into which category it should be included, based on whether testing and exploring these relationships would identify, statistically, if they are valued in different ways.

Respondents were asked to indicate the ease with which 16 working manoeuvres in the typical working dog can be trained: *cast*, *force*, *gathering*, *cover*, *backing*, *bark*, *bite*, *heading*, *hold*, *balance*, *drive*, *break*, *width*, *pull*, *lift* and *draw*. They answered using a semantic differential-type 5-point rating scale. Descriptive phrases 'extremely easy' and 'almost impossible' were used at either extreme of the scale. Respondents were advised to not provide a rating for any working manoeuvre terms that they were unfamiliar with.

The same 16 working manoeuvres were used again in the next question, which asked the respondents to indicate how valuable they considered these behaviours in livestock herding dogs. Respondents were asked to answer this question separately for up to four herding contexts in which they had experience. These included three types of work, namely, utility (generally known among trainers as all-round), mustering, and yard, and one competition context referred to as trial (i.e., working trials, generally known among trainers as arena or 3-sheep trials). Answer options included a semantic differential-type 5-point rating scale, ranging from 'no value' to 'highly valuable'. Respondents unfamiliar with any working manoeuvre terms were advised to not provide a rating.

The handlers were asked to score the value of 11 working attributes—*shows eye*, *control*, *initiative* (which we described as including working independently, *keenness* and willingness), *trainable* (including tractable), *intelligent* (including sagacious, brainy and clever), *calm*, *firmness* (including strength and power), *style of work* (including width), *physical suitability* (including stamina and durability), *anticipation* and *boldness*. These attributes' values were recorded for each working and competition environment within which the respondent had handled livestock herding dogs. Answer options included a semantic differential-type 5-point rating scale. Descriptive phrases 'no value' to 'extremely valuable' were used at either end of the scale. Respondents unfamiliar with any working attribute terms were advised to not provide a rating.

Respondents were asked to indicate the degree of expression of five general attributes they would expect to be present in the ideal dog for the working and competition environments in which they had experience. These attributes were *excitability*, *trainability*, *motivation and confidence*, *friendliness* and *cautiousness*. They were drawn, and slightly modified from, the "Big Five" personality traits identified by Ley et al. [32]. Respondents answered using a semantic differential-type 5-point rating scale ranging from 'none' to 'a very high degree' at each end point with 'a moderate degree' at the midway point. Respondents unfamiliar with any general attribute terms were advised to not provide a rating.

The final question asked handlers to indicate, on a 100-point visual analogue scale, the balance of shyness–boldness expression they would expect the ideal dog to exhibit for each herding context in which they had experience, where zero was maximum shyness and 100 was maximum boldness. Specifically, they were asked: *Please indicate, by moving the sliding scale/marking on the scale, the balance of shyness and boldness that the ideal dog would exhibit.* As a guide, the descriptive phrases 'extremely shy' and 'extremely bold' were used at either extreme of the scale. Respondents unfamiliar with the concept of shyness–boldness expression were advised to not provide a rating.

Genstat Version 16 (VSN International, Hemel Hempstead, UK) was used for statistical analysis of shyness–boldness expression results. REML analysis was performed using the variate means of shyness–boldness expression for each herding context: utility, mustering, yard and trial. This permitted assessment of whether differences in ideal shyness–boldness expression across the four contexts were significant.

Individual handler optima on the shyness–boldness expression in the ideal dog for each of the four herding contexts were gathered and descriptive statistics collated. The same approach was taken to the amount of five general attributes in the ideal dog; the 11 working attributes in their dogs; the 16 working manoeuvres and the training ease of each working manoeuvre.

To explore the influence of context on handler preferences, a hierarchical cluster analysis based on Gower distance (as all variables in this study were ordinal; this corresponded to Manhattan distance) was conducted using the hclust function of the R statistical and computing software [33].

A Pearson's chi-squared test was performed to assess the significance of the results.

3. Results

3.1. Respondent Demography

Of the 812 livestock herding dog handlers and breeders who completed the survey, 563 were male and 249 were female. Most respondents were aged 50–59 years (n = 213), followed by 60–70 years (n = 182), 40–49 years (n = 165), 30–39 years (n = 120), 20–29 years (n = 120), and over 70 years (n = 44). There were more respondents who did not breed dogs (non-breeders, n = 451) than who did (breeders, n = 361). More than half of respondents had acquired knowledge of dog training beyond 'on-the-job' experience (experience only, n = 302; further education, n = 510), but only 19 of these respondents had completed a certified course. The remainder reported having attended dog-training schools and/or read dog-training books.

3.2. Herding Context Experience

Among the 811 livestock herding dog handlers and breeders who selected having experience in one or more of the four herding contexts, the following contexts (and combinations of contexts) were selected: mustering/yard/utility (n = 241), mustering/yard (n = 169), utility (n = 125), mustering/yard/trial/utility (n = 96), mustering (n = 85), mustering/utility (n = 33), mustering/yard/trial (n = 14), yard (n = 13), mustering/trial (n = 10), mustering/trial/utility (n = 10), trial (n = 5), yard/utility (n = 4), trial/utility (n = 3), yard/trial (n = 2), and yard/trial/utility (n = 1). Totals for each herding context were: mustering (n = 658), yard (n = 540), utility (n = 513) and trial (n = 141).

3.3. Preferred Balance of Shyness–Boldness Expression in the Ideal Dog

The highest boldness expression in the ideal dog selected by respondents was in the yard context, followed by utility, mustering and trial (means of 79.12, 72.11, 69.77, and 65.43, respectively). Between contexts, differences in mean value were statistically significant between trial and mustering (p = 0.004), mustering and utility (p = 0.006), highly significant between utility and trial (p < 0.001) and yard and all other contexts (mustering/trial/utility, p < 0.001) (see Figure 1).

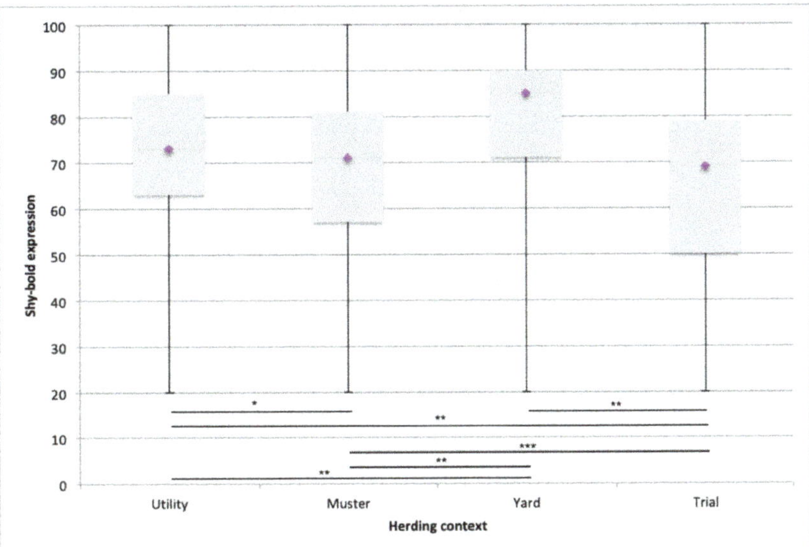

Figure 1. Boxplot of shyness–boldness expression in the ideal dog per herding context: the box spans the interquartile range of the values; middle 50% of the values lie within the box with a diamond indicating the mean. The whiskers extend beyond the box to represent the range of the data. Respondents marked shyness–boldness expression on a visual scale that used the descriptive phrases 'extremely shy' and 'extremely bold' at either end. * p = 0.006 ** p < 0.001 *** p = 0.004.

3.4. Amount in the Ideal Dog

3.4.1. Amount of General Attributes in the Ideal Dog for Utility Work

For the utility context, most respondents selected 'a very high degree' for the general attributes of being *trainable* (n = 351; out of 492) and *motivation and confidence* (n = 320; out of 491). Meanwhile, over half the respondents selected 'a moderate degree' (midway on the five-point scale) for *excitability* (n = 258; out of 491) and *cautiousness* (n = 245; out of 485) (see Figure 2).

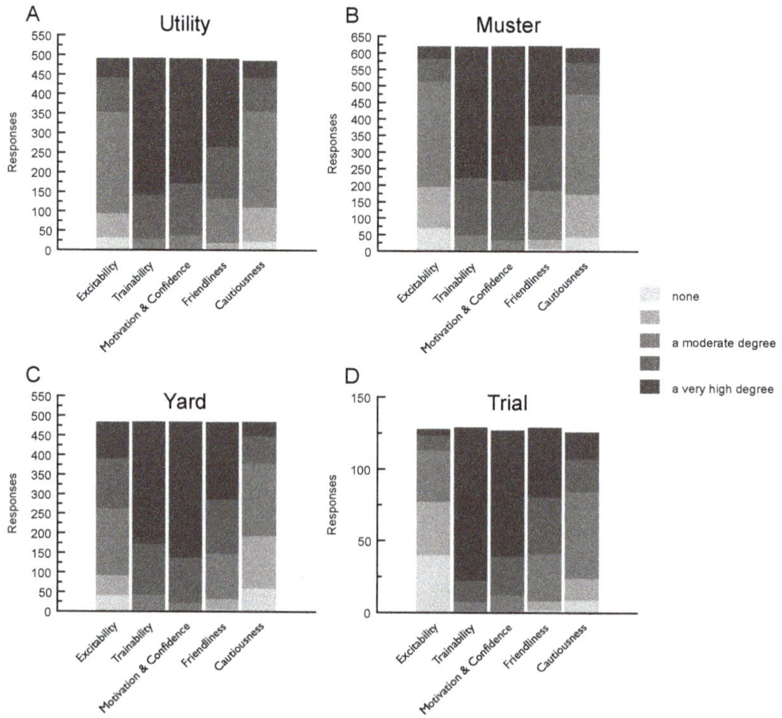

Figure 2. Amount of five (5) general attributes in the ideal dog across the four herding contexts. Respondents' ratings: **A**—utility; **B**—mustering; **C**—yard; **D**—trial.

3.4.2. Amount of General Attributes in the Ideal Dog for Mustering Work

For the mustering context, similar to the utility context, most respondents selected 'a very high degree' for *trainable* ($n = 397$; out of 619) and *motivation and confidence* ($n = 406$; out of 621). In comparison, for *excitability* ($n = 317$; out of 620) and *cautiousness* ($n = 302$; out of 616), no respondent selected more than 'a moderate degree' in the ideal dog (see Figure 2).

3.4.3. Amount of General Attributes in the Ideal Dog for Yard Work

Most respondents who supplied data for the yard context, similar to both utility and mustering contexts, selected 'a very high degree' for *trainable* ($n = 313$; out of 485) and *motivation and confidence* ($n = 348$; out of 485). Unlike the utility and mustering context results, respondents selected increased amounts of *excitability* and similar *cautiousness* in the ideal dog (see Figure 2).

3.4.4. Amount of General Attributes in the Ideal Dog for Herding Trials

In the trial context, 107 out of 129 respondents selected 'a very high degree' for *trainability*. Additionally, more than half of respondents selected 'a very high degree' for *motivation and confidence* ($n = 88$; out of 127). *Excitability* was considered least useful in the trial context with 'none' selected relatively more by respondents ($n = 40$; out of 128) than the working contexts: mustering ($n = 70$; out of 620), yard ($n = 41$; out of 484) and utility ($n = 32$; out of 491) (see Figure 2).

3.5. Value of Working Attributes

3.5.1. Value of Working Attributes—Utility

Respondents scored five working attributes as 'extremely valuable': *control* (n = 355; out of 490), *intelligent* (n = 350; out of 491), *initiative* (n = 349; out of 491), *trainable* (n = 347; out of 490), and *physical suitability* (n = 343; out of 490). *Style of work* received the fewest ratings of 'extremely valuable' overall (n = 186; out of 461). Most respondents selected one of the two highest ratings for *boldness* ('extremely valuable' n = 208, followed by the next point on the scale (unlabeled) n = 191; out of 485) (see Figure 3).

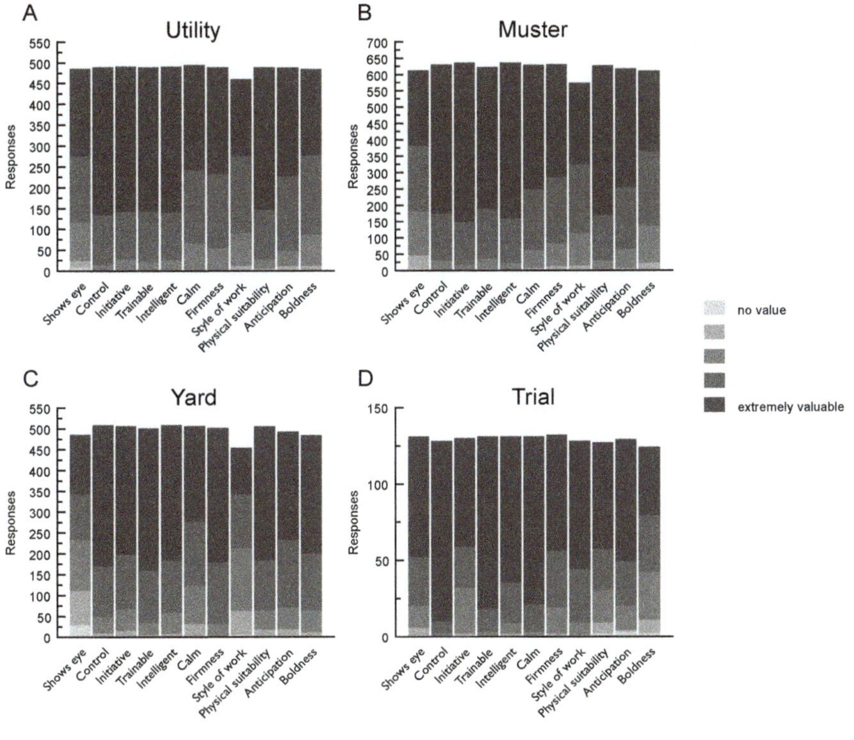

Figure 3. The value of eleven (11) working attributes across the four herding contexts. Respondents' ratings: **A**—utility; **B**—mustering; **C**—yard; **D**—trial.

3.5.2. Value of Working Attributes—Mustering

Initiative and *Intelligence* were considered 'extremely valuable' by 489 and 476 (both out of 636) respondents. Of the 611 respondents who assigned scores to this context for *boldness*, 475 selected the two highest ratings (see Figure 3).

3.5.3. Value of Working Attributes—Yard

Respondents considered the most valuable working attributes in the yard context *trainable* ('extremely valuable' n = 342; out of 501), *control* ('extremely valuable' n = 339; out of 509), *intelligence* ('extremely valuable' n = 326; out of 509), *firmness* ('extremely valuable' n = 323; out of 502) and *physical suitability* ('extremely valuable' n = 323; out of 506). They considered *eye* and *style of work* as being of least value, with responses more evenly spread across the ordinal scale compared to all other working attributes (see Figure 3).

3.5.4. Value of Working Attributes—Trial

Control (n = 118; out of 130), *trainable* (n = 113; out of 131) and *calm* (n = 110; out of 131) were scored as 'extremely valuable' by most respondents when describing the ideal trial dog. *Boldness* was rated the least valuable of the working attributes assessed (see Figure 3).

3.6. Value of Working Manoeuvres

3.6.1. Value of Working Manoeuvres—Utility

Respondents describing the ideal utility dog considered *cast* to be of highest value ('highly valuable' n = 365; out of 485 responses), and *bite* received the highest score for 'no value' ('no value' n = 177; out of 456). *Break, width, pull, lift* and *draw* received fewer than 300 responses, likely indicating that knowledge of these terms was limited to a sub-set of survey respondents (see Figure 4).

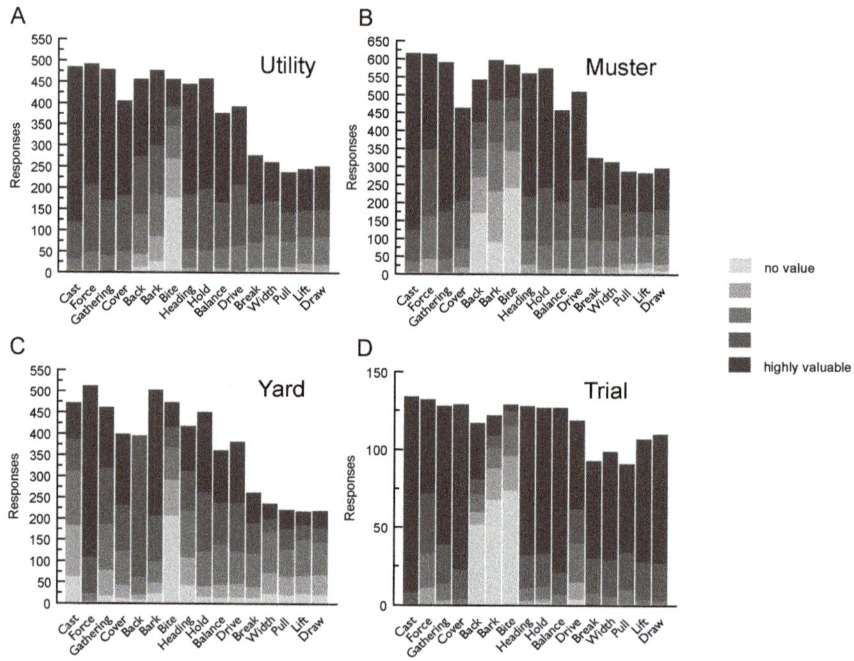

Figure 4. The value of sixteen (16) working manoeuvres across the four herding contexts: Respondent's ratings: **A**—utility; **B**—mustering; **C**—yard; **D**—trial.

3.6.2. Value of Working Manoeuvres—Mustering

Six working manoeuvres were considered 'highly valuable' by most respondents in the mustering context: *Cast* (n = 492; out of 616), *gathering* (n = 415; out of 591), *cover* (n = 259; out of 463), *heading* (n = 344; out of 561), *hold* (n = 333; out of 575) and *balance* (n = 255; out of 458). *Bite* was considered to be of 'no value' (n = 242) or of limited value (n = 102) by most respondents (total n = 584). Responses on the value of *backing* were spread across the five-point scale ('extremely valuable' n = 117, n = 77, n = 77, n = 100, 'no value' n = 172; out of 543) (see Figure 4).

3.6.3. Value of Working Manoeuvres—Yard

In the yard context, *force* and *bark* were considered 'highly valuable' (n = 404; out of 513, n = 295; out of 503) by most respondents. No one considered *back* 'highly valuable' (n = 0; out of 395). *Bite* was scored as having 'no value' by nearly half of respondents (n = 208; out of 474). Similar to the mustering context results, *break*, *width*, *pull*, *lift* and *draw* each received fewer than 262 responses, compared to more than 361 each for the other traits (see Figure 4).

3.6.4. Value of Working Manoeuvres—Trial

The highest value traits in the trial context were *cast* ('highly valuable' n = 126; out of 134), *balance* ('highly valuable' n = 106; out of 127), *cover* ('highly valuable' n = 106; out of 128), *heading* ('highly valuable' n = 96; out of 128) and *hold* ('highly valuable' n = 94; out of 127). Most respondents scored *bite* ('no value' n = 74; out of 129) and *bark* ('no value' n = 68; out of 122) as 'no value' (see Figure 4).

3.7. Trainability of Working Manoeuvres

Among the 16 working manoeuvres, all but two traits were scored by respondents at the midway point between 'extremely easy' and 'almost impossible'. These were *force* (first two points at the end of the scale including 'extremely easy' n = 387; out of 748) and *heading* (first two points at the end of the scale including 'extremely easy' n = 371; out of 684), indicating respondents' possible awareness of these traits being innate behaviours. *Break* (n = 435), *width* (n = 397), *pull* (n = 366), *lift* (n = 361) and *draw* (n = 363) received fewer than 435 responses. For all the other traits measured, there were at least 569 responses each: *Cast* (n = 741), *force* (n = 748), *gather* (n = 730), *cover* (n = 588), *back* (n = 679), *bark* (n = 691), *bite* (n = 604), *heading* (n = 684), *hold* (n = 708), *balance* (n = 569), *drive* (n = 650) (see Figure 5).

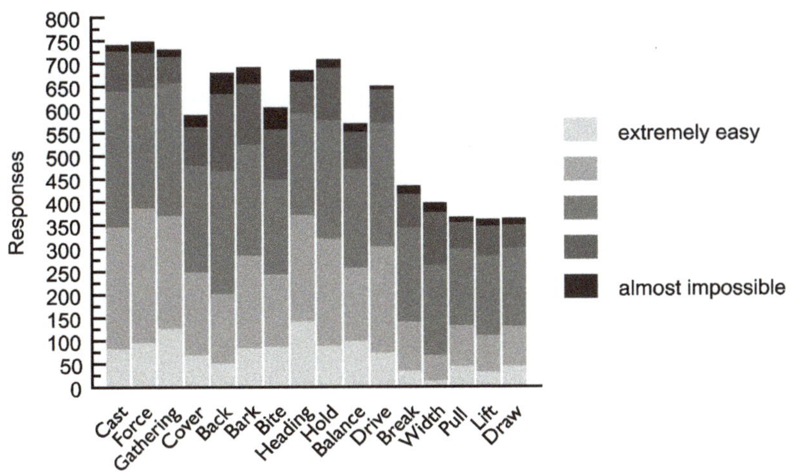

Figure 5. Reported ease of training for sixteen (16) working manoeuvres.

3.8. Cluster Analysis

The hierarchical cluster analysis resulted in three groups: Group Three (most common preference pattern), Group Two (the smallest preference pattern) and Group One (intermediate to Groups Two and Three) (see Figure 6).

Group One owners prioritised cast, force, gathering, cover, heading, hold, balance, firmness, calmness, intelligence, trainability, initiative, control, anticipation, physical suitability and confidence (see Figure 7).

Group Two owners prioritised *cast*, *gathering*, *cover*, *heading*, *control* and *intelligence*. *Back*, *bark* and *excitability* were clearly not preferred in this group (see Figure 7).

Group Three owners prioritised a more balanced approach to the value and amount of each trait. They were less concerned with *draw*, *lift*, *pull*, *width* and *break* (see Figure 7).

Across all three groups, *bite* was consistently less in demand than the other traits analysed.

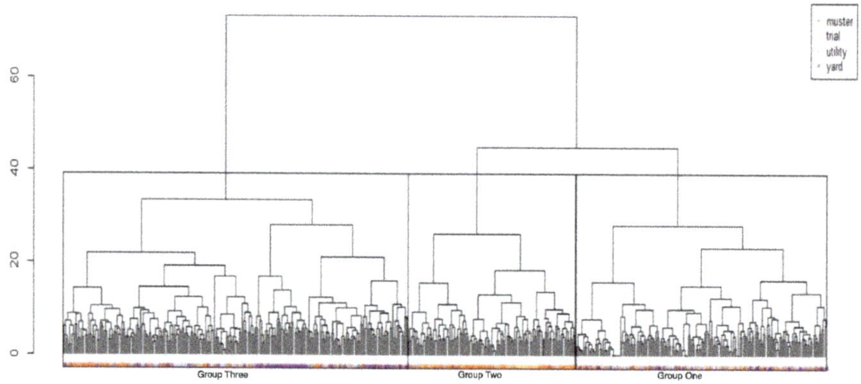

Figure 6. Cluster dendogram showing hierarchical, agglomerative clustering of Euclidean Distances of preference scores for analysed traits. The type of work or competition for which a trait was favoured by respondents is indicated by: Green = utility, Red = mustering, Blue = yard, Yellow = trial. Dendogram shows, from left to right, Group Three, Group Two, Group One.

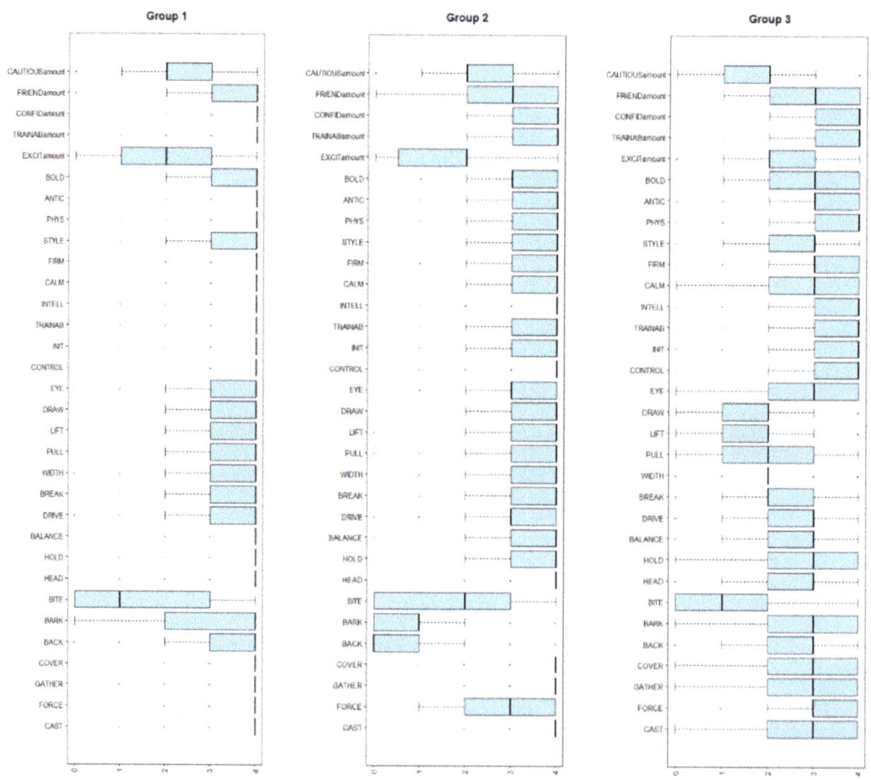

Figure 7. Cluster group preferences by trait: Boxplot demonstrating the different medians, interquartile ranges and whisker lengths of preferences for each trait in the three clusters of respondents shown in Figure 6.

When the three groups were analysed across the four herding contexts (see Figure 8), using Pearson's chi-square test (chi-squared = 136.21, df = 6), highly significant ($p < 0.01$) preferences were apparent for particular contexts. Group One's trait preferences were overrepresented for utility and underrepresented for mustering. Group Two's trait preferences were overrepresented for mustering and trial, while underrepresented for yard and utility. Group Three's preferences were overrepresented for yard and underrepresented for mustering and trial.

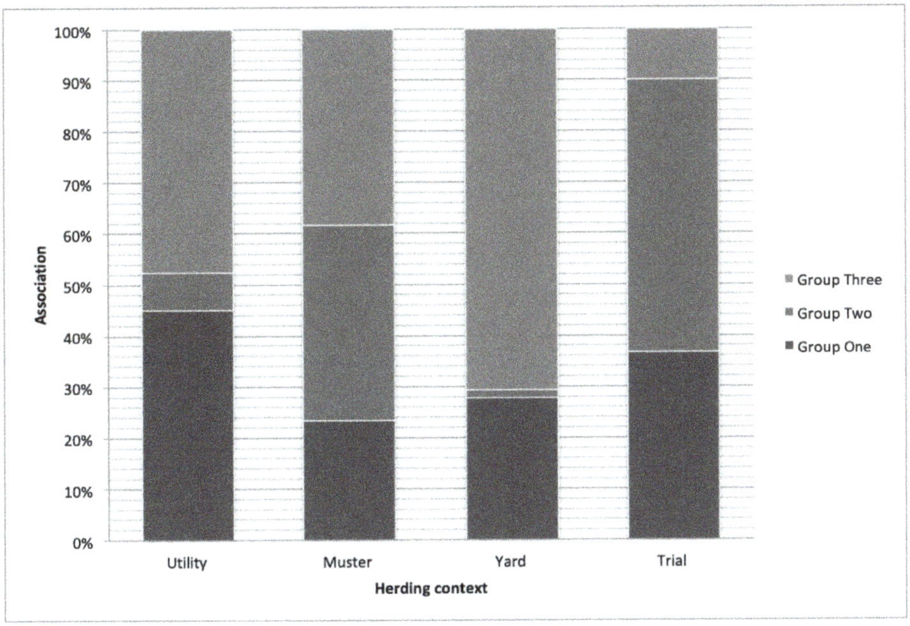

Figure 8. Associations of each cluster group with working and competition contexts. Each group would be expected to have equal representation across each herding context; the chart indicates which group's preferences were over or underrepresented for each herding context.

4. Discussion

Although surveys of owners and/or experts have been previously used to develop behavioural profiles of companion dog breeds [34,35], to our knowledge, this is the first to assess the relative value of personality and working traits in livestock herding dogs.

By identifying the most valuable working and personality traits across multiple herding contexts, the current results help to show how these traits influence successful movement of livestock while also identifying traits that could enhance both context-specific and general breeding programs. Genetic analysis from the hunting dog sector in Sweden has shown that, for at least six traits, if Best Linear Unbiased Prediction breeding values were used instead of phenotype, genetic gain would be 89% higher [36]. With the sequencing of the canine genome [37], molecular genetics provides the opportunity to identify suitable livestock herding dogs at an earlier age than behavioural assessment currently offers.

Of the four herding contexts surveyed, mustering was selected as the context in which most respondents had experience with handling livestock herding dogs. This was followed by yard and utility, with only a small number of respondents involved in the trial (competition) environment. These distributions reflect prevalent Australian working conditions in that livestock are often collected over large areas of farmland prior to being managed within yards. Whether this Australian distribution of potential respondents and the current resultant data are similar in other large-scale livestock-producing countries requires further investigation.

4.1. Shy–Bold

The ideal reported balance of shyness–boldness expression significantly differed across the four herding contexts. Livestock herding dogs working in the yard context were considered by respondents to require a higher level of *boldness* than in the utility and mustering contexts, followed by the trial context. However, the middle 50% of responses overlapped between each context. This finding reflects the necessary blend of confidence, motivation, composure and resilience that presumably make up the term boldness, the so-called super trait that yard dogs require to be successful in this close-up, threatening, high risk environment. The expression of *boldness* appears to be a simple way for respondents to differentiate between the type of dog required between each herding context.

4.2. General Attributes

For general attributes, handlers and breeders across all herding contexts reported that the ideal dog has a high degree of *trainability, motivation and confidence* and *friendliness*. The cluster analysis also identified similarities across the three groups for the amount preferred of these traits. The results for *excitability* highlighted one of the more interesting findings for the general attributes group. Across all contexts and the three cluster analysis groups, respondents identified this as the 'Goldilocks' trait—not too little, not too much, but 'just right'; the term Goldilocks principle or effect is referred to in other research fields including economics and education to describe contexts that seek balance [38,39]. Finding a balance for *excitability* is not unique to livestock herding dogs, for example it is also noted in guide dogs [40]. However, improvements in selective breeding for this trait may assist future research about behavioural tendencies being undertaken among service dogs [41,42].

Respondents' preference for a dog that is easy to work with, through long hours, and often as the handler's only companion in conducting their work has implications for the welfare of these dogs. Specifically, livestock herding dogs that do not meet handlers' expectations of being 'trainable, motivated, confident and friendly' could be at more risk of becoming so-called behavioural wastage (being discarded from the industry because of poor performance related to behaviour, rather than physical inadequacy) [43]. 'A moderate degree' of *excitability* and *cautiousness* was preferred, reflecting respondents' awareness of the potential for high levels of these traits to compromise the successful working performance of a dog. This preference for trait-specific expressions of certain qualities is a complex finding. A common difficulty in breeding or selection of livestock herding dogs is consistency between what one handler or breeder and another considers to be 'the right amount' of a trait expressed in a dog.

4.3. Working Attributes

As far as working attributes are concerned, the current study revealed both similarities and differences between the four herding contexts and confirmed, by the cluster analysis, the key working characteristics preferred and valued by handlers. Across all contexts, the high values assigned to *control* and *trainable* indicate the importance of these traits in allowing handlers and breeders to breed, rear and train the best dogs for each herding context. While dogs working at a distance from their handler are not unique to livestock herding, what is specialised to livestock herding dogs is the concurrent gathering and movement of livestock to the handler. The understanding of respondents that dogs in this context pose a risk to themselves (e.g., placing themselves in dangerous positions leading to injury) and the livestock (e.g., injuring livestock due to poor herding technique) should not be underestimated. For both mustering and utility contexts, respondents reported that dogs with *initiative, intelligence* and *physical suitability* were of most value to them. These traits reflect the complex and demanding nature of work in these contexts and the requirement for handlers to both direct their dogs, when needed, and rely on them to perform independently, as required. In the yard context, most respondents also assigned high value to *firmness* and *physical suitability*. *Boldness*, a personality trait often referred to in the peer-reviewed canine behaviour literature (e.g., [44,45]), was considered of less value than the other working attributes in the utility, mustering and trial contexts, yet was one of the highest value attributes in the yard context. These results reflected those obtained from the shyness–boldness expression question. Similarly, *shows eye*, a core attribute of herding dogs that has clear analogues in the predatory sequence [8], was considered of high value only in the trial context. These apparent anomalies suggest that the demands of herding work peak within the broad yard context, which often requires dogs to move fearful livestock at a close distance, through tight spaces, repeatedly.

4.4. Working Manoeuvres

The value of the 16 working manoeuvres across the four herding contexts revealed some key similarities between the four herding contexts. In the utility, mustering and trial contexts, *cast*, *force*, *gathering*, *cover*, *heading*, *hold*, *balance* and *drive* were consistently of high value to respondents whereas *force* and *bark* were of extreme value to respondents compared to the other manoeuvres. While these results are not surprising, they represent a selection of manoeuvres on which breeders or handlers may focus to ensure their dogs meet expectations when working. Notable exceptions were *bite* and *backing*. Across all four contexts, many respondents assigned 'no value' for *bite* while *backing* was assigned 'no value' by many respondents in the mustering and trial contexts, reflecting their relative unhelpfulness in these contexts. Additionally, throughout the cluster analysis, *bite*, being the trait of least value, revealed awareness among respondents that the ideal dog should show limited expression of this trait.

The current study clearly identified manoeuvres whose reported value differed from one context to the next. For example, *cast* was rated as the most valuable in utility, mustering and trial contexts, but as only of general value in the yard context. Many respondents consistently rated *bite* as having 'no value' across each context.

The cluster analysis provided further insight into respondents' thoughts on working manoeuvres. Among handlers, there appear to be two main groups who require all-rounder type dogs. The first group (Group One) were those who have a clear preference for a select group of traits they need to perform the work with a focus on the utility context. This group also appears to have a keen interest in the skills their dogs need to be successful. The second group (Group Three) were those handlers who want 'jack of all trades' dogs that possess a broad range of working manoeuvres and consider most manoeuvres to have only moderate value. This group reported primarily on the yard context. Further studies are required to differentiate between the first group, who may be experienced students of livestock herding dogs with well-developed ideas of the most valuable traits to have a successful dog, and the second group, who seem to represent the majority of yard-based Australian livestock herding dog handlers who are primarily outcomes-focused. These results have provided an additional layer to this survey's findings: that, while there appears to be a move towards a need for specialist livestock working dogs on the whole, generalities remain and there is still a need for those who primarily work in the yard context to have dogs with a broad, albeit average, skill set.

4.5. Ease of Training Working Manoeuvres

When respondents were asked how easy or difficult it is to train a group of working manoeuvres, the hypothesis was that respondents of the survey would select innate behaviours as being more difficult to train. The results indicate that, overall, the manoeuvres were neither easy nor difficult to train. This may reflect our use of scales that had insufficient granularity to detect real differences between ease of training. Two manoeuvres that appear to be elements of the predatory sequence, *force* and *hold*, were reported by respondents to be easier to train than the other manoeuvres. It is possible that some handlers with a good understanding of livestock herding dog behaviour may find innate behaviours easier to train or shape, while others may credit their own training skills for any apparent ease of training certain innate behaviours. For accomplished trainers, creating scenarios for a livestock herding dog to practice and trigger these innate behaviours may make them relatively easy to train by simply fine-tuning a natural behaviour. In Australia, mean failure rates of herding dogs in training have been estimated to be at least 20% [2]. Improving understanding of the practical application of behavioural science findings as they pertain to livestock herding dogs may boost training outcomes [46]. This approach has met with considerable success in horse training through the nascent discipline of equitation science [47]. Additionally, identifying consistently successful trainers with low failure rates, who can provide assistance to fellow trainers through sharing their knowledge, may also have merit.

While the current results provide some helpful information on the relative value of working and personality traits in the four herding contexts, they do contain some limitations. Among all three non-competition contexts, the manoeuvres *break*, *width*, *pull*, *lift* and *draw* were unfamiliar to many respondents in the working context. This is most probably because these terms are often used in competition and technical literature but rarely in the practical, working contexts. An additional limitation is that the current survey did not explore the specific reasons for why individual manoeuvres were either of high or low value to each respondent or why some respondents did not assign a rating or score to a given quality. While the terms presented in the current survey were identified as those most commonly used in relevant manuals [26], there is scope to further investigate why these working manoeuvres provide value within each herding context.

Furthermore, due to the large number of questions, it was apparent that respondents missed answering parts of questions, developed survey fatigue or were not familiar with some of the terms or traits presented. Improved survey design, including attempts to verify whether respondents deliberately failed to answer part of any given question, may assist in reducing missing data. A further limitation that we acknowledge as a potential source of lost detail among the data arose in the general attributes questions where, for consistency with Ley et al. [32], the personality dimensions *motivation* and *confidence* were grouped together. Because of this, it was not possible to determine whether respondents were selecting just one or both terms in their response. As such, this result should be interpreted with some caution. The question on training ease of the 16 working manoeuvres attempted to tease out respondents' understanding of the difference between innate and learnt behaviours in relation to livestock herding. While the results indicate some variability in understanding by respondents, a published analysis of a different section of results from the same survey, focusing on dogmanship, found a general lack of understanding of learning theory and training principles among the current respondents [27]. Further investigation into survey respondents' understanding of the behavioural underpinnings of these working manoeuvres may assist interpretation of this section of the survey.

There is limited research analysing the effect of sex on behavioural traits in livestock working dogs. While data on sex in relation to the behavioural traits in this study was not collected, *confidence* and *boldness* are examples where sex differences could be explored in future studies. Additionally, sex differences in training ease among livestock working dogs may provide interesting comparisons to other working dog sectors [48].

5. Conclusions

This survey identified preferred levels of boldness and those general attributes, working attributes and working manoeuvres of greatest value to most Australian handlers and breeders across four herding contexts. These results highlight similarities in the attributes and manoeuvres valued across these contexts but also the need for dogs working in an individual context to develop specialised skills. However, there was, among the respondents, a sub-group of handlers with a focus on the yard context, who need dogs with a broad range of skills that are easy to work with. The most valued general attributes were *trainable*, *friendly*, *motivation* and *confidence*. The most valued working attributes were *control* and *trainable* while the most valued working manoeuvres included *cast*, *force*, *cover* and *gathering*. Further investigation is required to explore handlers' understanding of the distinction between innate and learnt behaviours in training livestock herding dogs.

Author Contributions: Conceptualization, J.E., E.A., C.W. and P.M.; Data curation, J.E. and E.A.; Formal analysis, J.E., E.A., B.W., C.W. and P.M.; Funding acquisition C.W. and P.M.; Investigation, J.E., E.A., Claire Wade and Paul McGreevy; Methodology, J.E., E.A., B.W., C.W. and P.M.; Project administration, C.W. and P.M.; Resources, J.E., E.A., C.W. and P.M.; Supervision, C.W. and P.M.; Validation, J.E., E.A. and B.W.; Visualization, J.E. and B.W.; Writing—original draft, J.E.; Writing—review & editing, J.E., E.A., B.W., C.W. and P.M.

Funding: This study was supported by a grant from Agrifutures (previously the Rural Industries Research and Development Corporation) (Grant number, PRJ 007 806), including Meat and Livestock Australia.

Acknowledgments: In-kind support from the Working Kelpie Council of Australia. The authors also thank the 2013 Casterton Kelpie Auction Committee for their support of this study.

Conflicts of Interest: The authors declare no conflict of interest.

References and Notes

1. ACAC. *Contribution of the Pet Care Industry to the Australian Economy*, 7th ed.; Australian Companion Animal Council Inc.: North Sydney, Australia, 2010.
2. Arnott, E.R.; Early, J.B.; Wade, C.M.; McGreevy, P.D. Estimating the economic value of Australian stock herding dogs. *Anim. Welf.* **2014**, *23*, 189–197. [CrossRef]
3. Arnott, E.R.; Early, J.B.; Wade, C.M.; McGreevy, P.D. Environmental factors associated with success rates of Australian stock herding dogs. *PLoS ONE* **2014**, *9*, e104457. [CrossRef] [PubMed]
4. Butler, W. Using Stockdogs For Low Stress Livestock Handling. In *Proceedings: Grazing Behavior of Livestock and Wildlife*; University of Idaho: Moscow, ID, USA, 1999; pp. 23–24.
5. McConnell, P.B.; Baylis, J.R. Interspecific communication in cooperative herding: Acoustic and visual signals from human shepherds and herding dogs. *Z. Tierpsychol.* **1985**, *67*, 302–328. [CrossRef]
6. Coppinger, R.; Coppinger, L. *Dogs: A New Understanding of Canine Origin, Behavior and Evolution*; University of Chicago Press: Chicago, IL, USA, 2002.
7. Coppinger, R.; Schneider, R. Evolution of Working Dogs. In *The Domestic Dog: Its Evolution, Behaviour and Interactions with People*; Serpell, J.A., Ed.; Cambridge University Press: Cambridge, UK, 2008; pp. 21–47.
8. Burns, M. The mutual behaviour of sheep and sheep dogs in Ghana. *Trop. Agric. Trinidad* **1969**, *46*, 91–102.
9. Hiby, E.F.; Rooney, N.J.; Bradshaw, J. Dog training methods: Their use, effectiveness and interaction with behaviour and welfare. *Anim. Welf.* **2004**, *13*, 63–69.
10. Rooney, N.; Gaines, S.; Hiby, E. A practitioner's guide to working dog welfare. *J. Vet. Behav. Clin. Appl. Res.* **2009**, *4*, 127–134. [CrossRef]
11. Rooney, N.J.; Cowan, S. Training methods and owner–dog interactions: Links with dog behaviour and learning ability. *Appl. Anim. Behav. Sci.* **2011**, *132*, 169–177. [CrossRef]
12. Marschark, E.D.; Baenninger, R. Modification of instinctive herding dog behavior using reinforcement and punishment. *Anthrozoos Multidiscip. J. Inter. People Anim.* **2002**, *15*, 51–68. [CrossRef]
13. McGreevy, P.D.; Boakes, R.A. *Carrots and Sticks: Principles of Animal Training*; Darlington Press: Sydney, Australia, 2011.
14. Hoffmann, U.; Hamann, H.; Distl, O. Genetic analysis of markers of the performance test for herding dogs. 1. Performance traits. *Berl. Munch. Tierarztl. Wochenschr.* **2003**, *116*, 81–89.
15. Hoffmann, U.; Hamann, H.; Distl, O. Genetic analysis of markers of the working test for herding dogs 2. Undesired behavior traits. *Berl. Munch. Tierarztl. Wochenschr.* **2003**, *116*, 90–95.
16. Arvelius, P.; Malm, S.; Svartberg, K.; Strandberg, E. Measuring herding behavior in Border collie—effect of protocol structure on usefulness for selection. *J. Vet. Behav. Clin. Appl. Res.* **2013**, *8*, 9–18. [CrossRef]
17. Kelley, R.B. *Sheep Dogs: Their Breeding, Maintenance and Training*, 3rd ed.; Angus & Robertson: Sydney, Australia, 1949.
18. Williams, T. *Working Sheep Dogs*; Landlinks Press: Melbourne, Australia, 2008.
19. Parsons, A.D. *The Working Kelpie: The Origins and Breeding of a Dinkum Australian*, 1st ed.; Thomas Nelson: Melbourne, Australia, 1986.
20. Cavanagh, R. *Australian Sheep Dogs: Training and Handling Including a Complete Guide to Commands*; R. Cavanagh: Whittlesea, Australia, 1990.
21. Parsons, A.D. *The Kelpie: The Definitive Guide to the Australian Working Dog*; Penguin Group Australia: Camberwell, Australia, 2010.
22. Austin, T. *Dog Psychology: Thoughts on Developing the Man-Dog Bond*; T.G. Austin: Port Fairy, Australia, 1978.

23. Gedye, J., n.d. *Training Notes for farmers Re: Kelpies*. J. Gedye, Ballarat, Victoria.
24. Lithgow, S. *Training and Working Dogs: For Quiet Confident Control of Stock*; University of Queensland Press: St. Lucia, Australia, 1991.
25. Seis, C. *Working Dogs: Training for Sheep and Cattle*; Inkata Press: Sydney, Australia, 1995.
26. Early, J.B.; Arnott, E.R.; Wade, C.M.; McGreevy, P.D. Manual muster: A critical analysis of the use of common terms in Australian working dog manuals. *J. Vet. Behav. Clin. Appl. Res.* **2014**, *9*, 370–374. [CrossRef]
27. Payne, E.M.; Arnott, E.R.; Early, J.B.; Bennett, P.C.; McGreevy, P.D. Dogmanship on the farm: Analysis of personality dimensions and training styles of stock dog handlers in Australia. *J. Vet. Behav. Clin. Appl. Res.* **2015**, *10*, 471–478. [CrossRef]
28. The University of Sydney. Available online: https://sydney.edu.au/ (accessed on 1 May 2019).
29. Meat and Livestock Australia. Available online: https://www.mla.com.au (accessed on 1 May 2019).
30. The Working Kelpie Council of Australia. Available online: http://www.wkc.org.au (accessed on 1 May 2019).
31. Farm Dog Survey. Available online: https://www.dropbox.com/s/zozwd6h8iugu09b/Survey%20Hardcopy10.pdf?dl=0 (accessed on 1 May 2019).
32. Ley, J.; Bennett, P.; Coleman, G. Personality dimensions that emerge in companion canines. *Appl. Anim. Behav. Sci.* **2008**, *110*, 305–317. [CrossRef]
33. R Core Team. *R: A Language and Environment for Statistical Computing*; R Foundation for Statistical Computing: Vienna, Austria, 2018; Available online: https://www.R-project.org/ (accessed on 29 October 2018).
34. Hart, B.L.; Miller, M.F. Behavioral profiles of dog breeds. *J. Am. Vet. Med. Assoc.* **1985**, *186*, 1175–1180. [PubMed]
35. Gosling, S.D.; Kwan, V.S.Y.; John, O.P. A Dog's Got Personality: A Cross-Species Comparative Approach to Personality Judgments in Dogs and Humans. *J. Personal. Soc. Psychol.* **2003**, *85*, 1161–1169. [CrossRef]
36. Arvelius, P.; Klemetsdal, G. How Swedish breeders can substantially increase the genetic gain for the English Setter's hunting traits. *J. Anim. Breed. Genet.* **2013**, *130*, 142–453. [CrossRef]
37. Lindblad-Toh, K.; Wade, C.M.; Mikkelsen, T.S.; Karlsson, E.K.; Jaffe, D.B.; Kamal, M.; Clamp, M.; Chang, J.L.; Kulbokas, E.J.; Zody, M.C.; et al. Genome sequence, comparative analysis and haplotype structure of the domestic dog. *Nature* **2005**, *438*, 803–819. [CrossRef]
38. Kagan, D.M. Ways of Evaluating Teacher Cognition: Inferences Concerning the Goldilocks Principle. *Rev. Educ. Res.* **1990**, *60*, 419–469. [CrossRef]
39. Zak, P.J.; Park, K.W. Population Genetics and Economic Growth. *J. Bioecon.* **2002**, *4*, 1–37. [CrossRef]
40. Goddard, M.E.; Beilharz, R.G. A factor analysis of fearfulness in potential guide dogs. *Appl. Anim. Behav. Sci.* **1984**, *12*, 253–265. [CrossRef]
41. Tomkins, L.M.; Thomson, P.C.; McGreevy, P.D. Behavioral and physiological predictors of guide dog success. *J. Vet. Behav. Clin. Appl. Res.* **2011**, *6*, 178–187. [CrossRef]
42. Foyer, P.; Bjallerhag, N.; Wilsson, E.; Jensen, P. Behaviour and experiences of dogs during the first year of life predict the outcome in a later temperament test. *Appl. Anim. Behav. Sci.* **2014**, *155*, 93–100. [CrossRef]
43. Branson, N.J.; Cobb, M.L.; McGreevy, P.D. *Australian Working Dog Survey Report*; Australian Government Department of Agriculture, Fisheries and Forestry: Canberra, Australia, 2009.
44. Starling, M.J.; Branson, N.; Thomson, P.C.; McGreevy, P.D. Age, sex and reproductive status affect boldness in dogs. *Vet. J.* **2013**, *197*, 868–872. [CrossRef] [PubMed]
45. Starling, M.J.; Branson, N.; Thomson, P.C.; McGreevy, P.D. "Boldness" in the domestic dog differs among breeds and breed groups. *Behav. Process.* **2013**, *97*, 53–62. [CrossRef] [PubMed]
46. Cobb, M.; Branson, N.; McGreevy, P.; Lill, A.; Bennett, P. The advent of canine performance science: Offering a sustainable future for working dogs. *Behav. Process.* **2015**, *110*, 96–104. [CrossRef]
47. McGreevy, P.D. The advent of equitation science. *Vet. J.* **2007**, *174*, 492–500. [CrossRef]
48. Scandurra, A.; Alterisio, A.; Di Cosmo, A.; D'Aniello, B. Behavioral and Perceptual Differences between Sexes in Dogs: An Overview. *Animals* **2018**, *8*, 151. [CrossRef]

animals

MDPI

Article

Development of a Spatial Discount Task to Measure Impulsive Choices in Dogs

Paolo Mongillo [1], Anna Scandurra [1,2], Carla Jade Eatherington [1], Biagio D'Aniello [2] and Lieta Marinelli [1,*]

[1] Laboratorio di Etologia Applicata, Dipartimento di Biomedicina Comparata e Alimentazione, Università degli Studi di Padova. Viale dell'Università 16, 35020 Legnaro, Italy
[2] Dipartimento di Biologia, Università di Napoli Federico II, Via Cinthia, Edificio 7, 80126 Napoli, Italy
* Correspondence: lieta.marinelli@unipd.it

Received: 12 June 2019; Accepted: 13 July 2019; Published: 23 July 2019

Simple Summary: Impulsivity is believed to play a role in problematic behaviors in dogs. In this study, we developed a test to assess dogs' tendency to make impulsive choices, that is their preference for smaller immediate reward instead of larger, but harder to obtain ones. Dogs were first trained that a bowl presented on a certain side always contained a large food amount, whereas the one presented on the opposite side (although at the same distance from the dog) contained less food. Then, the bowl with less food was progressively placed closer to the dog. As expected, dogs' choices to feed from the bowl with less food increased as the distance of the latter decreased. Choices did not depend on factors that could interfere, such as dogs' level of motivation for food, training experience, or learning ability. This indicates that the test is likely to be actually assessing impulsivity, not other traits. Also, female dogs were more likely to make impulsive choices than males, in accordance with what is known in humans and rodents, supporting the validity of the test. The test was completed in less than 1 h, making it a valid option to assess impulsivity in dogs in various contexts.

Abstract: Impulsive choices reflect an individual's tendency to prefer a smaller immediate reward over a larger delayed one. Here, we have developed a behavioural test which can be easily applied to assess impulsive choices in dogs. Dogs were trained to associate one of two equidistant locations with a larger food amount when a smaller amount was presented in the other location, then the smaller amount was placed systematically closer to the dog. Choices of the smaller amount, as a function of distance, were considered a measure of the dog's tendency to make impulsive choices. All dogs (N = 48) passed the learning phase and completed the entire assessment in under 1 h. Choice of the smaller food amount increased as this was placed closer to the dog. Choices were independent from food motivation, past training, and speed of learning the training phase; supporting the specificity of the procedure. Females showed a higher probability of making impulsive choices, in agreement with analogue sex differences found in human and rodent studies, and supporting the external validity of our assessment. Overall, the findings support the practical applicability and represent a first indication of the validity of this method, making it suitable for investigations into impulsivity in dogs.

Keywords: dog; behavioral test; impulsivity; sex differences; learning; validation

1. Introduction

Impulsivity is generally referred to as the tendency to act prematurely, without forethought or consideration of the consequences [1], or as the failure to defer gratification [2]. In humans, impulsivity has been indicated as a vulnerability factor for a range of maladaptive behaviours, including substance abuse, gambling, or pathological conditions such as attention deficit and hyperactivity disorders [3,4]. Although impulsivity is sometimes measured as a single dimension of personality, it is best described as a multidimensional trait [4]. Many studies converge on the recognition of two broad classes of impulsive behaviour, namely impulsive actions and impulsive choices [5,6]. The former is regarded as the result of an inability to inhibit or stop a motor act in response to prepotent stimuli. Behavioural paradigms such as the go/no-go task, or the stop-signal reaction time task, analogue versions of which exist for humans and rodents, are designed to pinpoint this behavioural facet of impulsivity [7]. Impulsive choices instead reflect an individual's preference for smaller immediate gratifications over delayed ones of greater value or quantity [8]. This dimension of impulsive behaviour is typically assessed in delay-discounting tasks, which measure the maximum delay tolerated by individuals who are informed of (for humans) or trained to expect (for animal paradigms) the possibility to obtain higher value rewards if they can wait for sufficiently long time-intervals [7,9]. These tasks do not simply represent different measures of a single construct. There is evidence that these two measures are independent [10,11], and that they are underpinned by different neurobiological mechanisms [12–14]. Also, they are differently related to individual characteristics, such as sex and age. For instance, while a tendency to perform impulsive actions is blandly, if at all, associated with male sex, robust associations exist between sex and impulsive choices, where females discount more steeply than males in both humans and rodents [15]. Sex differences are believed to root in differential activation of the dopaminergic signalling system between sexes, which mediate subjects' sensitivity to rewards. Less clear is the interplay between these mechanisms and circulating gonadal hormones, the role of which impulsive choice behaviour still has to be clarified [15]. As regards age, evidence indicates a higher tendency to express impulsive choices during adolescence/young adulthood, than later in life [16].

The current knowledge about impulsivity comes mostly from studies in humans and rodents. However, the same construct has been tentatively applied to dogs, where high impulsivity is thought to be a correlate of different maladaptive behavioural manifestations or cognitive processes. For instance, impulsivity may play a role in aggression [17,18], and more generally in the expression of behavioural problems [19]. Furthermore, some evidence suggests that impulsivity is associated with lower problem-solving abilities [20,21]. As it occurs in the human literature, methods used to assess impulsivity in dogs vary in scope and methodology. A plethora of tasks that were proposed as assessments of dogs' impulsivity actually represent (tentative) measures of impulsive actions, including reversal learning tasks, the A-not-B task, the cylinder task the middle-cup task, the wait-for-treat task, and buzzer tasks [21–26]; although a thorough description of these paradigms fall outside the scope of this paper, all encompass the necessity to withhold a prepotent response, either spontaneous or learnt. A much smaller variety of tasks assess impulsive choices. Although with some variations in the nature or the source (social or non-social) of the reward, these methods are based on the same general paradigms which measure dogs' ability to tolerate temporal delays on the expectation of a larger/more valuable reward [17,25,27–29]. A common disadvantage of these delay-discounting tasks is that they generally require dogs to undergo a long training (in most cases lasting more than one day), which also makes it difficult to complete the test as proved by a relatively low success rate (e.g., 58.8% [19]; 51.4% [30]). This obviously represents a strong limit to the practical applicability of these tasks, and to the possibility of administering them routinely to large dog samples.

There is accumulating evidence that measures provided by these methods are in many cases uncorrelated [23,30]. Moreover, there is variability in terms of how the outcomes of these tasks relate to the broader, indirect assessment of impulsivity provided by owners' answers to a questionnaire (Dog Impulsivity Assessment Scale, DIAS [18]), which range from no correlation [29,30], to correlation with one of the DIAS subscales [19] or with the overall DIAS score [19,30]. Although this lack of consistency may reflect the complex, multidimensional nature of the construct, it nonetheless prompts us to question which of these tasks provide a valid and easy measure of impulsive behaviours in dogs. In only a few cases, attempts have been made to assess impulsivity as a function of external variables known to influence impulsivity measures in other species, such as sex or age (see for instance [30]). However, none of the aforementioned studies addressed problems of potential intervening variables, including dog's motivation for food, learning abilities, and previous experience, which may represent confounds in the outcome of the assessments, as highlighted by some of the very same authors [30].

Upon these premises, in this study we aimed to develop and validate a task to assess dogs' tendency to express impulsive choices. In view of the possibility to administer the task to large dog samples, one of the requisites of the task was to be successfully easily completed by most dogs, and within a reasonably short time (i.e., a single session, no longer than 1.5 h). To circumvent the difficulties associated with training the dogs to wait in the classical delay-discounting tasks, here the immediacy of the possibility to obtain the smaller reward was operationalized as a smaller space to travel, rather than as a shorter time to wait (although space differences inherently imply a time difference [31]). In the lack of a gold standard that could provide an external validation measure, we aimed at providing a first assessment of the tasks' validity, by (a) looking at the psychometric relationship between the task contingencies and dogs' performance, (b) excluding effects of other intervening factors, namely the dogs' previous training history, level of food motivation, and the learning requirements of the task, (c) assessing the tasks sensitivity to biological factors that are known to influence impulsive choices in other species, namely sex, reproductive status and age, and (d) looking at the relationship between the outcomes of the task and indirect measures of impulsivity provided by the DIAS questionnaire.

During the writing of this paper, results of the development of an analogous paradigm, independently developed by Brady and collaborators [32], came to our attention. Like the method described in the present paper, the task was a spatial version of the classical delay-discounting task. The procedure involved a single test session, preceded a short pre-training phase, and was completed by dogs in one day. The short time requirement, and a training success rate of 96% (24 out of 25 dogs), provide excellent indications in terms of feasibility of this kind of procedure. As far as validation was concerned, the primary means of validation reported in the study were the assessment of test-retest reliability and correlations with a score of the DIAS. On the other hand, the study did not look at factors included in our investigation, and highlighted by the very same authors as potential confounds in their results. In this sense, the results reported in the present study represent fundamental additional indications about the validity of this spatial-discounting task.

2. Materials and Methods

2.1. Subjects

Forty-eight pet dogs were recruited for this study through advertisement in veterinary clinics and the University of Padua. Apart from being healthy, no specific criteria for inclusion in the study were required. The sample included 15 mongrels of small (≤30 cm at the withers, N = 2), medium (>30 and ≤55 cm; N = 9) and large size (>55 cm; N = 4), and 33 pure breed dogs (N = 7 Border Collies, N = 4 Australian Shepherds, N = 3 Golden Retrievers, N = 2 Beagles, N = 2 Cocker Spaniels, N = 2 Labrador Retrievers, N = 1 American Staffordshire Terrier, N = 1 Bernese Mountain Dog, N = 1 Breton, N = 1 Czechoslovakian Wolfdog, N = 1 Dachshund, N = 1 German Shepherd, N = 1 Greyhound, N = 1 Hovawart, N = 1 Labradoodle, N = 1 Newfoundland, N = 1 Rhodesian Ridgeback, N = 1 Samoyed, N = 1 Siberian Husky). Recruitment was aimed at forming four groups of equal size based on the dogs' sex and reproductive status, namely: non-orchiectomized males (mean age ± SD: 4.4 ± 3.2 years, min = 1, max = 12), non-ovariectomized females in dioestrous or anoestrous phase (mean age ± SD: 4.7 ± 2.7 years, min = 1.5, max = 11), orchiectomized males (mean age ± SD: 4.8 ± 2.3 years, min = 1, max = 10) and ovariectomized females (mean age ± SD: 4.8 ± 1.5 years, min = 2, max = 9). Dogs of the last two groups had their gonads removed at least 6 months prior to participating in the study. The owners were asked to indicate if their dogs had any previous experience of training, choosing between four options (no training, basic training with no professional support, obedience training with a professional trainer, training to specific activities with a professional trainer). Finally, owners were asked to evaluate their dog's food motivation, as high (would always eat if given the chance, eats most types of food, never leaves food in the bowl, fights for food), medium (sometimes leaves food in the bowl, eats many, but not all types of food, does not fight for food), or low (always leaves some food in the bowl, only eats some specific types of food, never fights for food). The distribution of training history and food motivation within each of the four experimental groups is reported in Table 1. Owners were asked to not feed their dogs on the day of the experiment.

Table 1. Distribution of categories of Training history and Food motivation within groups of dogs of different sex and reproductive status.

Sex/Reproductive Status	Training History				Food Motivation		
	No	Home	Obedience	Work	Low	Medium	High
Intact males	1	5	3	3	0	7	5
Intact females	0	7	1	4	0	5	7
Orchiectomized males	2	5	4	1	0	5	7
Ovariectomized females	0	6	2	4	0	5	7

2.2. Impulsivity Evaluation Questionnaire

Owners were asked to fill out an Italian translation of the DIAS. This required owners to indicate their degree of agreement with the proposed statements, according to a score scale from 1 (complete disagreement) to 5 (complete agreement). For each dog, an Overall Questionnaire Score (OQS) was calculated as the average score obtained in all items. Moreover, for the sake of comparison with other studies, average scores were calculated for three sub-scales corresponding, in terms of item composition, to the three factors described by Wright and collaborators [18]. However, a factor analysis performed on the data collected in the current study resulted in a very different factorial structure (data not reported), thus the sub-scales used in this study could not be described using the same names adopted elsewhere.

2.3. Experimental Setting

Tests were conducted at the Laboratory of Applied Ethology (Department of Comparative Biomedicine and Food Science, University of Padua) in a room of approximatively 5 × 5 m, equipped with a chair behind a curtain (140 cm high and 160 cm wide) and two plastic panels (24 × 38 cm), placed vertically at a maximum distance of 360 cm from the chair (the actual distance depended on the experimental phase, as detailed below) and 80 cm apart (Figure 1). The panels represented placeholders for positioning food bowls (circular metal bowls, 20 cm in diameter) during the experiment and concealed the bowls from the dog's view, while the curtain served to temporarily conceal the actions of the experimenter from the dog's view during the experimental procedures.

2.4. General Procedure

The test was based on a two-alternative forced choice, between two different quantities of food, in a ratio of 1 to 7. Prior to beginning the test, the dog and owner were taken into the room, and the dog was left free to explore and familiarize itself with the experimental setting and the experimenter for approximately 5 min. During this time, an experimenter explained the procedure to the owner. Then, the owner was invited to attach the leash to the dog and sit on the chair, and the experimental procedure began.

The experiment comprised a Pre-training phase, a Training phase, and a Test phase. All phases were composed of a number of consecutive trials following a similar procedure: the owner sat on the chair behind the curtain, holding the dog next to him/her. In a separate room, the experimenter baited the bowls with 7 food pieces (each being 1/4 of a ring of Frolic®, a commercial semi-humid dog food) of in one (S+) and 1 piece in the other (S−). Then, she entered the experimental room, placed the bowl(s) behind each plastic panel, walked towards the curtain and opened it, allowing the dog to see the two plastic panels. At that point, the experimenter walked behind the owner and placed a hand on the owner's shoulder, which signalled that the dog could be released. The dog was allowed to reach only one of the two bowls, so as soon as the dog approached one bowl, the experimenter removed the other bowl, preventing the dog from eating its content. As soon as the dog ate the food, the selected bowl was also removed. Finally, the owner took the dog back to the starting position, and the curtain was again lowered, the experimenter went into the separate room to prepare the bowls for the next trials. If the dog did not make a choice within one minute, bowls were removed and the trial was considered null.

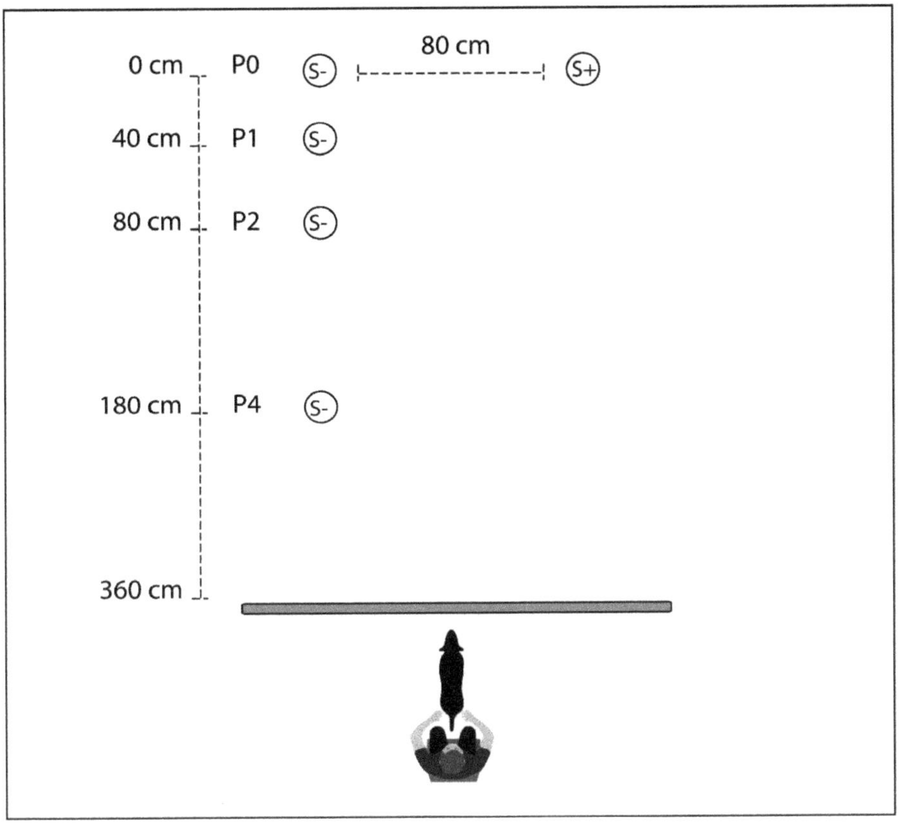

Figure 1. Experimental setting. Representation of the experimental setting, illustrating the owner's and dog's position behind the curtain (large horizontal grey bar) at the start of presentations, and the position of the bowls containing the larger (S+) and the smaller amount of food (S−) during training (P0) and test trials (P0 for S+, P0 to P4 for S−).

2.5. Pre-Training Phase

The aim of the Pre-training phase was to allow the dog to familiarize with the experimental procedure and experience that bowls in different location contained different amounts of food. This phase consisted of 6 trials, which followed the procedure described above, with the difference that only one food bowl was presented in each trial (S+ was presented on 3 trials, and S− on the other 3). In this phase, the food bowls were placed at the distance of 350 cm from the dog. For any given dog, S+ was always presented on the same side thorough the test, and S− on the opposite side. The side of presentation varied between subjects, and was counterbalanced within each of the four experimental groups. To be admitted to the training phase, dogs needed to promptly eat the food from the presented bowl in each of the 6 trials.

2.6. Training Phase

This phase was meant to teach dogs to choose the bowl containing the larger amount of food when both S+ and S− were presented simultaneously. Both S+ and S− bowls were placed at the same distance (P0, 350 cm from the dog). For each dog, S+ and S− were placed on the same side as in the Pre-training phase. A maximum of 30 trials were presented, and the criterion for passing this phase was to choose S+ in 6 consecutive trials. If a dog did not reach the learning criterion within the 30 trials, it was excluded from further testing. Before the test phase began, the owner was allowed to walk outdoors with her/his dog for 10 min.

2.7. Test Phase

The test phase was aimed at verifying the effect of increasing proximity of the smaller amount of food on dogs' choice. The rationale for the test was that lower levels of impulsivity would result in dogs' higher ability to choose the larger amount of food, despite the progressively higher proximity of the smaller food amount. The test phase consisted of 14 trials, which followed the general procedure, with the exception that, while S+ was always placed at the distance of 350 cm, the proximity of S− from dogs was systematically increased along a geometric progression. Specifically, there were three levels of increasing proximity: P1 (proximity increased by 40 cm compared to P0; distance of S− from the dog: 310 cm); P2 (proximity increased by 80 cm; 270 cm from the dog); P4 (proximity increased by 160 cm; 190 cm from the dog). Each of these three levels was presented three times among the fourteen trials; in the remaining 5 trials the distance of S+ and S− from the dog was the same (P0, 350 cm from the dog) as in the Training phase. The trials were randomly presented, with the constrain that S− could not be presented at the same distance in consecutive trials.

2.8. Data Collection and Analysis

All experiments were recorded by two ceiling-mounted cameras and coded with the Observer XT software (Ver.12.5, Noldus, Gröeningen, The Netherlands). In the Training and the test phases the dog's choices were codified as S+, S−, or null.

The analysis of dogs' choices in the Test phase aimed to provide an indication regarding the validity of the procedure. To this aim, the analysis was meant to verify that dogs' ability to choose the larger food amount decreased as a function of the proximity of the smaller food amount. In addition, to obtain an indication about the specificity of the measure, the analysis was meant to exclude that the dogs' performance reflected non-impulsivity related factors, such as different levels of motivation towards food, the dogs' learning ability in acquiring the initial discrimination task, or the dogs' training level. Finally, the analysis was aimed at highlighting possible differences in performance linked to the dogs' age, sex and/or reproductive status, in accordance with associations between these factors and impulsivity reported in the literature, as an indication of the external validity of the procedure.

Training history was unevenly distributed across groups of different sex and reproductive status, making it impossible to include the variable in the model described below. To achieve a better distribution, we recoded the variable using the following two levels: "non professionally trained dogs", which included untrained dogs, and dogs trained without support of a professional trainer, and "professionally trained dogs", which included all other dogs. Prior to such recoding, we ascertained that training history had no main effect on dogs' probability to make impulsive choices. To this aim, a Generalized Linear Mixed Model (GLMM) was used, which included the dog's choice of S+ or S− as a binary dependent variable, the dogs' ID as a random variable accounting for the repeated measurement within each dog, and training history as a four-level factor. As the GLMM revealed no significant main effects of training ($p = 0.217$), the variable was recoded which was used in the analysis described below.

To ascertain the specificity and external validity of our task, a GLMM was used, which included the dog's choice of S+ or S− as a binary dependent variable, and the dogs' ID as a random variable accounting for the repeated measurement within each dog. Separate models were run to investigate the effect of sex and reproductive status: one model was run on data collected from non-gonadectomized males and females and included the dog's sex as fixed factor; the other two models were run on data collected respectively from females and males and included the reproductive status as a fixed factor. In addition to sex or reproductive status, the model included distance of S−, the dogs' training history, and food motivation, as two-level fixed factors, and the dog's age and number of trials to reach the learning criterion in the Training phase as covariates. First-order interactions between S− distance and each of the other fixed factors were also included in the model. A stepwise backwards elimination procedure was used to eliminate non-significant interactions. Post-hoc comparisons were run between factor levels when a significant effect was found for a factor, applying a sequential Bonferroni correction.

As the analysis revealed a significant effect of sex on dogs' choices of S+ in the Test phase (see Results), a one-way ANOVA was performed to ascertain that there were no differences between dogs of different sex or reproductive status on their ability to acquire the initial Training phase, as measured by the number of errors made and in the number of trials needed to reach the learning criterion in such phase.

In order to further exclude that the dogs' performance in the test reflected their learning ability in initial discrimination training, Pearson's correlations coefficients were calculated between the number of trials needed to reach the learning criterion in the Training phase and the percentage of choices of S+, both across the entire test and at each different distance of S−.

Finally, as a further way to assess the relationship between the measure provided by the proximity test and other putative measures of impulsivity, Pearson's correlation coefficients were calculated between the DIAS OQS, and the DIAS sub-scales scores, and the number of trials to reach criteria in the Training phase, the percentage of choices of S+ in the Test phase.

All analysis was run with SPSS (ver. 23, IMB, Armonk, NY, USA). A value of 0.05 was adopted as threshold for statistical significance.

3. Results

3.1. Pre-Training and Training Phases

All dogs involved in the study successfully passed the Pre-training and the Training phase. In the latter, dogs reached the training criterion with an average (± SD) of 10.8 ± 6.6 trials and made an average of 7.7 ± 3.0 choices of S+ and 3.1 ± 4.7 choices of S−. The mean ± SD number of errors (choices of S−) and of trials required to reach the learning criterion by dogs split by sex and reproductive status is reported in Table 2; the one-way ANOVA indicated that there were no differences between sexes in the number of errors ($F = 1.09$; $p = 0.36$) or trials required to reach the learning criterion in this phase ($F = 0.56$; $p = 0.64$). No null trials (i.e., a dog not approaching any of the two bowls) were observed by any dog. The Training phase was completed in an average of 14.6 ± 6.5 min (min: 5.5; max: 28.1).

Table 2. Mean ± SD number of errors (choices of S−) and trials required to reach the learning criterion (TTC) in the Training phase by dogs of different sex and reproductive status.

Parameter	Intact Males	Intact Females	Orchiectomized Males	Ovariectomized Females
Errors	1.2 ± 1.6	4.4 ± 6.9	3.1 ± 4.0	3.7 ± 4.3
TTC	8.7 ± 3.6	11.7 ± 8.4	11.3 ± 7.3	11.6 ± 6.6

3.2. Test Phase

Overall, dogs chose S+ in 75.8% of trials (mean N of trials ± SD: 3.79 ± 1.47 out of 5) when S− was presented at distance P0, 61.8% of trials (1.85 ± 1.15 out of 3) at P1, 35.4% (1.06 ± 1.25 out of 3) at P2, and 25.7% (0.77 ± 1.17 out of 3) at P4. The average of S+ choices for each of the four groups of different sex and reproductive status, at the different S− distances are summarized in Table 3. No null trials were observed in this phase. The Test phase was completed in an average of 15.4 ± 2.7 min (min:10.6; max: 24.4).

Table 3. Mean ± SD number of S+ choices in the Test phase by dogs of different sex and reproductive status. In brackets: mean ± SD percentage of S+ choices on the total number of trials for each distance (i.e., 5 for P0, 3 for P1, P2, and P4).

Distance	Intact Males	Intact Females	Orchiectomized Males	Ovariectomized Females
P0	4.3 ± 1.4 (85 ± 27%)	3.5± 1.6 (70 ± 32%)	3.4 ± 1.6 (68 ± 31%)	4.0 ± 1.4 (80 ± 27%)
P1	2.3 ± 0.9 (78 ± 30%)	1.4 ± 1.3 (47 ± 43%)	1.8 ± 1.0 (61 ± 34%)	1.8 ± 1.3 (61 ± 42%)
P2	1.1 ± 1.2 (36 ± 39%)	1.1 ± 1.2 (36 ± 39%)	1.0 ± 1.4 (33 ± 45%)	1.1 ± 1.4 (36 ± 48%)
P4	0.5 ± 1.0 (17 ± 33%)	0.8 ± 1.0 (28 ± 34%)	0.8 ± 1.2 (25 ± 41%)	1.0 ± 1.5 (33 ± 49%)
Overall	8.25 ± 3.2 (58 ± 23%)	6.8 ± 4.4 (48 ± 31%)	7.1 ± 4.4 (50 ± 30%)	7.9 ± 4.6 (56 ± 33%)

Table 4 summarizes the results of the three GLMM models, investigating the effect of S− distance, speed of acquisition of the Training phase, and dogs' age and training history, food motivation and sex/reproductive status, on dog's probability of choosing S+.

Table 4. Results of the Generalized Linear Mixed Models investigating the effect of the distance of the bowl with the smaller amount of food, the dog's sex or reproductive status (investigated in separate models, and with different data subsets), age, food motivation, type of training received, and number of trials needed to reach the learning criterion in the Training phase (TTC). Only significant first-order interactions between distance and other factors are reported. IF = intact females, IM = intact males, OF = ovariectomized females, OM = orchiectomized males. Subscript numbers indicate the numerator and denominator degrees of freedom, respectively.

Factor	Experimental Groups Which Data Were Analysed in the Model		
	IF and IM	**OF and IF**	**OM and IM**
Distance	$F_{3,303} = 5.17; p = 0.002$	$F_{3,303} = 3.26; p = 0.022$	$F_{3,303} = 2.93; p = 0.034$
Sex	$F_{1,303} = 0.061; p = 0.805$	-	-
Reproductive status	-	$F_{1,303} = 0.77; p = 0.380$	$F_{1,303} = 0.61; p = 0.433$
Age	$F_{1,303} = 2.67; p = 0.197$	$F_{1,303} = 2.10; p = 0.149$	$F_{1,303} = 0.43; p = 0.512$
TTC	$F_{1,303} = 0.55; p = 0.457$	$F_{1,303} = 1.42; p = 0.234$	$F_{1,303} = 0.13; p = 0.723$
Food motivation	$F_{1,303} = 0.88; p = 0.353$	$F_{1,303} = 0.92; p = 0.337$	$F_{1,303} = 1.33; p = 0.250$
Training history	$F_{1,303} = 1.50; p = 0.222$	$F_{1,303} = 0.030; p = 0.863$	$F_{1,303} = 0.71; p = 0.401$
S− distance × Sex	$F_{3,303} = 2.70; p = 0.045$	-	-

All models evidenced an effect of the distance of S− on dog's probability of choosing S+, which generally decreased as S− was placed closer to the dog. When data from intact dogs were analysed, an effect of the interaction between the distance of S− and the dog's sex was found (Figure 2). Post-hoc analysis revealed a significant difference between males and females in the probability of choosing S+ at distance P1 and, while in females the probability already decreased when S− was moved from P0 to P1, in males the first significant drop in probability was only observed when S− was moved from P1 to P2.

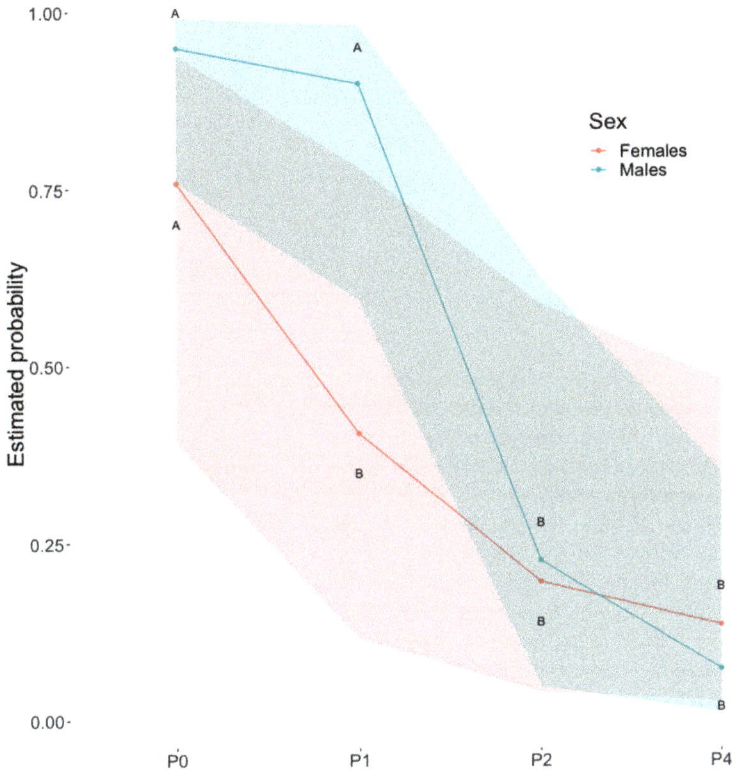

Figure 2. Choices of S− as a function of sex and distance. Generalized Linear Mixed Model (GLMM) mean estimates of the probability of choosing S+ as a function of the distance of S−, by intact male and female dogs. Shaded areas represent the lower and upper 95% confidence intervals. Different capital letters indicate significantly different probabilities between sexes and different levels of proximity of S− ($p < 0.05$) after sequential Bonferroni corrected post-hoc comparisons.

Models using data from the whole group of male dogs, and from the whole group of female dogs, revealed no effect of reproductive status on the probability of choosing S+ as a function of the distance of S−. None of the three models found any effect of the dog's age, training history, food motivation, or speed of acquisition of the Training phase.

3.3. *Correlations of Test Outcomes with Training Phase Performance and DIAS Scores*

The DIAS questionnaire resulted in a mean ± SD of 0.51 ± 0.10 (range: 0.31–0.77) for the OQS, 0.48 ± 0.14 (0.28–0.88) for Factor 1, 0.45 ± 0.09 (0.28–0.68) for Factor 2 and 0.57 ± 0.09 (0.36–0.80) for Factor 3. Results of the correlation analysis between choices of S+ in the Test phase and both the speed of learning of the Training phase and the DIAS scores are reported in Table 5. No correlation was found between any of these variables. However, the number of trials to reach the criterion in the Training phase correlated positively with the DIAS OQS (Pearson's correlation: 0.42, $p < 0.01$) and its score for Factor 1 ($r = 0.44$, $p < 0.01$) and Factor 2 ($r = 0.40$, $p < 0.01$), but not Factor 3 ($r = 012$, $p = 0.44$).

Table 5. Pearson's correlations coefficients between the percentage of choices of S+ in the test phase, both at different S− distances (P0 = 350 cm, P1 = 310 cm, P2 = 270 cm, P4 = 190 cm) and across the whole test, and the number of trials needed to reach the learning criterion in the Training phase (TTC), the DIAS overall score (OQS), the score of the DIAS' Factor 1, Factor 2 and Factor 3.

	P0	P1	P2	P4	Overall
TTC	0.026	−0.063	−0.160	−0.255	−0.108
DIAS OQS	0.243	−0.167	−0.074	−0.025	0.010
DIAS Factor 1	0.295	−0.135	−0.017	0.040	0.075
DIAS Factor 2	−0.216	−0.140	−0.142	−0.203	−0.221
DIAS Factor 3	−0.220	−0.035	−0.100	−0.011	0.036

4. Discussion

In this study, we devised a behavioral test for the assessment of dogs' tendency to make impulsive choices, which was conceived as a spatial implementation of the conventional delay discounting paradigm. All dogs who participated in the study successfully achieved the initial training, which required them to consistently select the larger of two food quantities presented at the same distance. In the subsequent test phase, as expected, dogs expressed a higher probability to choose the smaller amount of food, as the latter was positioned increasingly closer to the dog. The entire assessment procedure was completed in less than approximately 1 h. Overall, the findings represent a good indication of the feasibility of the paradigm, and its better suitability for the assessment of impulsive choices in dogs, compared with lengthier and harder-to-complete delay discounting tasks. A spatial discounting test analogous to the one presented in this study was independently developed and recently presented by Brady and collaborators [32]. This study also reports a high success rate, and an outcome which conformed to expectations (i.e., choices of the larger food amount dependent on the relative distance). Thus, in agreement with this study, we converge on this paradigm's ease of application, which makes it a good candidate for the assessment of impulsivity in large dog samples.

Besides evaluating the feasibility of the procedure, we aimed at providing a first validation of the task as a measure of impulsive choices, by assessing its specificity and its external validity. To the first aim, we ascertained that dog's performance in the spatial discounting task could not be explained by factors different from impulsivity. As our task was based on the acquisition of food, one of our first concerns was to exclude that the dogs' performance did not reflect their motivation towards food rather than their impulsivity. The interplay between impulsive behaviour and motivation to obtain food is certainly a complex one [33]. In fact, while impulsivity and sensitivity towards food are independent traits, they interact to determine food-related behavioral outcomes in humans and rats [33,34]. To the best of our knowledge, the role of food motivation was seldom taken into account in dogs' impulsivity studies. Brucks and collaborators [30] report that varying the quantity and the quality of food-rewards affects dogs' ability to tolerate delays in a delay of gratification task. The same authors highlight the potential confounds represented by food motivational factors on impulsivity measures in dogs. Therefore, the finding that food motivation was not a significant predictor of the dogs' performance in our task provides a first indication in favour of the tasks' specificity. One caveat in the interpretation of these finding is that no dog was present with low levels of food motivation, restricting the validity of this claim to dogs with medium to high food motivation levels.

Past training was another factor that could potentially interfere with dogs' performance in our tasks; for instance, dogs with experience of prolonged training may be more accustomed to sustained work and be less susceptible to mental fatigue, thereby performing better than untrained dogs in the test phase of our procedure. The finding that training had no effect in explaining dogs' choices of the larger food amount was therefore another indication in favour of the tasks' specificity as a measure of impulsivity. Importantly, while this result indicates that our assessment is unaffected by differences in training history, it does not negate that some forms of training may improve dogs' ability to exert self-control. Recent findings suggest that specific forms of training can improve some measures of impulsive control, such as impulsive actions [26]. Moreover, deliberate training for self-control can lead to a generalized increased ability in different forms of impulsivity in humans, while extensive and specific training brings improvement in impulsive choice in animals [35]. However, data about the dogs' training history is seldom reported in previous studies on dog impulsive choices, thus it is difficult to make conclusions about the role of such experience on this facet of impulsivity. The availability of an easily applicable procedure for the assessment of impulsive choices, like the one presented in the present study, will allow to study the role of specific training history on canine impulsivity.

A related finding was that the number of errors (S− choices) made by dogs before reaching the learning criterion in the training phase did not explain their choices in the test. Previous research highlighted how the task's learning requirement may represent a confound in measures of dog impulsivity. In fact, the idea that alleged measures of impulsivity may actually reflect the dogs' learning ability was presented as a potential explanation for the lack of consistency across tasks [29]. In view of such concerns, the finding that dogs' performance in our assessment was not affected by the dogs' ability to learn the initial tasks represents an important indication of specificity. Another concern that relates to the learning requirements of delay discounting tasks, is that the necessary initial training is often achieved only by a fraction of dogs, producing an inherent bias in the selection of dogs who undergo the actual assessment. This does not seem to apply to spatial discounting tasks, as the training phase was acquired by all dogs who participated in our study, as well as by nearly all those who took part in the task developed by Brady and collaborators [30].

As an indication of external validity, we investigated how dogs' performance in our task was affected by age, sex, and reproductive status. Age had no effect on dogs' probability to make impulsive choices. Considering the majority of our dogs were adults, the result is in line with human studies, where evidence indicates a stabilization of impulsive choice behavior after adolescence/young adulthood [16]. The performance of the dogs in our task showed a clear dimorphic pattern: females discounted more steeply, as their probability to choose the larger food amount decreased significantly as soon as the bowl with the smaller amount was moved closer to the dog. Many sex related behaviors have been described in dogs [36]. In the present study, analysis of sex differences was undertaken to provide an indication of the tasks' goodness as a measure of impulsive behavior. In fact, our results conform to the what is reported in both humans and rodents, where steeper discount curves are generally found in females than in males [15]. No difference in performance was found between our intact and gonadectomized females. On the one hand this suggests that the main contribution to the observed sex difference is due to organizational effects of sex hormones, rather than by these hormone's circulating levels. On the other hand, as our intact female dogs were in the anœtrous phase (based on the report of the owners on the date of their last manifestations of œstrous) it cannot be excluded that the performance of intact female dogs may have been different, had females been tested in other phases of the oestrous cycle, as seen in other species [37,38]. Our current data cannot elucidate the mechanisms underlying the observed differences, Thus, we cannot tell whether dopamine transmission is involved in these differences, as suggested for other species. To the best of our knowledge there is no data about sex differences or the role of ovarian hormones in dopaminergic transmission in dogs. However, it is worth noting that sex differences are consistently found in dogs' spatial learning tasks [39–41], where dopamine plays a crucial role [42]. Regardless of the mechanism, our results indicate that the phenomenon our task is measuring is subject to the same biological influence seen in other species, providing an indication of the tasks' external validity.

Finally, no correlation was found between our dogs' performance in the test, and the score obtained by dogs in a putative assessment of impulsivity made through the DIAS questionnaire, either in terms of its overall score or the score of its subscales (calculated as described in the validation study by Wright and collaborators [18]). On the one hand, the finding clashes with the significant correlations between the DIAS score and the measures of impulsivity obtained in the spatial discounting task presented by Brady and collaborators [32], or in a delayed reward paradigm [19]. On the other hand, several other studies on dogs' impulsive choices report no association with the DIAS score [25,29], or correlations in opposite directions than expected [30]. Although the reason of these discrepancies is not immediately clear, it must be considered that the DIAS was developed to assess impulsivity as a generic personality trait, rather than to pinpoint a specific facet of the phenomenon. As already highlighted by others [30], expressions of impulsivity are highly context specific and it is possible that the questionnaire and our task are assessing different facets of the same phenomenon. Alternatively, it is possible that they assess completely independent traits. In fact, our finding of a positive correlation between dogs' speed of learning of the initial training phase and the questionnaire scores suggests that the latter reflects the dogs' learning ability rather than their impulsivity. Moreover, questionnaires are based on indirect evaluations of the animals' behavior made by their owners, which incorporates a considerable degree of subjectivity in the assessment. Such individual variability could be further amplified by cultural differences, and translation-related nuances. In fact, while significant correlations between the DIAS and impulsivity measures were reported by studies conducted in the UK, the opposite was generally true for studies made in non-Anglo-Saxon countries, either using the original English version (e.g., [25]) or a translated version of the questionnaire [29,30], as in the current one.

5. Conclusions

In this study we presented a spatial discount task, aimed at assessing impulsive choices in dogs. A similar task was independently developed by Brady and collaborators [32] at around the same time. Both studies converge on the ease of application of the task, which advocates the procedure as a good candidate for larger-scale studies on impulsivity. We ascertained the lack of effect of several factors which may have interfered with the dogs' measure, thereby providing indications of the procedure's specificity. In addition, we provided indications about its external validity by showing a susceptibility of the assessment to sex differences, similar to those already observed in humans and rodents. Overall, the task seems to be promising as a valid, easily applicable procedure for the assessment of impulsive choices.

However, although these findings, together with those of Brady and collaborators [32], provide indications about the goodness of this assessment, other steps would be needed to provide conclusive evidence of its validity, as well as to fine-tune the procedure. For instance, it would be important to determine how the present assessment relates to the outcome of other procedures, that are assumed to measure other facets of impulsivity, such as tasks assessing dogs' tendency to express impulsive actions. Moreover, in view of a potential application in large-scale or cross-cultural studies, it would be important to extend the assessment to larger representation of size and age than those included in this study, as well as to ascertain the reproducibility of the assessment across different laboratories. Considering the ease of administration of the procedure, it is foreseeable that the same would be applied as a screening/selection tool in clinical (e.g., for the identification of pathological impulsivity) or other professional contexts (e.g., for the selection of dogs to be trained for specific activities); to this aim, evaluation of the applicability of the procedure in non-experimental settings and of its predictive validity for expected outcomes, would be required. Finally, considering the known interplay between training and impulsivity, the procedure could be used to assess the efficacy of specific forms of training, including its applications as a therapeutic intervention, in reducing impulsive behavior.

Author Contributions: The study was conceived by L.M. and P.M.; the experiments were conducted by A.S.; statistical analysis was performed by P.M.; drafting was done by P.M., A.S., and C.J.E.; interpretation of results was done by P.M., A.S., B.D., and L.M.; revision and editing were done by B.D. and L.M. All authors have approved the submitted version and agree to be personally accountable for the author's own contributions and for ensuring that questions related to the accuracy or integrity of any part of the work, even ones in which the author was not personally involved, are appropriately investigated, resolved, and documented in the literature.

Funding: This research was supported by a research grant from Università degli Studi di Padova (to P.M., Grant Nr. CPDA144871/14), two postdoc grants awarded by Università degli Studi di Padova (to A.S., Grant Nr. CPDR148844/14; to C.E., Grant Nr. BIRD178748/17). The APC was funded by Università degli Studi di Padova, Dipartimento di Biomedicina Comparata e Alimentazione (Grant Nr. DOR1809748/18) and Università degli Studi di Napoli Federico II, Dipartimento di Zoologia (Grant Nr. PRD2019).

Acknowledgments: We are very grateful to Carlo Poltronieri for his technical assistance and to all the dogs' owners for volunteering to take part.

Conflicts of Interest: The authors declare no conflict of interest.

Ethical Statement: The study was conducted in accordance with relevant legislation about research involving animals, and, for the type of procedures involved, no formal ethical approval was required.

References

1. Dalley, J.W.; Everitt, B.J.; Robbins, T.W. Impulsivity, compulsivity, and top-down cognitive control. *Neuron* **2011**, *69*, 680–694. [CrossRef] [PubMed]
2. Monterosso, J.; Ainslie, G. Beyond discounting: Possible experimental models of impulse control. *Psychopharmacology* **1999**, *146*, 339–347. [CrossRef] [PubMed]
3. Dussault, F.; Brendgen, M.; Vitaro, F.; Wanner, B.; Tremblay, R.E. Longitudinal links between impulsivity, gambling problems and depressive symptoms: A transactional model from adolescence to early adulthood. *J. Child. Psychol. Psychiatry* **2011**, *52*, 130–138. [CrossRef] [PubMed]
4. Robbins, T.W.; Gillan, C.M.; Smith, D.G.; De Wit, S.; Ersche, K.D. Neurocognitive endophenotypes of impulsivity and compulsivity: Towards dimensional psychiatry. *Trends Cogn. Sci.* **2012**, *16*, 81–91. [CrossRef] [PubMed]
5. Evenden, J.L. Varieties of impulsivity. *Psychopharmacology* **1999**, *146*, 348–361. [CrossRef] [PubMed]
6. Bechara, A. The neurology of social cognition. *Brain* **2002**, *125*, 1673–1675. [CrossRef]
7. Broos, N.; Schmaal, L.; Wiskerke, J.; Kostelijk, L.; Lam, T.; Stoop, N.; Weierink, L.; Ham, J.; De Geus, E.J.C.; Schoffelmeer, A.N.M.; et al. The relationship between impulsive choice and impulsive action: A cross-species translational study. *PLoS ONE* **2012**, *7*, e36781. [CrossRef]
8. Ainslie, G. Specious reward: A behavioral theory of impulsiveness and impulse control. *Psychol. Bull.* **1975**, *82*, 463–498. [CrossRef]
9. Oas, P. The psychological assessment of impulsivity: A review. *J. Psychoeduc. Assess.* **1985**, *3*, 141–156. [CrossRef]
10. Van den Bergh, F.; Spronk, M.; Ferreira, L.; Bloemarts, E.; Groenink, L.; Olivier, B.; Oosting, R. Relationship of delay aversion and response inhibition to extinction learning, aggression, and sexual behaviour. *Behav. Brain Res.* **2006**, *175*, 75–81. [CrossRef]
11. MacKillop, J.; Weafer, J.; Gray, J.; Oshri, A.; Palmer, A.; de Wit, H. The latent structure of impulsivity: Impulsive choice, impulsive action, and impulsive personality traits. *Psychopharmacology* **2016**, *233*, 3361–3370. [CrossRef] [PubMed]
12. Lovic, V.; Keen, D.; Fletcher, P.J.; Fleming, A.S. Early-life maternal separation and social isolation produce an increase in impulsive action but not impulsive choice. *Behav. Neurosci.* **2011**, *125*, 481. [CrossRef] [PubMed]
13. Wiskerke, J.; Stoop, N.; Schetters, D.; Schoffelmeer, A.N.; Pattij, T. Cannabinoid CB1 receptor activation mediates the opposing effects of amphetamine on impulsive action and impulsive choice. *PLoS ONE* **2011**, *6*, e25856. [CrossRef] [PubMed]
14. Sun, H.; Cocker, P.J.; Zeeb, F.D.; Winstanley, C.A. Chronic atomoxetine treatment during adolescence decreases impulsive choice, but not impulsive action, in adult rats and alters markers of synaptic plasticity in the orbitofrontal cortex. *Psychopharmacology* **2012**, *219*, 285–301. [CrossRef] [PubMed]
15. Orsini, C.A.; Setlow, B. Sex Differences in Animal Models of Decision Making. *J. Neurosci. Res.* **2017**, *95*, 260–269. [CrossRef] [PubMed]
16. Green, L.; Myerson, J.; Lichtman, D.; Rosen, S.; Fry, A. Temporal discounting in choice between delayed rewards: The role of age and income. *Psychol. Aging* **1996**, *11*, 79–84. [CrossRef] [PubMed]
17. Amat, M.; Manteca, X.; Mariotti, V.M.; Ruiz de la Torre, J.L.; Fatjó, J. Aggressive behavior in the English cocker spaniel. *J. Vet. Behav. Clin. Appl. Res.* **2009**, *4*, 111–117. [CrossRef]
18. Wright, H.F.; Mills, D.S.; Pollux, P.M. Development and Validation of a Psychometric Tool for Assessing Impulsivity in the Domestic Dog (*Canis familiaris*). *Int. J. Comp. Psychol.* **2011**, *24*, 210–225.
19. Wright, H.F.; Mills, D.S.; Pollux, P.M. Behavioural and physiological correlates of impulsivity in the domestic dog (*Canis familiaris*). *Physiol. Behav.* **2012**, *105*, 676–682. [CrossRef]
20. Bray, E.E.; MacLean, E.L.; Hare, B.A. Increasing arousal enhances inhibitory control in calm but not excitable dogs. *Anim. Cogn.* **2015**, *18*, 1317–1329. [CrossRef]

21. Müller, C.A.; Riemer, S.; Virányi, Z.; Huber, L.; Range, F. Inhibitory control, but not prolonged object-related experience appears to affect physical problem-solving performance of pet dogs. *PLoS ONE* **2016**, *11*, 1–19. [CrossRef] [PubMed]
22. Tapp, P.D.; Siwak, C.T.; Estrada, J.; Head, E.; Muggenburg, B.A.; Cotman, C.W.; Milgram, N.W. Size and reversal learning in the beagle dog as a measure of executive function and inhibitory control in aging. *Learn. Mem.* **2003**, *10*, 64–73. [CrossRef] [PubMed]
23. Bray, E.E.; MacLean, E.L.; Hare, B.A. Context specificity of inhibitory control in dogs. *Anim. Cogn.* **2014**, *17*, 15–31. [CrossRef] [PubMed]
24. Marshall-Pescini, S.; Virányi, Z.; Range, F. The effect of domestication on inhibitory control: Wolves and dogs compared. *PLoS ONE* **2015**, *10*, 1–16. [CrossRef] [PubMed]
25. Brucks, D.; Soliani, M.; Range, F.; Marshall-Pescini, S. Reward type and behavioural patterns predict dogs' success in a delay of gratification paradigm. *Sci. Rep.* **2017**, *7*, 42459. [CrossRef] [PubMed]
26. Barrera, G.; Alterisio, A.; Scandurra, A.; Bentosela, M.; D'Aniello, B. Training improves inhibitory control in water rescue dogs. *Anim. Cogn.* **2019**, *22*, 127–131. [CrossRef] [PubMed]
27. Leonardi, R.J.; Vick, S.J.; Dufour, V. Waiting for more: The performance of domestic dogs (Canis familiaris) on exchange tasks. *Anim. Cogn.* **2012**, *15*, 107–120. [CrossRef] [PubMed]
28. Riemer, S.; Mills, D.S.; Wright, H. Impulsive for life? The nature of long-term impulsivity in domestic dogs. *Anim. Cogn.* **2014**, *17*, 815–819. [CrossRef]
29. Fagnani, J.; Barrera, G.; Carballo, F.; Bentosela, M. Tolerance to delayed reward tasks in social and non-social contexts. *Behav. Process.* **2016**, *130*, 19–30. [CrossRef]
30. Brucks, D.; Marshall-Pescini, S.; Wallis, L.J.; Huber, L.; Range, F. Measures of dogs' inhibitory control abilities do not correlate across tasks. *Front. Psychol.* **2017**, *8*, 1–17. [CrossRef]
31. Mühlhoff, N.; Stevens, J.R.; Reader, S.M. Spatial discounting of food and social rewards in guppies (*Poecilia reticulata*). *Front. Psychol.* **2011**, *2*, 68. [CrossRef] [PubMed]
32. Brady, K.; Hewison, L.; Wright, H.; Zulch, H.; Cracknell, N.; Mills, D. A spatial discounting test to assess impulsivity in dogs. *Appl. Anim. Behav. Sci.* **2018**, *202*, 77–84. [CrossRef]
33. Velázquez-Sánchez, C.; Ferragud, A.; Moore, C.F.; Everitt, B.J.; Sabino, V.; Cottone, P. High trait impulsivity predicts food addiction-like behavior in the rat. *Neuropsychopharmacology* **2014**, *39*, 2463. [CrossRef] [PubMed]
34. Appelhans, B.M.; Woolf, K.; Pagoto, S.L.; Schneider, K.L.; Whited, M.C.; Liebman, R. Inhibiting food reward: Delay discounting, food reward sensitivity, and palatable food intake in overweight and obese women. *Obesity* **2011**, *19*, 2175–2182. [CrossRef] [PubMed]
35. Renda, C.R.; Madden, G.J. Impulsive choice and pre-exposure to delays: III. Four-month test-retest outcomes in male wistar rats. *Behav. Process.* **2016**, *126*, 108–112. [CrossRef] [PubMed]
36. Scandurra, A.; Alterisio, A.; Di Cosmo, A.; D'Aniello, B. Behavioral and perceptual differences between sexes in dogs: An overview. *Animals* **2018**, *8*, 151. [CrossRef]
37. Carroll, M.E.; Anker, J.J. Sex differences and ovarian hormones in animal models of drug dependence. *Horm. Behav.* **2010**, *58*, 44–56. [CrossRef]
38. Smith, C.T.; Sierra, Y.; Oppler, S.H.; Boettiger, C.A. Ovarian cycle effects on immediate reward selection bias in humans: A role for estradiol. *J. Neurosci.* **2014**, *34*, 5468–5476. [CrossRef]
39. Fugazza, C.; Mongillo, P.; Marinelli, L. Sex differences in dogs' social learning of spatial information. *Anim. Cogn.* **2017**, *20*, 789–794. [CrossRef]
40. Mongillo, P.; Scandurra, A.; D'Aniello, B.; Marinelli, L. Effect of sex and gonadectomy on dogs' spatial performance. *Appl. Anim. Behav. Sci.* **2017**, *191*, 84–89. [CrossRef]
41. Scandurra, A.; Marinelli, L.; Lõoke, M.; D'Aniello, B.; Mongillo, P. The effect of age, sex and gonadectomy on dogs' use of spatial navigation strategies. *Appl. Anim. Behav. Sci.* **2018**, *205*, 89–97. [CrossRef]
42. Kempadoo, K.; Mosharov, E.V.; Choi, S.J.; Sulzer, D.; Kandel, E.R. Dopamine release from the locus coeruleus to the dorsal hippocampus promotes spatial learning and memory. *Proc. Natl. Acad. Sci. USA* **2016**, *113*, 14835–14840. [CrossRef] [PubMed]

animals

Article

The Impacts of a Reading-to-Dog Programme on Attending and Reading of Nine Children with Autism Spectrum Disorders

Stefania Uccheddu [1,*], Mariangela Albertini [2], Ludovica Pierantoni [3], Sara Fantino [3] and Federica Pirrone [2]

1 Vet Ethology, Leemveldstraat 44, 3090 Overijse, Belgium
2 Department of Veterinary Medicine, University of Milan, Via Celoria, 10, 20133 Milan, Italy
3 CAN (Comportamento Animale Napoli) S.S.D.R.L., Via Camaldolilli, 79, 80128 Naples, Italy
4 L'altra via, Psychology and Animal Assisted Intervention, Via Petrarca, 4, 09010 Pula, Italy
* Correspondence: uccheddus@gmail.com; Tel.: +39-347-8533239

Received: 29 May 2019; Accepted: 23 July 2019; Published: 26 July 2019

Simple Summary: The purpose of this research was to compare reading motivation and attitude, as well as reading and cognitive skills, of school-age children diagnosed with Autism Spectrum Disorders (ASD) who attended a 10 session reading programme with and without the presence of a dog. Children who read to a dog had 100% attendance at sessions over the course of the programme versus 75% (range 25–100%) of children attending reading sessions without a dog. In addition, after the programme, they were significantly more motivated and willing to read at home, as perceived by their parents. However, there were no significant differences in scores on reading and cognitive tests either within each group or between groups. Based on these results, we can conclude that reading to a dog can have positive effects on an ASD child's motivation and attitude toward reading. More research is needed to better understand if it can also have positive effects on children with ASD's overall reading and cognitive abilities.

Abstract: Poor knowledge is available on the effectiveness of reading to dogs in educational settings, particularly in children with Autism Spectrum Disorders (ASD). In this study, we test the hypothesis that reading to a dog improves propensity towards books and motivation to read after the end of the programme, as well as reading and cognitive skills in children with ASD. The study is a prospective, randomized controlled trial, consisting of testing and re-testing after a 10 sessions reading programme with and without the presence of a dog. Nine Children with ASD (6–11 years old) were randomly assigned to a control (CG, reading without a dog, n. 4) or experimental group (EG, reading to a dog, n. 5). Children's attendance at reading sessions was recorded at each session. Parents' perceptions were evaluated at the end of the programme to detect changes in children's attitudes and motivation toward reading. Psychologist-administered validated reading (Cornoldi's MT2 reading test; test of reading comprehension, TORC; metaphonological competence test, MCF) and cognitive tests (Wechsler intelligence scale for children Wisc IV, Vineland) to all children, at baseline and at the end of the reading programme. Compared with CG children, children in the EG group participated more frequently in the reading sessions, and they were reported to be more motivated readers at home after the programme. However, there were no differences on reading and cognitive tests' scores either within each group of children or between groups. Further studies are warranted in order to understand whether and how incorporating dogs into a reading programme is beneficial to Children with ASD at the socio-emotional and cognitive level.

Keywords: dogs; children; Autism Spectrum Disorders; cognition; reading-to-dog programme

1. Introduction

Launched in 1999 by Intermountain Therapy Animals, the Reading Education Assistance Dogs (READ) is the first programme and still one of the most comprehensive involving animals to strengthen children reading skills [1]. The Reading Education Assistance Dogs® programme improves children's reading and communication skills by employing a powerful method: reading to a dog (R.E.A.D. webpage)

Recent work focused on children has shown that reading-to-dog programmes might reduce blood pressure and offer a nonjudgmental, safe environment in which to practice reading [2,3]. Dogs seem able to offer children a unique type of emotional support in the education setting because they are fully capable of being active, supportive listeners, but are also unable to verbally criticize or comment upon a child's reading abilities [4]. In the wider literature [1], measurements of reading skills have included improved scores on test of reading comprehension (TORC), measures of academic progress (MAP), reading rate, and reading ability [1]. According to Pillow-Price et al. [5], all reading scores for children participating in a reading programme improved significantly. Sorin et al. [6] noted improvements in reading, behavior, confidence, self-esteem, and school attendance with special education students who worked on literacy skills with dogs. Changes in reading motivation may reflect a better reading performance [1]. In Guthrie and Cox [7], engaged and motivated children who opened a book more frequently were also highly achieving in reading abilities because cognitive functioning was powerfully facilitated through interest and motivation [8].

The presence of a dog has already been suggested to reduce physiological parameters of stress (decreased blood pressure [9] and cortisol [10]) in children with autism. A child with autism spectrum disorder improved on the Dynamic Indicators of Early Literacy Skills (DIBELS) and Elementary Reading Attitudes Scale (ERAS) after completing a reading-to-dog programme [11]. Based on the final version of DSM-5 [12], autism is currently counted in one general term, Autism Spectrum Disorders (ASD), with three different levels, level 1 (requiring support), level 2 (requiring substantial support), and level 3 (requiring very substantial support). ASD is characterized by delays in the development of multiple basic functioning including socialization and communication and behavioral challenges (such as rituals and repetitive behaviors) [12]. In the clinical setting, anxiety-related concerns are among the most common presenting problems for school-age children and adolescents with ASD [13]. Recently, one study developed an educational setting in which three Children with ASD read social stories in the presence of therapy dogs [14]. The authors aimed to test the hypothesis that the presence of a therapy dog improves the effectiveness of Social Story method, which is used to communicate clear and detailed information to autistic children on a context, skill, achievement, or concept [15]. Although improvements in these children's indicators of social skills were reported (e.g., increased frequency of the initiations of social interactions and decreased level of prompt needed to provide the expected social response), interpretation of these indicators can be difficult, and the quality of the evidence is still unclear, also due to the small sample size used [1].

Here, we decided to apply recent and innovative psychological approaches to detect potential improvements in reading abilities and changes in behavioral and emotional processes in ASD school-age children's reading in the presence of a dog compared with children reading without a dog. Moreover, we compared attendance at sessions of children and parents' perception of their reading motivation and willingness to read. The main hypothesis was that a social environment enriched by the presence of a dog strengthens the effectiveness of a reading programme in enhancing both reading and cognitive abilities in Children with ASD.

2. Materials and Methods

2.1. Participants

A total of nine children in the age range of 6 to 11 years (mean 7 ± 0.45 SE), seven boys and two girls, were recruited from the CTR Esperienze ONLUS (Comunicazione Territorio Relazioni) Cagliari, Sardinia, Italy, where the reading sessions took place.

Informed consent was obtained from parents of all children, who were previously advised by the facility staff members of an experimenter's presence for the videotaping procedure. In signing the consent, parents ensured that there was a clear understanding of the information given to them, and also that they agreed with that and with the disclosure of their personal details. Besides favorable opinion from a clinical psychologist or neuropsychiatry holding also a certification as Board Certified Behavior Analyst® (BCBA®), to be eligible for participation in the programme, children were required to: (a) be diagnosed according to the diagnostic tools described in the DSM-5 and in the guidelines elaborated by the Italian Ministry of Health guidelines. The diagnosis was determined by a multidisciplinary equipe composed of a child neuropsychiatrist, a psychologist specialized in child development, and a pedagogist. The DSM-5 diagnosis also includes new guidelines for categorizing autism by level. There are three levels, each reflecting a different level of support each child needs (from level 1: little support, to level 3: higher support); (b) show lack of initiation of appropriate social response in a given social situation during therapy or free-time activities; (c) have some reading prerequisites, such as the ability to open and browse through a book; (d) be willing to interact with dogs, as evaluated in a preintervention screening; (e) possess basic speaking skills, and (f) immunocompetency. Fear of dogs was considered an exclusion criterion. Diagnosis and severity level have been established by the neuropsychiatry according to [12].

2.2. Reading Session

Children with ASD were randomly divided into two groups according to demographics characteristics and severity levels expressed in the diagnosis: (1) the experimental group (EG, n. 5: four boys and one girl; mean 7.60 ± 2.30 SE) read a book with a dog present, and (2) the control group (CG, n. 4: three boys and one girl; mean 8.25 ± 1.73 SE) read a book without a dog. Details on age, gender, and level of severity indicated in the diagnosis are reported in Table 1. Groups were homogeneous in terms of mean age and diagnosis. Both EG (experimental group) and CG (control group) were involved in 10 weekly group sessions, run over a period of 70 days. Each session was approximately 30 min in length, without pauses, during which children read a book one-on-one, upon request by the psychologist. A book was selected by the psychologist at the beginning of the programme. Both EG and CG children read the same book. A copy of the book was available for each child. The same psychologist was present for all the sessions for both EC and CG groups. The psychologist, before starting the reading session, reminded the children of the rule of the session. The rules were presented, if necessary, more times during the session, only in oral form (Now we are going to read to the dog. Her name is Bella/Lilli. The dog is pleased to listen to our reading, but we need to respect some specific rules: please, do not be loud, do not run, do not touch the dog since this is going to make the dog fearful. We are not going to pass through the benches during the reading session. We cannot touch the dog during the session, but we can talk with her).

Sessions were performed in the afternoon in order to exclude parental factors/obligations that could impact the child's attendance.

As for the experimental group, two dogs (both neutered females, mixed-breed, 2 and 8 years old) participated in the sessions, one at a time, on alternate weeks. Two dogs were chosen by a team composed of two veterinarians expert in behavior and welfare and a psychologist specialized in animal-assisted intervention. Inclusion criteria considered their kindness and cooperation when handled by children, their interest in people, and absence of any signs of anxiety, fearfulness, reactivity, or aggression. The dogs, both neutered females, mixed-breed dogs, were 2 and 8 years old (mean

5.0 ± 3.0 SE) and weighed between 3 and 18 kg (mean 10.5 ± 7.5 SE) at the time of the sampling period. Dogs were recruited from the local nonprofit organization "Effetto Palla ONLUS", with the aim of enhancing their socialization and adoption rates [16]. Dogs were subjected to regular health screening and behavioral monitoring by a veterinarian with expertise in animal behavior and welfare. In order to be eligible for participation in the reading programme, the dogs were required to be in perfect clinical health (i.e., free from pain, external and internal parasites, and immunized). These dogs' characteristics, behavior, and welfare during the reading sessions have been described also in more detailed in [17]. Child–animal interaction was limited to verbal contact: no child-initiated contacts with dogs were allowed. Children could only talk to the dog, and they did so by praising her or asking whether she enjoyed the story or was getting bored.

Table 1. Age, gender, and diagnosis of Children with ASD involved in the project.

Experimental Group			Control Group		
Age (Years)	Gender	Severity Level [1]	Age (Years)	Gender	Severity Level [1]
11	M	1	7	M	1
9	M	1	10	M	1
6	M	2	10	F	1
6	F	1	6	M	2
6	M	1	-	-	

M = Male; F = Female; ASD = Autistic spectrum disorder. [1] According to [12].

2.3. Setting Room

The two dogs were handled by a female veterinarian expert in animal welfare and behavior, who was familiar with them and was always present during the sessions to guarantee their well-being.

Sessions were performed in a 6 × 5 m carpeted room at the facility, where children were also involved in other activities, in the presence of a psychologist. In more detail, at reading sessions, one visiting dog, one dog handler/veterinarian, one psychologist, and one experimenter were always present. The room temperature ranged between 20 °C and 24 °C. Two 30 cm high benches were placed to separate the room into two identical spaces, one for the dog and one for the children [10].

2.4. Test

At baseline (T0) and at the end of the 10 sessions programme (T1), the psychologist administered validated reading and cognitive tests to all the children (Table 2). Pre- and posttest in both groups followed the same order. Reading tests such as Cornoldi reading test (MT2) [18], test of reading comprehension (TORC) [19], metaphonological competence (MCF) [20] and cognitive tests, Wechsler intelligence scale for children (Wisc IV) [21], and Vineland [22].

Session attendance was recorded in both groups. A short self-report questionnaire was prepared by the psychologist by reviewing similar literature [23] in order to collect parents perceptions after the 10 reading sessions. The questionnaire, presented in Table 3, was composed of seven yes/no closed questions. The questions focused on the perception of the parents about: (1) reading motivation, (2) motivation to follow the programme, (3) social skills, and (4) attention towards dogs.

2.5. Statistical Analysis

Data was analyzed with SPSS, version 25.0 (SPSS Inc, Chicago, IL, USA) through nonparametric statistics as they did not follow a normal distribution (Shapiro–Wilk normality tests, all $p > 0.05$). The Mann–Whitney U test was used to compare differences between the two groups, while the Wilcoxon signed-rank test was used for paired data. Due to the multiple comparisons, Benjamini–Hochberg multiple testing correction [24] was applied. Fisher's exact test was used to investigate associations between the presence of the dog and both children's attendance and parents' answers in the questionnaire. Values of $p < 0.05$ were considered statistically significant.

Table 2. Different tests administered to children by psychologist at T0 and T1.

Area of Interest	Test	Details	Domains	Items	Scores
Reading	Cornoldi Reading Test (MT2) [18]	assesses reading literacy	Fluency (speed and accuracy)	2	Speed: syllabus in one second;Accuracy: number of auto-correction
Reading	Test of Reading Comprehension (TORC) [19]	measures a child's abilities in reading comprehension	Reading comprehension	1	0 to 100 (0 low level, 100 higher level)
Reading	Metaphonological Competence (MCF) [20]	measures the child ability to talk about a topic and explain his or her use or understanding of the phonological awareness skill	Recognition, Fluidity, Phonemic, Segmentation, Letter deletion, Final deletion	5	0 to 100 (0 low level, 100 higher level)
Cognitive	Wechsler Intelligence Scale for Children (Wisc IV) [21]	measures a child's abilities in some cognitive domains. It generates a Full-Scale IQ (formerly known as an intelligence quotient or IQ score) that represents a child's general intellectual ability	Intelligence Quotient, Fluid Reasoning Index, Processing Speed Index, Verbal Comprehension Index, Working Memory Index	5	70 to 130 (70 low level, 130 higher level)
Adaptive behavior	Vineland [22]	measures the personal and social skills of individuals from birth through adulthood	Compressive Results, Communication, Daily Living Skills, Socialization, and Motor Skills	5	Specific for the age range: 34 to 144 (34 low level, 144 higher level)

Table 3. Parent-completed questionnaire.

Questions	EG Parents Answers		CG Parents Answers		p (Fisher's Exact Test)
At the End of the Reading Programme	**Yes**	**No**	**Yes**	**No**	
(1) Was the child pleased to read?	4	1	1	3	$p > 0.05$
(2) Was the child looking for any book autonomously or in presence of an adult?	4	1	0	4	$p = 0.04$
(3) Was the child more motivated and enthusiastic to read a book?	4	1	0	4	$p = 0.04$
(4) Was the child able to pay more attention to dogs in daily routine?	4	1	2	2	$p > 0.05$
(5) Was the child able to keep a relationship with other children in the group?	1	4	1	3	$p > 0.05$
(6) Was the child more motivated in doing homework at home?	4	1	0	4	$p = 0.05$
(7) Was the child motivated to follow the sessions?	4	1	2	2	$p > 0.05$

EG = Experimental group; CG = Control group.

3. Results

3.1. Session Attendance

EG children achieved 100% attendance in each reading session, which was statistically higher than the 75% of CG children (range 25–100%, U = 11.0, z = −3.468, $p = 0.002$, Figure 1). In particular, in CG children, attendance was significantly different on day 9 (Fisher's exact test $p = 0.04$) and day 10 (Fisher's exact test $p = 0.05$) compared with the other days.

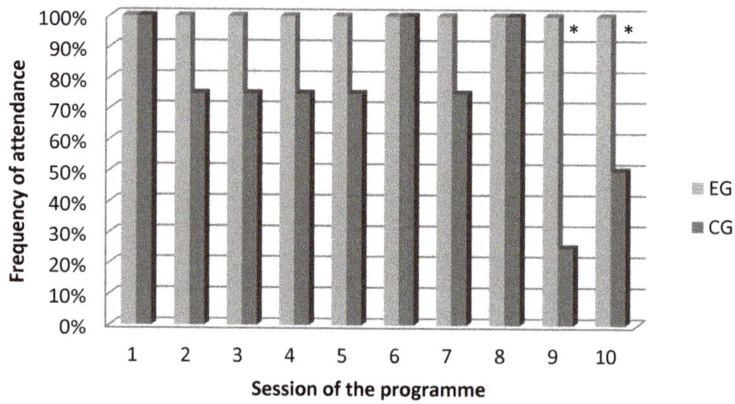

Figure 1. Children's attendance. EG and CG children. EG = Experimental group; CG = Control Group; * $p < 0.05$.

3.2. *Reading Tests*

We explored the two domains of MT2, namely speed (S) and accuracy (A), the reading comprehension (RC) for the TOR test, and the five domains [18] for the MCF test: Recognition (RE) Fluidity (F), Phonemic (FO), Segmentation (SG), Letter deletion (LD). On all reading tests, no significant differences were found between T0 and T1 within each group (Mann–Whitney U test, $p > 0.05$) or even between the EG and CG groups at each time point (Wilcoxon signed-rank test, $p > 0.05$) (Tables 4 and 5).

Table 4. Statistical results of the MT2 and TOR tests at baseline and at the end of the reading sessions ($p > 0.05$). Mean ± Standard Deviation is reported.

Group	Time Points	MT2-S	MT2-A	TOR-RC
Experimental	T0	1.5 ± 1.2	7.3 ± 2.3	53.5 ± 12.0
	T1	1.7 ± 1.5	7.3 ± 2.3	64.0 ± 26.8
Control	T0	2.7 ± 0.4	8.3 ± 2.9	76.5 ± 33.2
	T1	3.0 ± 0.3	8.3 ± 2.8	72.3 ± 24.5

MT2-S = Cornoldi reading test speed; MT2-A = Cornoldi reading test accuracy; TOR-RC = TOR test reading comprehension.

Table 5. Statistical results of the CMF tests at baseline and at the end of the reading sessions ($p > 0.05$). Mean ± Standard Deviation is reported.

Group	Time Points	RE	F	FO	SG	LD
Experimental	T0	30.0 ± 28.2	33.3 ± 14.4	35.0 ± 25.9	37.5 ± 17.6	21.2 ± 21.3
	T1	26.6 ± 22.5	28.3 ± 20.2	35.0 ± 25.9	35.0 ± 25.9	35.0 ± 2.9
Control	T0	30.0 ± 28.8	50.0 ± 0.0	27.5 ± 31.8	27.5 ± 31.8	27.5 ± 31.8
	T1	50.0 ± 0.0	50.0 ± 0.0	30.0 ± 28.2	30.0 ± 28.2	50.0 ± 0.0

RE = Recognition; F = Fluidity; FO = Phonemic; SG = Segmentation; LD = Letter deletion.

3.3. *Cognitive Test: WISC IV Test and Vineland Tests*

The five domains of the WISC IV test have been explored: Intelligence Quotient (IQ), Fluid Reasoning Index (RF), Processing Speed Index (PS), Verbal Comprehension Index (VC), Working Memory Index (WM). Vineland test's domains have been analyzed: Compressive Results (CO), Communication (CM), Daily Living Skills (DLS), Socialization (S), and Motor Skills (MS).

On all the cognitive tests, no significant differences were found between T0 and T1 within each group (Mann–Whitney U test, $p > 0.05$), as well as between the EG and CG groups at each time point (Wilcoxon signed-rank test, $p > 0.05$) (Tables 6 and 7).

Table 6. Statistical results of the WISC tests at baseline and the end of the reading sessions ($p > 0.05$). Mean ± Standard Deviation is reported.

Group	Time Points	IQ	RF	PS	VC	WM
Experimental	T0	75.2 ± 16.4	84.0 ± 18.7	79.0 ± 13.34	84.7 ± 5.6	75.5 ± 16.2
	T1	75.0 ± 16.6	82.5 ± 20.2	80.5 ± 11.4	84.7 ± 5.6	74.0 ± 17.4
Control	T0	108.2 ± 24.7	86.0 ± 33.1	106.5 ± 15.7	102.0 ± 26.9	133.0 ± 0.0
	T1	100.0 ± 25.2	83.0 ± 35.5	103.7 ± 10.3	96.0 ± 21.7	92.5 ± 44.5

IQ = Intelligence Quotient; RF = Fluid Reasoning Index; PS = Processing Speed Index; VC = Verbal Comprehension Index; WM = Working Memory Index.

Table 7. Statistical results of the Vineland tests at baseline and the end of the reading sessions ($p > 0.05$). Mean ± Standard Deviation is reported.

Group	Time Points	CO	CM	DLS	S	MS
Experimental	T0	53.7 ± 19.6	69.2 ± 25.8	45.0 ± 8.3	50.0 ± 17.1	46.5 ± 9.1
	T1	76.3 ± 29.2	97.0 ± 36.7	76.3 ± 29.6	62.6 ± 22.1	48.0 ± 0.0
Control	T0	63.4 ± 26.1	74.8 ± 29.8	50.4 ± 10.7	55.0 ± 19.0	40.0 ± 0.0
	T1	78.5 ± 34.6	99.0 ± 45.2	78.0 ± 36.8	65.5 ± 21.9	55.0 ± 0.0

CO = Compressive Results; CM = Communication; DLS = Daily Living Skills; S = Socialization; MS = Motor Skills.

3.4. Parents' Questionnaire

Questions and answers reported by the parents of the EG and CG children are shown in Table 3.

4. Discussion

In the present study, we evaluated a programme that aims to understand the impact of 10 weekly reading sessions with dogs on children with ASD to read. Attendance and parents' perceptions were evaluated. Validated reading and social tests were employed prior to the beginning and after the end of the programme in order to offer an evidence-based evaluation approach. To our knowledge, this is the first time that tests measuring reading and social skills have been applied to assess the effectiveness of a reading-to-dog programme in children with ASD. Being willing to interact with dogs has been considered as inclusion criteria: this makes it difficult to generalize the results to all children with ASD, although it might be applicable to other children with ASD who happen to like dogs (or at least not dislike them). However, this study wants to work as a pilot in the reading-to-dog programme field.

Motivation has been defined as "a psychological process in which personality traits (e.g., motives, reasons, skills, interests, expectations, and future perspectives) interact with perceived environmental characteristics" [25]. Thus, student motivation can be affected by changes in their learning environment. In our study, the reading-to-dog programme significantly increased the propensity of children to read at home and look autonomously for a book, as showed by EG parents scoring higher on the related questions of the survey compared with CG parents immediately after the end of the programme. This is in line with what is reported in reading studies [1], in which motivation is often discussed in terms of intrinsic motivation (motivated from internal factors; e.g., curiosity to read, enjoyment of the experience) and extrinsic (motivated by external factors; e.g., to get a good grade). Children in the EG group were also perceived by their parents as having a significantly higher motivation to follow reading sessions. EG children actually attended the sessions significantly more frequently (100% attendance) than those in the control group (25% to 100% attendance). According to Newman-Ford et al. [26] attendance is a measure of a student's motivation for learning, which is considered a galvanizing energy in the learning process. From this perspective, it is not surprising that EG children were significantly more motivated to do homework at home than controls, as reported by their parents. The dog might have acted as motivator for children to attend, which might be due to a dog's recognized ability to be an active, nonjudgmental listener [1]. As reported in [27], "The dogs 'listened' while the students were reading at their own pace. The dogs did not laugh, judge or criticize them, and therefore they were not embarrassed by their own mistakes". Moreover, in [28]), children with autism interacted most frequently and for the longest periods with a real dog in comparison with objects or a person. The presence of the dog assumed an important role during the session. The authors concluded that students reading in the presence of a dog were more likely to participate in reading-to-dog sessions, because that was an environment in which they could build their self-confidence [27]. Also. children with pervasive developmental disorders (including autism) were more playful in interaction with a live dog compared with toys, and also more aware of their social environment in the presence of the dog [29]. However, in the questionnaire, when parents were asked to answer to a specific question about "attention to dogs", no differences between EG and CG were reported. We wanted to ask this question in order to understand if the children with ASD were more aware of the social environment, as reported in literature [29].

In our study, children's engagement in social interactions with peers was not increased at the end of the programme in both groups, according to parents' perceptions. Similarly, Socialization Area results obtained on Vineland tests (for example, the score related to Plays with peer/s for 5 min under supervision, Plays with peer/s for 20 min under supervision, Asks others to play or spend time together) showed no improvements in social skills of children from both groups when the programme was over. This is in contrast with what was reported in the study by Grigore et al., [14], in which the author found improved social interactions in three preschool autistic children following a combined social story method and canine-assisted intervention. As far as we know, there are no other published researches conducted with children with ASD reporting results based on engagement in social interactions with peers. Paul and Serpell [30] found that normal families who obtained a dog, 1 month later engaged in more leisure activities together and their children were more often visited by friends. In a classroom of first-graders, the presence of a dog led to a better social integration among students, as documented via indirect psychometric indicators [31] as well as via direct behavior observation [32].

The possible role of the Oxytocin (OT) in these child–dog interactions during reading-to-dog sessions needs to be underlined too. Nagasawa et al. [33] assessed the effect of 30 min of interaction between dogs and their owners, particularly the duration of friendly gazes from the dogs to the owners. In a control condition lasting for 30 min, owners were instructed not to look at their dogs directly. In the normal interaction condition, longer gaze was linked to higher OT levels in the owner, while this was not the case in the control condition without eye contact. The interaction, even without direct contact, is related to OT increases that are strictly related to social interaction (see [34] for a detailed review). The release of OT via contact with animals may contribute to explain many of the effects of dog–human interactions.

As for both groups, we found no significant gains in children's reading test (MT2, TOR, MCF) scores after taking part in our reading-to-dog programme. In contrast, Konarski et al. [11] reported improved Dynamic Indicators of Early Literacy Skills (DIBELS) and Elementary Reading Attitudes Scale (ERAS) in a child with autism spectrum disorder after completing a reading-to-dog intervention. However, this was a case study, which did not use any control measures or include a case series, and therefore it does not allow to conclude that any change observed is due to the intervention being studied rather than to other factors. Several other authors described positive effects of reading-to-dog programmes in children (see [1] for review). For example, Fisher et al. [35] applied the Neale Analysis of Reading Ability [36] to test reading abilities in one child, before and after participating in a BaRK programme. BaRK is a free programme that involves reluctant readers in the middle-upper primary school classes. In this programme, a child was involved in eight weekly reading sessions with a dog. The results indicated a dramatic improvement between pretest and post-test scores for both reading accuracy and comprehension, with greater gains being made in comprehension skills. In [37], 26 children had higher scores after reading to a dog on the Gray Oral Reading Test (GORT-4), in which the child has to read aloud narrative passages (of medium length) and, for each passage, answer to multiple-choice comprehension questions read by the examiner. These results were supported by those collected by The Intermountain Therapy Animal [38] that indicated students' reading skills improved by two to four grade levels during a reading programme. However, again, failure to use appropriate controls makes it impossible to draw meaningful conclusions from these studies. Booten [39] and Petersen [40] included a control group in their investigation, and they did not report any differences between children who read to a dog and those who read without a dog. Conversely, Treat et al. [41] found improved reading fluency, accuracy, and comprehension after reading to a dog, while in the study by LeRoux et al. [27] children in the dog group scored higher on the Neale reading comprehension test compared with the control groups, and Kirnan [42] found an improvement in reading skills based on teachers perception. It should be noted that all these studies involved typical children, making it difficult to compare results with ours. In fact, a meta-analysis by Fuchs [43] revealed that the reading achievement of students with a learning disability is significantly different from that of typical students, even if low-achieving students are considered: students with learning disabilities have more severe reading problems than others [43]. Overall, children with ASD can be characterized by a triad of persistent impairments with core deficits in social interaction, language, and communication, as well as restrictive, repetitive thoughts, routines, and behavior patterns: ASD and learning disability are then co-associated. ASD is more likely to be present in individuals with a learning disability, impacting on all aspects of learning, especially among more severely affected individuals [44]. In our study, children with ASD had to follow important but easy rules related to the setting. The reason for this limitation is dual. As for dogs, this allowed activities to be predictable and controllable [17,45]. For children, it was a way to receive a simple but useful rule.

There are some limitations to our study, so the findings should be interpreted with caution. First, the programme involved a small sample size and did not control for the confounding effect of variables, including parenting styles but also comorbid outcomes such as anxiety, which makes it difficult to generalise to a wide population. Second, although parent-completed questionnaires are considered as accurate as developmental screening instruments (see [46] for example), parents were required to interpret their children's motivation and attitudes, inevitably resulting in a degree of subjectivity. In addition, it is possible that the parents' answers were influenced by perceptions of which answers would be deemed acceptable, even if the questionnaire was anonymous. Third, we implemented a short-term intervention, and future studies should examine interventions over a longer time (e.g., the entire school year), possibly analyzing academic performances. However, a standard programme for Children with ASD has not been developed and validated yet [42].

5. Conclusions

In conclusion, reading to a dog has the potential to bring significant improvements to typical children's social and reading abilities [1]. The results of the present pilot study suggest that such a programme can have specific effects on session attendance and literacy motivation at home in children with ASD, as perceived by their parents. Previous research demonstrated that increased engagement in reading is linked to improved academic performance [47,48]. Thus, the attendance at (and engagement in) reading sessions, enriched by the presence of a dog should be further examined, together with the critical aspects of literacy, including testing accuracy, fluency, and comprehension.

The spatial setting used in this pilot can be applied in following studies in order to create a perfect welfare area for dogs and to take the chance to teach a rule to the children with ASD. In order to evaluate the success of a reading-to-a-dog programme, not only validated tests but also percentage of attendance and parents' perceptions should be taken into account. The next step should include large-scale, randomized control trials with longitudinal examinations of effects, to provide more tangible and reliable findings not only for children with ASD but also for dogs.

A recent review [1] reported positive results based on implementation of a reading-to-dog programme. Unfortunately, these are mostly based on ad-hoc reports, without undergoing a peer-review process [42]. These studies did not randomly allocate children to intervention or control groups, and only small groups (or case study) were investigated. Although extensive generalization should be avoided, the results of our study provide some tentative support for the effectiveness of a reading-to-dog programme based on the use of objective assessments. Specific tests did not confirm any effect on children's social and literacy skills due to the presence of a dog. More research is therefore needed to understand the impact of this type of intervention, considering potential confounding variables, including individual factors or a different number of sessions.

Author Contributions: The idea for the paper was conceived by S.U. and F.P. The experimental protocol was designed by S.U. and F.P. and carried out by S.U. and S.F. The data was statistically analyzed by S.U., and discussed by all authors (S.U., M.A., L.P., S.F., F.P.). The paper was written by S.U. and F.P., and revised by all authors (S.U., M.A., L.P., S.F., F.P.).

Funding: This project was supported by the Nestlé Purina Sponsorship for Human Animal Bond Studies and by LINEA2FPIRR_2017 University of Milan (19703_PSR2015-1716MDI_G_M).

Acknowledgments: We are grateful to all the children and family who participated in the study. Thanks to Effetto Palla Onlus for providing dogs for the study and Giulia Schirosi for the help in video analysis.

Conflicts of Interest: The authors declare no conflict of interest. The funding sponsors had no role in the design of the study; in the collection, analyses, or interpretation of data; in the writing of the manuscript, and in the decision to publish the results.

Ethics Statement: The study was performed according to the Declaration of Helsinki. The ethics committee of the CTR Onlus approved the study, protocol number CR3/403-18, 2018. No invasive intervention or drug experimentation on the dogs was performed; therefore, the application of Legislative Decree No 26/2014 and Directive2010/63/EU for the protection of animals used in scientific and experimental studies were not required. The Effetto Palla Onlus provided consent for use of the dogs in the study.

References

1. Hall, S.S.; Gee, N.R.; Mills, D.S. Children Reading to Dogs: A Systematic Review of the Literature. *PLoS ONE* **2016**, *11*, e0149759. [CrossRef] [PubMed]
2. Shaw, D.M. Man's Best Friend as a Reading Facilitator. *Read. Teach.* **2013**, *66*, 365–371. [CrossRef]
3. McKinney, T. Tales of Joy R.E.A.D.® Program. 2014. Available online: https://www.talesofjoyread.com/ (accessed on 25 July 2019).
4. Mallon, G.P. Some of our best therapists are dogs. *Child Youth Care Forum* **1994**, *23*, 89–101. [CrossRef]
5. Pillow-Price, K.; Yonts, N.; Stinson, L. Improving Literacy Skills Using Dogs! Celebrating the Writing Process Early Use of Technology Outdoor Learning. 2014. Available online: https://docplayer.net/66124185-Improving-literacy-skills-using-dogs-celebrating-the-writing-process-early-use-of-technology-volume-42-number-1-2014.html (accessed on 25 July 2019).
6. Sorin, R. The Impact of the Classroom Canines Program on Children's Reading, Social and Emotional Skills and Motivation to Attend School. *Intern. J. Lit.* **2015**, *22*, 23–35. [CrossRef]
7. Guthrie, J.T.; Cox, K.E. Classroom Conditions for Motivation and Engagement in Reading. *Educ. Psychol. Rev.* **2001**, *13*, 283–302. [CrossRef]
8. Hidi, S.; Harackiewicz, J.M. Motivating the Academically Unmotivated: A Critical Issue for the 21st Century. *Rev. Educ. Res.* **2000**, *70*, 151–179. [CrossRef]
9. Friedmann, E.; Katcher, A.H.; Thomas, S.A.; Lynch, J.J.; Messent, P.R. Social interaction and blood pressure. Influence of animal companions. *J. Nerv. Ment. Dis.* **1983**, *171*, 461–465. [CrossRef]
10. Viau, R.; Arsenault-Lapierre, G.; Fecteau, S.; Champagne, N.; Walker, C.-D.; Lupien, S. Effect of service dogs on salivary cortisol secretion in autistic children. *Psychoneuroendocrinology* **2010**, *35*, 1187–1193. [CrossRef]
11. Konarski, J. More Than Just A Class Pet: Reading Therapy Dogs and Young Readers With Autism. Available online: http://jkonarski.cmswiki.wikispaces.net/file/view/More+Than+Just+A+Class+Pet-+Reading+Therapy+Dogs+And+Young+Readers+With+Autism.rtf (accessed on 25 July 2019).
12. American Psychiatric Association. *Diagnostic and Statistical Manual of Mental Disorders (DSM-5®)*; American Psychiratric Publishing, Inc.: Arlington, VA, USA, 2013.

13. Ghaziuddin, M.; Ghaziuddin, N.; Greden, J. Depression in Persons with Autism: Implications for Research and Clinical Care. *J. Autism Dev. Disord.* **2002**, *32*, 299–306. [CrossRef]
14. Grigore, A.A.; Rusu, A.S. Interaction with a therapy dog enhances the effects of social story method in autistic children. *Soc. Anim.* **2014**, *22*, 241–261. [CrossRef]
15. Gray, C.A.; Garand, J.D. Social Stories: Improving Responses of Students with Autism with Accurate Social Information. *Focus Autistic Behav.* **1993**, *8*, 1–10. [CrossRef]
16. Hatch, A. The view from all fours: A look at an animal-assisted activity program from the animals' perspective. *Anthrozoos* **2007**. [CrossRef]
17. Uccheddu, S.; Albertini, M.; Pierantoni, L.; Fantino, S.; Pirrone, F. Assessing behaviour and stress in two dogs during sessions of a reading-to-a-dog program for children with pervasive developmental disorders. *Dog Behav.* **2018**, *4*, 3.
18. Cornoldi, C.; Oakhill, J. *Reading Comprehension Difficulties: Processes and Intervention*; L. Erlbaum Associates: Mahwah, NJ, USA, 1996; ISBN 0805818456.
19. Levorato, M.C.; Roch, M. *TOR 3-8. Valutazione Della Comprensione del Testo Orale*; Giunti Editori: Firenze, Italy, 2007.
20. Marotta, L.; Trasciani, M.; Vicari, S. *Test CMF. Valutazione Delle Competenze Metafonologiche. Con CD-ROM*; Erickson Editori: Trento, Italy, 2008.
21. Weiss, L.G.; Keith, T.Z.; Zhu, J.; Chen, H. WISC-IV and Clinical Validation of the Four- and Five-Factor Interpretative Approaches. *J. Psychoeduc. Assess.* **2013**, *31*, 114–131. [CrossRef]
22. Sparrow, S.S. Vineland Adaptive Behavior Scales. In *Encyclopedia of Clinical Neuropsychology*; Springer: New York, NY, USA, 2011; pp. 2618–2621.
23. Park, H. Home literacy environments and children's reading performance: A comparative study of 25 countries. *Educ. Res. Eval.* **2008**, *14*, 489–505. [CrossRef]
24. Benjamini, Y.; Hochberg, Y. Controlling the False Discovery Rate: A Practical and Powerful Approach to Multiple Testing. *J. R. Stat. Soc. Ser. B* **1995**, *57*, 289–300. [CrossRef]
25. Willy, L.; Matos, L.; Vansteenkiste, M. *Educação.*; Pontifícia Universidade Católica do Rio Grande do Sul: Porto Alegre, RS, Brazil, 2008; Volume 31.
26. Newman-Ford, L.; Fitzgibbon, K.; Lloyd, S.; Thomas, S. A large-scale investigation into the relationship between attendance and attainment: A study using an innovative, electronic attendance monitoring system. *Stud. High. Educ.* **2008**, *33*, 699–717. [CrossRef]
27. Le Roux, M.C.; Swartz, L.; Swart, E. The Effect of an Animal-Assisted Reading Program on the Reading Rate, Accuracy and Comprehension of Grade 3 Students: A Randomized Control Study. *Child Youth Care Forum* **2014**, *43*, 655–673. [CrossRef]
28. Prothmann, A.; Ettrich, C.; Prothmann, S. Preference for, and Responsiveness to, People, Dogs and Objects in Children with Autism. *Anthrozoos* **2009**, *22*, 161–171. [CrossRef]
29. Martin, F.; Farnum, J. Animal-assisted therapy for children with pervasive developmental disorders. *West J. Nurs. Res.* **2002**, *24*, 657–660. [CrossRef]
30. Paul, E.S.; Serpell, J.A. Obtaining a new pet dog: Effects on middle childhood children and their families. *Appl. Anim. Behav. Sci.* **1996**, *47*, 17–29. [CrossRef]
31. Hergovich, A.; Monshi, B.; Semmler, G.; Zieglmayer, V. The effects of the presence of a dog in the classroom. *Anthrozoos* **2002**, *15*, 37–50. [CrossRef]
32. Kotrschal, K.; Ortbauer, B. Behavioral effects of the presence of a dog in a classroom. *Anthrozoos* **2003**, *16*, 147–159. [CrossRef]
33. Nagasawa, M.; Kikusui, T.; Onakab, T.; Ohtaa, M. Dog's gaze at its owner increases owner's urinary oxytocin during social interaction. *Horm. Behav.* **2009**, *55*, 434–441. [CrossRef] [PubMed]
34. Beetz, A.; Julius, H.; Turner, D.; Kotrschal, K. Effects of Social Support by a Dog on Stress Modulation in Male Children with Insecure Attachment. *Front. Psychol.* **2012**, *3*, 352. [CrossRef] [PubMed]
35. Fisher, B.J.; Cozens, M. The BaRK (Building Reading Confidence for Kids) Canine Assisted Reading Program: One Child' s Experience. *Lit. Learn. Middle Years* **2014**, *22*, 70.
36. Neale, M. *Neale Analysis of Reading Ability*; AceReader: Grand Junction, CO, USA, 1999.
37. Smith, K.A. Impact of Animal Assisted Therapy Reading Instruction on Reading Performance of Homeschooled Students. Ph.D. Thesis, Northcentral University, San Diego, CA, USA, 2010.

38. Intermountain Therapy Animals R.E.A.D. Available online: http://www.therapyanimals.org/R.E.A.D.html (accessed on 14 October 2018).
39. Elizabeth Booten, A. Effects of Animal-Assisted Therapy on Behavior and Reading in the Classroom. Digital Scholar Theses, Marshall University, Hungtington, WV, USA, 2011.
40. Petersen, A. The Effect of Experiences with Animals on the reading Comprehension Skills of Students in the Seventh Grade. Ph.D. Thesis, Alliant International University, Alhambra, CA, USA, 2008.
41. Treat, W.A. Animal-Assisted Literacy Instruction for Students with Identified Learning Disabilities: Examining the Effects of Incorporating a Therapy Dog into Guided Oral Reading Sessions. Ph.D. Thesis, University of California, Santa Cruz, CA, USA, 2013.
42. Kirnan, J.; Siminerio, S.; Wong, Z. The Impact of a Therapy Dog Program on Children's Reading Skills and Attitudes toward Reading. *Early Child. Educ. J.* **2016**, *44*, 637–651. [CrossRef]
43. Fuchs, L.S.; Fuchs, D. Mathematical Problem-Solving Profiles of Students with Mathematics Disabilities With and Without Comorbid Reading Disabilities. *J. Learn. Disabil.* **2002**, *35*, 564–574. [CrossRef]
44. O'Brien, G. Autism and learning disability. *Autism* **2004**, *8*, 125–140. [CrossRef]
45. Pirrone, F.; Ripamonti, A.; Garoni, E.C.; Stradiotti, S.; Albertini, M. Measuring social synchrony and stress in the handler-dog dyad during animal-assisted activities: A pilot study. *J. Vet. Behav. Clin. Appl. Res.* **2017**, *21*, 45–52. [CrossRef]
46. Glascoe, F.P.; Dworkin, P.H. The Role of Parents in the Detection of Developmental and Behavioral Problems. *Pediatrics* **1995**, *95*, 829–836. [PubMed]
47. Klem, A.M.; Connell, J.P. Relationships Matter: Linking Teacher Support to Student Engagement and Achievement. *J. Sch. Health* **2004**, *74*, 262–273. [CrossRef] [PubMed]
48. Marks, M.; Zaccaro, S.; Mathieu, J. Performance implications of leader briefings and team-interaction training for team adaptation to novel environments. *J. Appl. Psychol.* **2000**, *85*, 971–986. [CrossRef] [PubMed]

animals

MDPI

Article

Thyroid and Lipid Status in Guide Dogs During Training: Effects of Dietary Protein and Fat Content

Biagina Chiofalo *, Esterina Fazio, Salvatore Cucinotta and Cristina Cravana

Department of Veterinary Sciences, University of Messina, Polo Universitario Annunziata, 98168 Messina, Italy
* Correspondence: biagina.chiofalo@unime.it; Tel.: +39-0906766833

Received: 18 July 2019; Accepted: 19 August 2019; Published: 23 August 2019

Simple Summary: Nutrition is one of the main causes of thyroid response and energetic metabolism. Presently, there is a lack of information on the physiological effect of moderate activities in canines, particularly in guide dogs. Aim was to compare the effect of diet on thyroid and lipid status in guide dogs, during a 12-weeks training, fed two commercial diets, one, HPF, characterized by low-carbohydrate/high-protein/high-fat (29:39:19% as fed) and the other, LPF, characterized by high-carbohydrate/low-protein/low-fat (50:24:12% as fed) content. Our hypothesis was that the intake of a diet rich in fat and protein would have given a better response than the carbohydrate-rich diet for thyroid and lipid homeostasis to cope with the increased energy demands of dogs subjected to the training period. Results evidenced that the consumption of a diet rich in fat and protein appears the nutritional plan most suitable to support moderate exercise for guide dogs during the training work; this diet supports maintenance of body weight, Body Condition Score (BCS), and serum baseline thyroid and lipid profiles, offering potential improvements in dogs' performances. However, the chronic ingestion of diets that are extreme in their composition of either fat or carbohydrate should be always approached with caution.

Abstract: Nutrition plays a leading role that most influences thyroid response and energetic metabolism. Aim was to compare the effect of diet on thyroid and lipid status in guide dogs during a 12-weeks training period. Eight Labrador Retrievers were divided into two groups homogeneous for sex, age, body weight, and Body Condition Score (BCS) and fed two commercial diets one, HPF, characterized by low-carbohydrate/high-protein/high-fat (29%:39%:19% as-fed) and the other, LPF, by high-carbohydrate/low-protein/low-fat (50%:24%:12% as-fed) content. The serum thriiodothyronine (T_3), thyroxine (T_4), cholesterol (CHOL), triglycerides (TAGs) and non-esterified fatty acids (NEFA) were determined at Day 0, 28, 56, and 84, before the daily training. Statistical model included the effects of Diet (HPF vs. LPF) and Time (Day 0 to Day 84), and their interaction. In the HPF group, Diet significantly ($p < 0.01$) increased T_4, CHOL, and TAGs and decreased NEFA. In both groups, Time significantly ($p < 0.05$) increased T_4 and TAGs, CHOL at Day 28, and NEFA at Day 56. The interaction did not influence serum hormones and lipid pattern. The adjustments in thyroid and lipid responses to moderate exercise in HPF group were driven mainly by the nutrient composition of the diet in relation to the involvement of metabolic homeostasis.

Keywords: guide dog; dietary protein/fat ratio; training; thyroid hormones; lipid panel

1. Introduction

Thyroid hormones are known to play a pivotal role in growth regulation, cognitive issues, cellular function and metabolic implication [1,2]. It was shown that the hypothalamic-pituitary-thyroid (HPT) axis activity decreases in response to food restriction, which is frequently interpreted to be an energy-saving mechanism [3]. It has also been speculated that energy signaling, like obesity and energy restriction, alters thyroid homeostasis in dogs, with significant higher T_3 and T_4 concentrations in obese dogs than lean dogs [4]. Moreover, the dietary protein quality and quantity content could change HPT axis activity; in particular, the long-term low-protein diet affects the thyroid axis activity, with the effect similar to that caused by starvation [5]. Some studies have shown a correlation between T_3 concentrations and resting metabolic requirements [6,7]; nevertheless, no correlation between body weight and serum T_3 concentrations was observed [4]. Moreover, serum concentrations of adiponectin were significantly and negatively associated with T_4 concentrations, and positively correlated with cholesterol [8]. It is interesting to note that there is also an evidence that thyroid hormones and lipoprotein alterations may have a role in susceptibility of dogs to infectious diseases [9]. Thus, thyroid hormones in dogs may be involved in the regulation of fatty acid delta-6-desaturase activity [10]. The background of the thyroid signal transduction [11] and lipid metabolism in dogs exhibits some unique characteristics compared to other species, and hyperlipidemia is common in dogs [12].

The primary plasma changes, that are needed to support long-lasting activity, are related to lipid metabolism and, for this reason, the lipid pattern is frequently assayed during clinical evaluation [13]. Fatty acids are an important source of energy for skeletal contraction, particularly during exercise of mild-moderate intensity, prolonged duration and in the fasting state [14]. Plasma free fatty acids (FFAs) transported from remote adipose tissue stores and triglycerides (TAGs) contained within skeletal muscle fibers are the major sources of these fatty acids. The relative contribution of each source is dependent on the mode, intensity and duration of exercise and on training status.

Cholesterol arrives in the small intestine from both the diet and bile. The liver –not the diet– is therefore the primary source of cholesterol available for intestinal absorption, a point that is often underappreciated [15].

The relative contribution of fat and carbohydrate changes according to intensity and duration of exercise, the physical training state and the environmental conditions in which the animal is working [16]. With light prolonged exercise, there is a progressively greater use of fat until it can contribute up to 80% of the caloric expenditure. Consumption of a diet rich in fat and protein produces a shift toward a greater use of fat, with a concomitant reduction of both the intensity and duration of effort that can be sustained. Conversely, ingestion of a carbohydrate-rich diet increases the percentage of carbohydrate used and increases endurance [17].

To date, canine athlete physiology studies have primarily focused on endurance sled dog racing and high intensity short duration Greyhound racing. There is a lack of information on the physiological and biochemical changes of low intensity endurance activities in canines, particularly in the guide dog [18]; this information is important in the determination of fitness level, detection of exercise-induced injury and improvement of success, with the development of more specific training programs [19].

Our hypothesis was that the intake of a diet rich in fat and protein would have given a better response than the carbohydrate-rich diet for thyroid and lipid homeostasis to cope with the increased energy demands of dogs subjected to the training protocol.

The aim of this study was to compare the dietary effect of two different concentrations of protein/fat/carbohydrates ratio on total iodothyronines and lipid panel in guide dogs for the blinds (GDB) during a 12-weeks training period.

2. Materials and Methods

Operative procedures and animal care were carried out in compliance with guidelines of Good Clinical Practices [20] and European regulation [21]. On the basis of the Italian regulation on animal experimentation and ethics [22], the research received the institutional approval by the Ethical Animal Care and Use Committee of the Department of Veterinary Science, of the University of Messina on 19 October 2016, Codex 006/2016 bis.

2.1. Animals and Diets

The study was carried out on eight neutered adult Labrador Retriever dogs, clinically health, housed at the Regional Centre Helen Keller (www.centrohelenkeller.it)of the Italian Blind and Guide Dog School Union, in Messina (Italy), during the training work to guide service for the blind. The Centre is unique in Italy and it is a member of the International Guide Dog Federation (IGDF), and as such accredited to the highest international standards.

The trial was carried out on dogs without a history of diabetes or hypo- or hyperthyroidism. Dogs admitted to the study were divided into two groups homogeneous for sex (2 male, 2 females), age (HPF: 18.3 months; LPF: 17.5 months), initial BW-body weight (HPF: 26.9 kg; LPF: 26.5 kg) and Body Condition Score-BCS (HPF: 4.33; LPF: 4.5, score 1–9). The first group, called HPF group, received a "performance" diet, characterized by low-carbohydrate/high-protein and fat diet (29%:39%:19% as-fed), whereas, the second group, called LPF group, received a "normal maintenance" diet characterized by high-carbohydrate/ low-protein and fat diet (50%:24%:12% as-fed).

Dogs were individually housed in pens of six square meters, adjacent to a large outdoor space where they could access during rest, and food was administered two times a day, in an individual bowl.

The trial was preceded by 7 days of adaptation period to the experimental diets; anonymous was a normal maintenance diet usually used in the Centre along training work. The adaptation diet was constituted by the mixture of the feeds; the HPF group received a mixture of anonymous with "performance diet" and the LPF group a mixture of anonymous and "normal maintenance diet"; during the 7 days, the anonymous was progressively replaced by the experimental diets. The quantity of administered diet was the same previously adopted by the breeder.

During the trial, the Company sent three lots of feed. Each lot was separately sampled and analyzed, as described by Chiofalo et al. [23].

Both the experimental diets of the Farmina Pet Foods line contained lamb meal as main protein source and, from a qualitative point of view, the same ingredients, analytical compounds, nutritional additives and antioxidants (tocopherol-rich extracts of natural origin). The information on the chemical composition of "performance" and "normal maintenance" diets is reported in the Table 1.

The amount of feed daily administered to each dog was calculated on the ratio between the calculated metabolizable energy requirements, as proposed by Hand et al. [24], for dogs that perform work, characterized by a moderate duration and frequency and the caloric density of metabolizable energy (ME) reported in the label [25], of each diet (HPF and LPF).

Table 1. Chemical composition and metabolizable energy of the diets [1].

Diet [2]	Anonymous [3]	HPF [4]		LPF [4]	
Moisture, g/100g as-fed	9.0	5.42	0.48	6.12	0.57
CP, g/100g as-fed	26	39.24	0.84	24.40	0.32
Fat, g/100g as-fed	15.50	18.69	0.51	11.78	0.29
OM, g/100g as-fed	ND	86.83	0.27	86.50	1.20
TDF, g/100g as-fed	2.80	11.59	1.13	13.03	1.46
Ash, g/100g as-fed	4.9	7.91	0.23	7.51	0.55
ME [3], kcal/kg as-fed	3900	4330	-	3423	-

CP = Crude Protein; OM = Organic Matter; TDF = Total Dietary Fibre; ME = Metabolizable Energy. ND = Not Determined; [1] Values are means ± standard deviation; [2] Anonymous was the normal maintenance diet usually used in the Centre along training work; HPF was the "performance" diet with low-carbohydrate/high-protein/high-fat diet and LPF was the "normal maintenance" diet with high-carbohydrate/low-protein/low-fat diet administered during the trial; [3] Values reported on the label; [4] Values determined analytically.

2.2. Conditioning Protocol

All dogs were conditioned for training program activities one month prior to the dietary study starts. All dogs were between 1 and 2 years of age and had been conditioned for the training program to be guide dogs in the area described below (see Section 2.3) and they fed a typical maintenance ration (see Section 2.1). Conditioning and training protocols remained the same for all dogs during the dietary trials each of which lasted 12-weeks; thus, each dog served as its own control during the dietary trial. Moreover, the dogs were accustomed to the blood collection since, before the beginning of the trial, hematological and biochemical analyses were carried out on each subject to evaluate their health status. Moreover, before the beginning of the trial, in order to assess the clinical status, all dogs were also submitted to a physical examination [26].

2.3. Training Program

The method of participant recruitment is described by Lloyd et al. [27].

The training consisted of a various phases program in which the dog gradually learned more guide work. This included leading a person in a straight line, stopping at any change in ground elevation as well as overhead obstacles and obstacle avoidance. Feed rewards were used in the guide dogs for the blind training program as a powerful motivation and reinforcement tool for learning and maintaining desired behavior. During each training session (at Day 0, day 28; Day 56 and Day 84), dogs were introduced to specific guide-work behaviors:

- Stopping at streets, regardless of the type of curb or wheelchair ramp;
- Clearing the space around the handler on the right and left sides as well as above the dog's head;
- Crossing streets on a line that efficiently reaches the up curb on the other side;
- Maintaining consistent pace and drive with the verbal cue "forward";
- Responding to the various uses of the 'hop-up' verbal cue e resuming or increasing pace; moving closer to a stopping point; or re-focusing;
- Stopping and standing calmly after the verbal cue "halt";
- Leading the handler in a 90_turn to the right and picking up the new travel line on "right";
- Leading the handler in a 90_turn to the left and picking up the new travel line on "left".

The guide dogs for the blind were trained 3 times a week. Each training session lasted approximately 60 min [23].

2.4. Physical Examination

To evaluate the performance of the studied dogs, from the Day 0 (start of the administration of the new food) to the Day 84, all dogs weekly underwent to physical examination [26] including: level of consciousness; posture and gait; hydration status; rectal temperature (°C); pulse rate; respiratory rate and breath character; perfusion indicators.

At the same time, on each animal, the BW and BCS were evaluated.

The determination of BW was measured on fasted animals, in the morning at 8:00 am, by using a digital scale.

BCS was evaluated by assigning a rating scale that ranged from 1 (too thin) to 9 (too heavy) using the table proposed by Nestle Purina [28–30].

2.5. Measurements of Hormonal and Lipid Patterns

In order to evaluate hormonal and lipid patterns in the fasting dogs, blood samples were monthly withdrawn, at Day 0, Day 28, Day 56, and Day 84 before the exercise (8:00 am).

Before the trial, the dogs were accustomed for the blood collection procedure (see paragraph *Conditioning protocol*). All samples were collected by the same operator into evacuated tubes (Venoject, Terumo®, Shibuya, Tokyo, Japan) and were immediately refrigerated at 4 °C after collection; the samples were subsequently (within 1 h) centrifuged for 15 min at 1500× *g* and collected and stored at −20 °C until their analyses. Serum total iodothyronine concentrations were analyzed in duplicate using commercial immunoenzymatic assays (RADIM, Rome, Italy). The method is based on a competitive immunoenzymatic assay and the reagents were prepared as described in the manufacturer's protocol. Total iodothyronines (T_3 and T_4) in the sample competed with T_3 and T_4 conjugated with horseradish peroxidase (conjugate) for binding to specific antibody sites of anti-T_3 and anti-T_4 coated on the wells. At the end of the incubation, all unbound material was removed by aspiration and washing. The enzyme activity which was bound to the solid phase would be inversely proportional to the concentration of T_3 and T_4 in calibrators and samples, and this was evidenced by incubating the wells with a chromogen solution (tetramethylbenzidine) in substrate buffer. Colorimetric readings were taken using a spectrophotometer at 450 nm (Sirio S, Radim/Seac Co., Rome, Florence, Italy). The sensitivities of the assays were as follows: 0 to12.3 nmol/L for T_3, and 0 to 512 nmol/L for T_4. The lower detection limits for T_3 and T_4 were 0.15 nmol/L and 12.8 nmol/L, respectively. The intraassay and interassay variance coefficients were 5.5% and 6.1% for T_3 and 4.9% and 8.4% for T_4, respectively.

Serum was analyzed for triglycerides (TAGs) using the enzymatic colorimetric method (GPO-PAP, glycerol-3-phosphate oxidase-p-aminophenazone) of McGowan et al. [31], for cholesterol (CHOL) using a modified Abell-Kendall/Levey-Brodie method [32] and for not esterified fatty acids (NEFA) by a coupled enzymatic reaction system (ACS-ACOD Method). First, Acyl CoA Synthetase (ACS) catalyzes fatty acid acylation of coenzyme A. Next, the acyl-CoA product is oxidized by Acyl CoA Oxidase (ACOD), producing hydrogen peroxide which reacts with the kit's Colorimetric Probe. The colorimetric reading was taken using a spectrophotometer at 570 nm.

2.6. Statistical Analyses

To account for the study design, a mixed model analysis of variance [33] with the fixed effects of Time (Day 0, Day 28; Day 56 and Day 84) and Diet (HPF vs. LPF) was applied. The interaction (Diet × Time) was forced into every model. Random effects in the model were individual dog. Residuals were examined for normality; in each case residuals were normally distributed. Least Squares Means (LSM) and standard error of the mean (SEM) were calculated. The comparison between LSM were performed using the Tukey test. Differences were considered significant for $p < 0.05$.

3. Results

The effect of environmental temperature is unlikely to play a significant role in this population's energy requirement. It is known that the temperatures outside of the thermoneutral zone of 20 to 30 °C will increase the energy requirements by 1 to 5 kcal • $BW^{0.75}$ per °C per day when above or below this zone [25]. During the 3 months of the study (1 March to 24 May), the dogs spent their time in thermoneutral zone (23 ± 2 °C); considering the kennel's geographic location, it was unlikely that the temperatures at night dropped below the thermoneutral zone.

3.1. Physical Examination and Body Weight

During the trial, the dogs presented adequate hydration status and rectal temperature within the physiological ranges (38.4 °C ± 0.32). The evaluation of pulse at femoral artery showed physiological characteristics about strength and quality, and the pulse rate within reference ranges for dogs (92 bpm ± 14) [26]. The mean of respiratory rate, determined visually or by auscultation as count either inspirations or expirations, was within the physiological ranges (18–29 ± 3) [26]. Mucous membrane color was pink, capillary refill time was less than 2 s.

The results of the present study regarding dog's performance were published by Chiofalo et al. [23]. Briefly, the diet influenced the animal performances (Table 2) in relation to their different protein, fat and carbohydrate contents, showing a significantly higher BW in the HPF group than those of the LPF group, as well as a significantly higher BCS in the HPF group than those of the LPF group.

Table 2. Effect of the diets on BW and BCS of the trial [1].

Items	Groups [2]				*p*-Value [3]
	HPF	**SEM**	**LPF**	**SEM**	
BW, kg	25.40 [a]	0.21	23.44 [b]	0.26	<0.001
BCS, score 1–9	4.64 [a]	0.12	4.01 [b]	0.15	0.003

BW = Body weight; BCS = Body Condition Score; [1] Values are means (LSM) ± standard error of the mean (SEM); [2] HPF = low-carbohydrate/ high-protein and fat diet; LPF = high-carbohydrate/low-protein and fat diet; [3] Probability values for the effects of Diet; [a, b] Within a row, means with different superscript letter were significantly different ($p < 0.05$).

The BW and the BCS of dogs were monitored weekly during the whole time of the 12-week feeding period. As observed by Chiofalo et al. [23], no significant differences of BW in the dogs of the HPF group from Day 0 (26.9 kg) to Day 84 (25.40 kg) were observed, whereas, the BW in the dogs of the LPF group was affected by the time showing a significant ($p < 0.05$) decrease from the beginning to the end of the trial (Day 0: 26.5 kg; Day 84: 23.44 kg). The interaction Diet × Time showed no significant differences ($p = 0.270$). This could be due to the high variability of the BW in each group during the trial. The trend of the BCS showed no significant differences in relation to the Time ($p = 0.997$) and to the interaction Diet × Time ($p = 0.991$) for the whole trial period.

3.2. Hormonal Response

As regards the trend of iodothyronine concentrations in relation to the diet (Table 3), after the 12-weeks diet intervention, T_3 concentration was significantly not influenced whereas, T_4 concentration showed significant higher mean level in HPF group than that observed in LPF group.

Table 3. Effect of the diet on serum hormonal and lipid panel concentrations for the whole trial period [1].

Items	Groups [2]				*p*-Value [3]
	HPF	**SEM**	**LPF**	**SEM**	
T_3 (nmol/L)	2.64	0.83	2.48	0.83	0.617
T_4 (nmol/L)	33.19 [a]	1.12	29.10 [b]	1.17	0.003
CHOL (mg/dL)	175.82 [a]	3.34	141.43 [b]	4.98	0.001
TAGs (mg/dL)	53.98 [a]	2.33	42.06 [b]	2.44	0.006
NEFA (mg/dL)	0.50 [b]	0.04	0.64 [a]	0.04	<0.001

T_3 = Thriiodothyronine; T_4 = Thyroxine; CHOL= Cholesterol; TAGs= Tryglicerides; NEFA= Non Esterificated Fatty Acids; [1] Values are means (LSM) ± standard error of the mean (SEM); [2] HPF = low-carbohydrate/high-protein and fat diet; LPF = high-carbohydrate/low-protein and fat diet; [3] Probability values for the effects of Diet; [a, b] Within a row, means with different superscript letter were significantly different ($p < 0.05$).

In relation to the variable Time (Table 4), the T_3 response was not significantly influenced during the trial, whereas T_4 concentration was significantly influenced, showing significant lower values at the Day 28 than those observed at Day 0, 56, and 84.

The interaction Diet × Time showed no significant differences for T_3 as well as for T_4 concentrations.

Table 4. Profile of serum hormonal and lipid panel concentrations during the trial [1].

	Groups [2]	Time [3]				SEM	p-Value [4]		
		0	28	56	84		Diet	Time	D × T
T_3 (nmol/L)	HPF	2.58	2.69	2.65	2.65	0.39	0.617	0.963	0.872
	LPF	2.42	2.54	2.50	2.50	0.38			
T_4 (nmol/L)	HPF	32.06 ^{a}	28.69 ^{b}	36.44 ^{a,x}	34.30 ^{a,x}	1.19	0.003	0.023	0.274
	LPF	33.91 ^{a}	27.77 ^{b}	29.23 ^{b,y}	29.47 ^{b,y}	1.16			
CHOL (mg/dL)	HPF	172.60 ^{b}	182.40 ^{a}	168.00 ^{b,x}	180.60 ^{a,x}	3.82	0.001	0.031	0.449
	LPF	161.50 ^{b}	176.50 ^{a}	131.50 ^{by}	154.00 ^{b,y}	4.54			
TAGs (mg/dL)	HPF	64.30 ^{a}	60.30 ^{a,x}	42.33 ^{b}	48.00 ^{b}	2.12	0.006	0.002	0.373
	LPF	54.00 ^{a}	46.00 ^{a,b,y}	42.00 ^{b}	32.00 ^{b}	2.34			
NEFA (mg/dL)	HPF	0.26 ^{b}	0.54 ^{b,y}	0.70 ^{b}	0.53 ^{b,y}	0.03	<0.001	<0.001	0.126
	LPF	0.36 ^{b}	0.62 ^{b,x}	0.82 ^{b}	0.74 ^{a,x}	0.03			

T_3 = Thriiodothyronine; T_4 = Thyroxine; CHOL= Cholesterol; TAGs= Tryglicerides; NEFA= non-esterified fatty acids. [1] Values are means (LSM) ± standard error of the mean (SEM); [2] HPF = low-carbohydrate/ high-protein and fat diet; LPF = high-carbohydrate/low-protein and fat diet; [3] Blood sampling at Day 0, Day 28, Day 56, and Day 84, before the exercise; [4] Probability values for the effects of Diet, Time, and Diet × Time; [a,b] Within row, means with different superscripts letter were significantly different ($p < 0.05$) due to time; [x,y] Within column, means with different superscript letter were significantly different ($p < 0.05$) due to diet.

3.3. *Lipid Pattern*

As regards the trend of lipid pattern in relation to the Diet (Table 3), after the 12-weeks diet intervention, CHOL and TAGs concentrations showed significant higher mean levels, whereas NEFA levels showed significant lower values in HPF group than those observed in LPF group.

CHOL, TAGs, and NEFA levels were significantly influenced by the Time (Table 4), showing in both groups the highest values at the Day 28 for CHOL and at Day 0 and 28 for TAGs and at Day 56 for NEFA.

The interaction Diet × Time showed no significant differences for CHOL, TAGs, as well as for NEFA concentrations.

4. Discussion

The primary objective of this study was to examine how normal-weight Labrador dogs respond to HPF and LPF diets during a 12-week GDB training programs and to examine the relationships with the potential changes in circulating THs and lipid panel concentrations. It is reasonable to assume that the single components of diet may induce the metabolic changes, according to workload and performance quality. On these bases, the shift of energy metabolism in a catabolic or anabolic direction during training programs and exercise is characterized by a wide range of metabolic hormones changes, such as total THs, according to lipid parameters. Our hypothesis was that the HPF diet would have given a better answer than the LPF diet for thyroid homeostasis to cope with the increased energy demands of dogs subjected to the training protocol.

Guide dogs for the blind have a great social impact because of their invaluable aid in providing independent mobility to people with visual impairment; their service comes at high cost (approximately 25,000 euros) due to the large amount of resources, housing, husbandry and training, required to train such animals [34]. Furthermore, success rates ranged between 50% and 56% for dogs in training [35] contribute to large production costs. Although the most important skills to train in these dogs are obedience, they also must have an appropriate nutritional plan, in order to support physical fitness. Moreover, also the stress derived from the changes of life style (work and kennels condition) may negatively affect food intake and live weight [24], causing metabolic disorders and some significant modifications in laboratory parameters.

For guide dogs, a normal maintenance diet (crude protein = 20–23%; crude fat = 10%–12%) does not meet the requirements during their training work and the use of large amounts of feed is not recommended. The consumption of the "performance" diet, characterized by low-carbohydrate/

high-protein and fat diet (29%:39%:19% as-fed), seems to be more appropriated for light prolonged exercise than the ingestion of a normal maintenance diet rich in carbohydrates [17], limiting the weight loss in the HPF group, as observed for the dogs of the LPF group (−18%). Nevertheless, all the animals during the trial lost weight; this could be due to the training work for the service guide for the blind. Weight loss is normal in guide dogs during the training, according to the exercise and the life in kennel [23]. Moreover, considering then the Labrador retrievers may be genetically predisposed to obesity and consequently to the osteoarticular diseases [36], and considering the important role of GDB, they always maintain moderate body weight during the training program.

Major depots of fat accumulation are present under the skin (as subcutaneous fat) and they can be readily observed and evaluated in dogs by using a BCS scale as indicator of the fat mass. If dietary energy intake is less than energy need, fat mass and BCS decrease. Conversely, if intake exceeds requirements, fat mass and BCS increase. This could explain our observation regarding a better BCS mean value of HPF group that that observed in LPF group. On the whole, all the animals showed a BCS within the ideal range (score 4 and 5). As reported by Hand et al. [24], a BCS of 2.5–3.5 (on a scale 1−5) is normal for more pets and for many canine athletes; the same authors observed that a much leaner body composition is desirable for some canine athletes. Even small excesses of body fat may represent an unnecessary handicap for working dogs.

In the present study, the circulating T_3 and T_4, CHOL, TAGs, and NEFA concentrations are reported for the first time in clinically healthy Labrador Retriever guide dogs during training. The comparison of hormonal data with published ranges for dogs revealed that T_3 and T_4 concentrations were in agreement with physiological wide ranges reported in literature [37,38]. The results of the lipid pattern ranged within the reference values [39] in agility dogs undergo during exercise [19] and in dogs during low intensity endurance activity [18].

Korhonen [40] monitored the levels of THs, total lipids and urea of adult farmed raccoon dogs, and compared these parameters with BW and feed consumption during intense, maintenance and restricted fasting feeding. He observed a marked adjustment of thyroid hormones as the result of changes in subcutaneous fat reserves. This could explain the significant differences observed in our trial for T_4 and BCS between HPF and LPF groups, according also to Eshratkhah et al. [41] that reported an influence of THs on lipid metabolism, through increasing lipolysis in adipose tissue and stimulating lipogenesis, by increasing the activities of some enzymes. In fact, the suitable function of thyroid gland is essential to metabolic regulations and for maintenance of the energy balance of body [42]. THs appear to contribute in the body energy balance, modulating the basal metabolic rate, primarily through actions in brain, heart kidney, liver, adipose tissue, and skeletal muscle [43]. The significant differences observed for T_4 between the HPF and LPF groups confirm that thyroid hormones were influenced by altering feed intake, changes in subcutaneous fat reserves [40], such as by different diets in relation to the quantity and quality of nutrient contents [44].

The consistent tendency to decrease of the T_4 concentrations in LPF diet group could be probably due to the continued weight loss showed by all dogs during the trial. This result confirmed previous studies related to a decrease in T_3 and T_4 concentrations in dogs undergoing a weight loss protocol [4]. It is well established that thyroid hormone status correlates with BW and energy expenditure [41]. This could be an energy-saving mechanism related to a down regulation of HPT axis activity in condition of caloric deprivation [3]. In human and rats, it's been previously reported that during food or caloric restriction, total body energy expenditure can slow down with an adaptive decrease resulting in a fall in circulating THs [6,7]. Iodothyronines modulate the fat metabolism, and alterations in T_4 may reflect increased lipolysis to offset reduced feed intake [2]. The significant decrease in T_4 and no change in T_3, in response to reducing energy intake, were unexpected. However, our data confirmed other studies that have examined the HPT axis response to starvation with a decrease in Thyroid Releasing Hormone, Thyroid Stimulating Hormone-β gene expression and circulating THs in rodents and human [3]; the decreased T_3 may be primarily due to diminished thyroidal secretion of T_4 or by increased Desiodinase (D_3) activity in the liver, kidney, and muscles of starved rats [45].

A particular note should be done on the thyroid response to variable diet. Only at the last of the trial (Day 56 to Day 84), data showed a significant higher basal concentration of T_4 in HPF group than the mean value of LPF group. The positive value trend might relate to a reduced catabolism resulting from decrease lipoprotein lipase activity [46], which determined a positive energetic balance of the animals fed the "performance" diet rich in fat and protein.

These results could indicate that adjustments in thyroid function and related consistent increase of circulating T_4 concentrations in HPF group, that were driven mainly by the nutrient composition of the diet in relation to the involvement of THs in the synthesis, mobilization, and degradation of lipids [47].

As TAGs are the most important type of fat in the diet and the body's primary for stored energy, during prolonged exercise and when energy intake is insufficient, they are metabolized in FFAs determining an increase of NEFA in the blood which became a primary energy source for long-lasting exercise [48]. Plasma FFA oxidation is directly related to the rate of lipolysis in adipose tissue [14]. Their oxidation can contribute 50 to 60 per cent of the energy expenditure during a bout of low intensity exercise of long duration [49]. This could confirm the better BCS and the lower NEFA levels recorded in the HPF than LPF groups related to the more adequate energy content of the "performances" diet than that of the "normal maintenance" carbohydrate-rich diet. Moreover, the results of this study indicate that the diet induces significant changes in TAGs concentrations. Circulating TAGs could be a potential source of fatty acids for ß-oxidation in working muscle, especially in animals in the fed state [50]. The rise in TAGs concentration after exercise depends on the intensity of exercise and the activity of lipolysis, although FFA concentrations are considered to be a better indicator of lipid metabolism [51]. The "normal maintenance" diet, characterized by high-carbohydrate/low-protein and fat diet (50%:24%:12% as-fed), increased blood TAGs, as effect of lipolysis stimulation inducted from the high request of energy during the metabolic adaptations that occur in skeletal muscle and adipose tissue, and that facilitate a greater delivery and oxidation of fatty acids during exercise. Our results are in accordance with Askew [49] and Kaciuba-Uścilko et al. [52] that observed a markedly enhancement of FFA mobilization modulated by the thyroid hormones, in relation to a decreased feed intake.

Total CHOL concentration is routinely measured during health checks in small animal clinics [13]. The total serum concentration of CHOL has been recognized as a potential biomarker for various processes related to lipoproteins metabolism [15]. Fialkovičová et al. [53] reported that THs have catabolic effects on muscle and adipose tissue and regulate CHOL synthesis and degradation; they are essential for an appropriate degree of metabolic activity, including generation and release of energy. It is possible presume that the higher CHOL levels of the HPF than LPF groups, would be probably correlated to a greater intestinal absorption of medium and long chain fatty acids, which would be esterified in situ and introduced again as lipoproteins and chylomicrons into the blood, testifying to an improvement in the intestinal absorption of the nutrients of HPF group.

Our data are not in accordance with Bruss' observations [54] where serum CHOL level generally varies inversely with thyroid activity. However, there are some contradictory findings regarding the relation between serum THs, CHOL, and TAGs; the concentrations of THs were not correlated with CHOL levels in some other animals [41,55,56]. Although the role of thyroid hormones is well known in many species, there is little evidence describing the relationship between thyroid hormones status and serum profiles of CHOL, TAGs, and NEFA in dogs [57].

This could probably due to the daily rhythmicity of total lipids, total CHOL, phospholipids, and TAGs that occurs in some animals and that vanished when dogs were food-deprived, indicating that these rhythms are driven by the digestive process [58].

Finally, literature data [54,57] report that changes in concentrations of THs in some animal species are due to the effect of temperature and season. Our trial was carried out in spring and the dogs spent their time in thermoneutral zone therefore, we think the environmental temperature could not have influenced the energy requirements, the functional activity of the thyroid gland and the concentration of THs [57].

On the whole, the significant higher values of BW, BCS, TAGs, and CHOL, together with T_4 concentrations, and the significant lower NEFA concentrations in the HPF group, testified a better physical fitness of the animals fed the "performance" diet [59].

5. Conclusions

Clinical biochemistry parameters are of major interest in canine sport medicine to assess health status and fitness level, as well to monitor the mental and physical stress imposed by exercise. Presently, there is a lack of information on the physiological effect of moderate activities in canines, particularly in the guide dog.

Furthermore, for working dogs, the lifetime cost of feed, even if specially formulated, represents a trivial fraction of the monetary investment in training. It is worth noting that guide dogs for the blind are expensive to train, as well as being expensive in personal terms for all concerning if the post-qualification period is unsuccessful; thus, this research is intrinsically leading for the guide dog trade in several ways. The knowledge of metabolic changes is essential in order to design specific and individual training protocols, for an early diagnosis of poor performances, to assess the impact of different feeding or supplementation strategies and to minimize the risk of exercise-linked disease. Results evidenced that the consumption of a diet rich in fat and protein appears the nutritional plan most suitable to support moderate exercise for guide dogs during their training work, supports maintenance of BW, BCS, and serum baseline thyroid and lipid profiles, contrasting the mobilization of subcutaneous fat reserves, and offers potential improvements in challenging work situations. However, the chronic ingestion of diets that are extreme in their composition of either fat or carbohydrate should be always approached with caution.

Author Contributions: Conceptualization, B.C.; Data curation, E.F. and C.C.; Formal analysis, S.C. and C.C.; Funding acquisition, B.C.; Methodology, S.C. and C.C.; Project administration, B.C.; Writing—review and editing, B.C. and E.F.

Funding: This research was funded by the Farmina Pet Foods/Russo Mangimi S.p.A. (Via Nazionale delle Puglie, 80035 Nola-NA, Italy) grant, 2016.

Acknowledgments: Authors want to thank for the continuing support the President and the Guide Dog Trainers of the Regional Centre Helen Keller of the Italian Blind and Guide Dog School Union in Messina.

Conflicts of Interest: The authors declare no conflict of interest. The funder had no role in the design of the study; in the collection, analyses, or interpretation of data; in the writing of the manuscript, or in the decision to publish the results.

References

1. Dodman, N.H.; Aronson, L.; Cottam, N.; Dodds, J.W. The effect of thyroid replacement in dogs with suboptimal thyroid function on owner-directed aggression: A randomized, double-blind, placebo-controlled clinical trial. *J. Vet. Behav.* **2013**, *8*, 225–230. [CrossRef]
2. Nieminen, P.; Saarela, S.; Pyykönen, T.; Asikainen, J.; Mononen, J.; Mustonen, A.M. Endocrine response to fasting in the overwintering captive raccoon dog (Nyctereutesprocyonoides). *J. Exper. Zool.* **2004**, *301A*, 919–929. [CrossRef] [PubMed]
3. Boelen, A.; Wiersinga, W.M.; Fliers, E. Fasting-induced changes in the hypothalamus-pituitary-thyroid axis. *Thyroid* **2008**, *18*, 123–129. [CrossRef] [PubMed]
4. Daminet, S.; Jeusette, I.; Ducateau, L.; Diez, M.; Van de Maele, I.; De Rick, A. Evaluation of thyroid function in obese dogs and in dogs undergoing a weight loss protocol. *J. Vet. Med. A Physiol. Pathol. Clin. Med.* **2003**, *50*, 213–218. [CrossRef] [PubMed]
5. Shi, Z.X.; Levy, A.; Lightman, S.L. The effect of dietary protein on thyrotropin-releasing hormone and thyrotropin gene expression. *Brain Res.* **1993**, *606*, 1–4. [CrossRef]
6. Rosenbaum, M.; Goldsmith, R.; Bloomfield, D.; Magnano, A.; Weimer, L.; Heymsfield, S.; Gallagher, D.; Mayer, L.; Murphy, E.; Leibel, R.L. Low-dose leptin reverses skeletal muscle, autonomic, and neuroendocrine adaptations to maintenance of reduced weight. *J. Clin. Investig.* **2005**, *115*, 3579–3586. [CrossRef]

7. Vella, K.R.; Ramadoss, P.; Lam, F.S.; Harris, J.C.; Ye, F.D.; Same, P.D.; O'Neill, N.F.; Maratos-Flier, E.; Hollenberg, A.N. NPY and MC4R signaling regulate thyroid hormone levels during fasting through both central and peripheral pathways. *Cell. Metab.* **2011**, *14*, 780–790. [CrossRef]
8. Mazaki-Tovi, M.; Abood, S.K.; Kol, A.; Farkas, A.; Schenck, P.A. Increased serum concentrations of adiponectin in canine hypothyroidism. *Vet. J.* **2014**, *203*, 253–255. [CrossRef]
9. Khaleghi Einakchi, M.; Sedaghat Sharifi, N.; Khoshnegah, J.; Heidarpour, M. Canine visceral leishmaniosis: The relationship of blood serum thyroid hormones, lipids, and lipoproteins with clinical status. *Parasitol. Res.* **2018**, *117*, 3761–3765. [CrossRef]
10. Campbell, K.L.; Davis, C.A. Effects of thyroid hormones on serum and cutaneous fatty acid concentrations in dogs. *Am. J. Vet. Res.* **1990**, *51*, 752–756.
11. Song, Y.; Massart, C.; Chico-Galdo, V.; Jin, L.; De Maertelaer, V.; Decoster, C.; Dumont, J.E.; Van Sande, J. Species specific thyroid signal transduction: Conserved physiology, divergent mechanisms. *Mol. Cell. Endocrinol.* **2010**, *19*, 56–62. [CrossRef] [PubMed]
12. Xenoulis, P.G.; Steiner, J.M. Lipid metabolism and hyperlipidemia in dogs. *Vet. J.* **2010**, *183*, 12–21. [CrossRef] [PubMed]
13. Kawasumi, K.; Suzuki, T.; Fujiwara, M.; Mori, N.; Yamamoto, I.; Arai, T. New criteria for canine metabolic syndrome in Japan. *J. Anim. Vet. Adv.* **2012**, *11*, 4005–4007. [CrossRef]
14. Martin, W.H. Effects of acute and chronic exercise on fat metabolism. *Exerc. Sports Sci. Rev.* **1996**, *24*, 203–231. [CrossRef]
15. Postle, A.D. Lipidomics. *Curr. Opin. Clin. Nutr. Metab. Care* **2012**, *15*, 127–133. [CrossRef] [PubMed]
16. Case, L.P.; Carey, D.P.; Hirakawa, D.A.; Daristotle, L. *Canine and Feline Nutrition*, 2nd ed.; Mosby, Inc.: St. Louis, MS, USA, 2000; ISBN-13 978-0323004435.
17. Gollnick, P.D. Metabolism of substrates: Energy substrate metabolism during exercise and as modified by training. *Fed. Proc.* **1985**, *44*, 353–357.
18. Huntingford, J.L.; Levine, C.B.; Mustacich, D.J.; Corrigan, D.; Downey, R.L.; Wakshlag, J.J. The effects of low intensity endurance activity on various physiological parameters and exercise induced oxidative stress in dogs. *Open J. Vet. Med.* **2014**, *4*, 134–144. [CrossRef]
19. Rovira, S.; Munoz, A.; Benito, M. Hematologic and biochemical changes during canine agility competitions. *Vet. Clin. Pathol.* **2007**, *36*, 30–35. [CrossRef]
20. EMEA (European Agency for the Evaluation of Medicinal Products). Guideline on good clinical practices. In *CVMP/VICH/595/98-FINAL*; EMEA: London, UK, 2000.
21. Directive 2010/63/EU of the European Parliament and of the Council of 22 September 2010. Protection of Animals Used for Scientific Purposes. *OJL* **2010**, *276*, 33–79. Available online: http://data.europa.eu/eli/dir/2010/63/oj (accessed on 20 July 2019).
22. Legislative Decree 26/2014. GU Serie Generale n. 61 del 14 March, 2014. Available online: https://www.gazzettaufficiale.it/eli/id/2014/03/14/14G00036/sg (accessed on 20 July 2019).
23. Chiofalo, B.; De Vita, G.; Lo Presti, V.; Cucinotta, S.; Gaglio, G.; Leone, F.; Di Rosa, A.R. Grain free diets for utility dogs during training work: Evaluation of the nutrient digestibility and faecal characteristics. *Anim. Nutr.* **2019**. [CrossRef]
24. Hand, M.S.; Thatcher, C.D.; Remillard, R.L.; Roudebush, P.; Novotny, B.J. *Small Animal Clinical Nutrition*, 5th ed.; Mark Morris Institute: Topeka, Kansas, 2010; ISBN 0-615297.
25. NRC—National Research Council (U.S.). Ad Hoc Committee on Dog and Cat Nutrition. In *Nutrient Requirements of Dogs and Cats*; National Academies Press: Washington, DC, USA, 2006. [CrossRef]
26. Ciaramella, P. *Semiologia Clinica Veterinaria*; Poletto Editore: Milano, Italy, 2014; ISBN 9788895033570.
27. Lloyd, J.K.F.; La Grow, S.J.; Stafford, K.J.; Budge, R.C. The guide dog as a mobility aid part 1: Perceived effectiveness on travel performance. *Int. J. Orientat. Mobil.* **2008**, *1*, 17–33. [CrossRef]
28. Kealy, R.D.; Lawler, D.F.; Ballam, J.M.; Mantz, S.L.; Biery, D.N.; Greeley, E.H.; Lust, G.; Segre, M.; Smith, G.K.; Stowe, H.D. Effects of diet restriction on life span and age-related changes in dogs. *J. Am. Vet. Med. Assoc.* **2002**, *220*, 1315–1320. [CrossRef] [PubMed]
29. Laflamme, D.P. Development and validation of Body Condition Score System for dogs. *Canine Pract.* **1997**, *22*, 10–15.

30. Mawby, D.; Bartges, J.W.; Moyers, T.; D'Avignon, A.; Laflamme, D.P.; Moyers, T.D.; Cottrell, T. Comparison of body fat estimates by dual-energy X-ray absorptiometry and by deuterium oxide dilution. *Compend. Contin. Educ. Pract. Vet.* **2001**, *23*, 70. [CrossRef]
31. McGowan, M.W.; Artiss, J.D.; Strandbergh, D. A peroxidase-coupled method for the colorimetric determination of serum triglycerides. *Clin. Chem.* **1983**, *29*, 538–542.
32. Burtis, C.A.; Ashwood, E.R. *Tietz Textbook of Clinical Chemistry*; WB Saunders: Philadelphia, PA, USA, 1994; pp. 1002–1093. ISBN 9781455777112.
33. XLSTAT. *Data Analysis and Statistical Solution for Microsoft Excel*; Addinsoft: Paris, France, 2014.
34. Guide Dogs NSW/ACT. Guide Dogs NSW/ACT February Newsletter. Hillfields, Burghfield Common, Reading Berkshire RG7 3YG United Kingdom. Registered in England Company No 2376162. 2010. Available online: www.igdf.org.uk (accessed on 20 July 2019).
35. Ennik, I.; Liinamo, A.-E.; Leighton, E.; van Arendonk, J. Suitability for field service in 4 breeds of guide dogs. *J. Vet. Behav.* **2006**, *1*, 67–74. [CrossRef]
36. Raffan, E.; Dennis, R.J.; O'Donovan, C.J.; Becker, J.M.; Scott, R.A.; Smith, S.P.; Withers, D.J.; Wood, C.J.; Conci, E.; Clements, D.N.; et al. Deletion in the Canine POMC Gene Is Associated with Weight and Appetite in Obesity-Prone Labrador Retriever Dogs. *Cell. Metab.* **2016**, *23*, 893–900. [CrossRef]
37. Kaneko, J.; Harney, J.W.; Bress, M.L. Blood analyte reference values in small and same laboratory animals (Appendix IX). In *Clinical Biochemistry of Domestic Animals*; Kaneko, J., Harney, J.W., Bress, M.L., Eds.; Academic Press: San Diego, CA, USA, 2010; p. 896. ISBN 9780080568829.
38. Köhler, B.; Stengel, C.; Neiger, R. Dietary hyperthyroidism in dogs. *J. Small Anim. Pract.* **2012**, *53*, 182–184. [CrossRef]
39. Thrall, M.A.; Weiser, G.; Allison, R.W.; Campbell, T. *Veterinary Hematology and Clinical Chemistry*, 2nd ed.; Wiley-Blackwell: West Sussex, UK, 2012; ISBN 978-0-813-81027-0.
40. Korhonen, H. Relationship between seasonal energy economy and thyroid activity in farm-raised raccoon dogs. *Comp. Biochem. Physiol. A Comp. Physiol.* **1987**, *87*, 983–988. [CrossRef]
41. Eshratkhah, B.; Sadaghian, M.; Khajeye, M.; Ahmadi, H.; Mostafavi, H. Evaluation of non-electrolytes normal values in blood of Makuei sheep bred. *J. Anim. Vet. Adv.* **2008**, *7*, 327–329.
42. Andersen, S.; Bruun, N.H.; Pedersen, K.M.; Laurberg, P. Biologic variation is important for interpretation of thyroid function tests. *Thyroid* **2003**, *13*, 1069–1078. [CrossRef] [PubMed]
43. Warner, A.; Mittag, J. Thyroid hormone and the central control of homeostasis. *J. Mol. Endocrinol.* **2012**, *49*, 29–35. [CrossRef] [PubMed]
44. Lartey, L.J.; Werneck-de-Castro, J.P.; O-Sullivan, I.; Unterman, T.G.; Bianco, A.C. Coupling between nutrient availability and thyroid hormone activation. *J. Biol. Chem.* **2015**, *290*, 30551–30561. [CrossRef] [PubMed]
45. Galton, V.A.; Hernandez, A.; St Germain, D.L. The 5′-deiodinases are not essential for the fasting-induced decrease in circulating thyroid hormone levels in male mice: Possible roles for the type 3′-deiodinase and tissue sequestration of hormone. *Endocrinology* **2014**, *155*, 3172–3181. [CrossRef] [PubMed]
46. da Luz, P.; Favarato, D.; Faria-Neto, J.J.; Lemos, P.; Chagas, A.C.P. High ratio of triglycerides to HDL-cholesterol predicts extensive coronary disease. *Clinics* **2008**, *63*, 427–432. [CrossRef] [PubMed]
47. Dixon, R.M.; Mooney, C.T. Evaluation of serum free thyroxine and thyrotropin concentrations in the diagnosis of canine hypothyroidism. *J. Small. Anim. Pract.* **1999**, *40*, 72–78. [CrossRef] [PubMed]
48. Webber, J.; Taylor, J.; Greathead, H.; Dawson, J.; Buttery, P.J.; Macdonald, I.A. Effects of fasting on fatty acid kinetics and on the cardiovascular, thermogenic and metabolic responses to the glucose clamp. *Clin. Sci.* **1994**, *87*, 697–706. [CrossRef] [PubMed]
49. Askew, E.W. Role of fat metabolism in exercise. *Clin. Sports Med.* **1984**, *3*, 605–621.
50. Terjung, R.L.; Kaciuba-Uscilko, H. Lipid metabolism during exercise: Influence of training. *Diabetes/Metab. Rev.* **1986**, *2*, 35–51. [CrossRef]
51. Pösö, R.A.; Viljanen-Tarifa, E.; Soveri, T.; Oksanen, H.E. Exercise-induced transient hyperlipidemia in the Racehorse. *Transbound. Emerg. Dis.* **1989**, *36*, 603–611. [CrossRef]
52. Kaciuba-Uścilko, H.; Greenleaf, J.E.; Kozlowski, S.; Brzezinska, Z.; Nzar, K.; Ziemba, M. Thyroid hormone-indiced changes in body temperature and metabolism during exercise in dogs. *Am. J. Physiol.* **1975**, *229*, 260–264. [CrossRef] [PubMed]

53. Fialkovičová, M.; Mardzinová, S.; Benková, M.; Mojžišová, J.; Gaálová, M.; Sesztáková, E. Seasonal influence on the thyroid gland in healthy dogs of various breeds in different weights. *Acta Vet. Brno.* **2012**, *81*, 183–188. [CrossRef]
54. Bruss, M.L. Lipids and Ketones. In *Clinical Biochemistry of Domestic Animals*, 6th ed.; Kaneko, J.J., Harvey, J.W., Bruss, M.L., Eds.; Academic Press Inc.: New York, NY, USA, 2008; pp. 81–116. ISBN 9780080568829.
55. Nazifi, S.; Gheisari, H.R.; Abbasali Poorkabir, M.; Saadatfar, S. Serum lipids and lipoproteins in clinically healthy male camels (Camelusdromedarius). *Vet. Res. Commun.* **2000**, *24*, 524–531. [CrossRef]
56. Nazifi, S.; Saeb, M.; Rowghani, E.; Hasankhani, M.; Hasankhani, F.; Ghafari, N. Studies on the relationship between thyroid hormones, serum lipid profile and erythrocyte antioxidant enzymes in clinically healthy Iranian fat-tailed sheep. *Bulg. J. Vet. Med.* **2007**, *10*, 161–167.
57. Mosallanejad, B.; Ghadiri, A.R.; Avizeh, R.; Pourmahdi, M.; Rajabalipour, M. Serum concentrations of lipids and lipoproteins and their correlations with thyroid hormones in clinically healthy German shepherd dogs: Effects of season, sex and age. *Iran. J. Vet. Sci. Technol.* **2015**, *6*, 53–61. [CrossRef]
58. Bertollucci, C.; Fazio, F.; Piccione, G. Daily rhythms of serum lipids in dogs: Influence of lighting and fasting cycles. *Comp. Med.* **2008**, *58*, 485–489.
59. Menchetti, L.; Guelfi, G.; Speranza, R.; Carotenuto, P.; Moscati, L.; Diverio, S. Benefits of dietary supplements on the physical fitness of German Shepherd dogs during a drug detection training course. *PLoS ONE* **2019**, *14*, e0218275. [CrossRef]

animals

Article

Local Dot Motion, Not Global Configuration, Determines Dogs' Preference for Point-Light Displays

Carla J. Eatherington [1], Lieta Marinelli [1,*], Miina Lõoke [1], Luca Battaglini [2] and Paolo Mongillo [1]

1 Laboratory of Applied Ethology, Department of Comparative Biomedicine and Food Science, University of Padua, Viale dell'Università 16, 35020 Legnaro, Italy

2 Department of General Psychology, University of Padua, Via Venezia 8, 35131 Padova, Italy

* Correspondence: lieta.marinelli@unipd.it; Tel.: +39-049-641219

Received: 26 July 2019; Accepted: 3 September 2019; Published: 6 September 2019

Simple Summary: Animal motion is characterised by predictable kinematics according to their body morphology and the laws of gravity. This pattern of movement, called biological motion, is traditionally studied using animated displays created by placing a small number of light dots on the major joints of living beings. Previous studies have shown that several animal species can reliably discriminate dot displays depicting an animal walking, and their performance is impeded when the display is turned upside-down and is variably affected when each dot is displaced to disrupt the global biological arrangement. In this study, we investigated this phenomenon in dogs during the presentation of dot displays depicting humans or dogs walking. Our findings showed that dogs preferred to view the display which depicted an upright dog, regardless of its global arrangement, and had no significant preferences when displays depicting humans were presented. This suggests that dogs' sensitivity to biological motion depends mainly on the presence of dot motion that moves in accordance with gravity. Also, our findings suggest that, despite dogs' extensive exposure to human motion, they are not sensitive to the bipedal motion presented in the human dot displays.

Abstract: Visual perception remains an understudied area of dog cognition, particularly the perception of biological motion where the small amount of previous research has created an unclear impression regarding dogs' visual preference towards different types of point-light displays. To date, no thorough investigation has been conducted regarding which aspects of the motion contained in point-light displays attract dogs. To test this, pet dogs (N = 48) were presented with pairs of point-light displays with systematic manipulation of motion features (i.e., upright or inverted orientation, coherent or scrambled configuration, human or dog species). Results revealed a significant effect of inversion, with dogs directing significantly longer looking time towards upright than inverted dog point-light displays; no effect was found for scrambling or the scrambling-inversion interaction. No looking time bias was found when dogs were presented with human point-light displays, regardless of their orientation or configuration. The results of the current study imply that dogs' visual preference is driven by the motion of individual dots in accordance with gravity, rather than the point-light display's global arrangement, regardless their long exposure to human motion.

Keywords: dog; biological motion; point-light display; visual perception; experience

---

1. Introduction

Animal motion is characterised by predictable kinematics according to their body morphology and the laws of gravity. Johansson [1] captured this movement by placing a small number of point-lights on the major joints of a human body and found that when viewed in isolation they still created the impression of a moving person—despite the lack of other visual information. Biological motion perception has been extensively researched in humans, the results of which demonstrate that people are able to extract a wealth of information from point-light displays, including gender [2,3], emotional state [4], familiarity [5,6] and action performed [7,8]. The perception of biological motion is also relevant to non-human animals, although instead of assessing their ability to infer specific information from point-light displays, research has tended to focus on demonstrating the relevance of biological motion cues to the species under investigation via conditioned discrimination (Baboons [9]; Bottlenose dolphins [10]; Cats [11]; Chimpanzees [12]; Pigeons [13,14]; Rats [15]) or spontaneous preference tasks (e.g., Dogs [16,17]; Chicks [18–21]; Medaka fish [22]; Marmosets [23]; Mice [24]).

In order to assess an individual's preference towards biological or non-biological motion, point-light displays are often presented in conjunction with manipulated ones. A common stimulus manipulation is to flip the point-light display along the horizontal axis. Inverting a point-light display preserves the spatial relationship between dots, but alters the movement of individual dots, which no longer conform to the laws of gravity. Studies conducted in human infants [25] and visually naive chicks [20] have revealed that this disruption of local dot motion reduces the attractiveness of point-light displays, so they are viewed for less time. It has also been shown that inverting a point-light display impaired cats' [11], marmosets' [23] and pigeons' [14] ability to discriminate biological motion. This influence is particularly apparent in dots representing wrists or ankles, which are therefore believed to represent crucial cues for detecting biological motion [26].

A different common manipulation is to scramble point-light displays, by moving individual dots to a different starting position. Scrambling a point-light display disrupts the spatial relationship between dots but maintains the trajectory and accordance with gravity of local dot motions. The impact of disrupting the display's global structure whilst preserving the local dots motion is less clear and may be dependent on the species under investigation. For instance, research in human infants [25] and chicks [19] revealed no visual preference towards coherent or scrambled point-light displays. On the other hand, mice [24], female marmosets [23] and female chicks [18] looked significantly less towards scrambled point-light displays than coherent displays. Blake [11] also found that scrambling a point-light display impaired cats' ability to discriminate biological motion. However, Parron and co-authors [9] found a higher rate of discrimination transfer between upright coherent point-light displays to scrambled point-light displays than from upright coherent point-light displays to inverted coherent point-light displays. Also, the finding that the inversion effect can still be detected when displays are scrambled [26] suggests that the global structure of a point-light display may be less important for biological motion perception than the motion of individual dots in accordance with gravity.

To date, limited investigation into biological motion perception in dogs has been conducted. The first to present animated point-light displays to dogs were Kovács and co-authors [16], who investigated the role of oxytocin on dogs' sensitivity to human motion. In their study, pairs of stimuli were presented comprising an upright coherent human and an inverted and scrambled human, with or without a background of random dots. The experiment revealed a significant effect of oxytocin in modulating dogs' looking preference, when the point-light displays were not masked by random dots, implying that reducing dogs' responsivity influenced their preference for different types of point-light displays. However, no direct comparison of dogs' visual preference to either stimulus was presented, thus leaving unanswered the question of whether dogs do show a visual preference bias towards human biological motion.

Ishikawa and co-authors [17] investigated the role of sociability on dogs' preference for viewing conspecific and human point-light displays. Several combinations of stimuli pairs were presented, varying in terms of manipulation (upright or inverted orientation), direction of movement (frontal or lateral) and species (dog or human). Dogs' level of sociability towards humans and dogs was measured via a questionnaire completed by the owner, which allowed researchers to categorise dogs into high or low sociability groups. A complex pattern of results revealed that overall dogs looked significantly more at human upright frontal point-light displays compared to their inverted control. However, they also found that high-sociability dogs preferentially viewed human inverted point-light displays when presented in the lateral orientation compared to its upright counterpart. And finally, that although low sociability dogs preferentially orientated towards upright dog displays presented laterally compared with its frontally orientated control, high-sociability dogs exhibited the diametrically opposite pattern of results.

In summary, the two previous studies into biological motion perception in dogs [16,17] were not able to clearly answer to what types of point-light displays dogs are preferentially attracted. A possible reason for this is because different types of point-light displays contain one or more different motion features (e.g., upright, inverted, coherent, scrambled). Consequently, the aim of the current experiment was to better understand what features of point-light displays dogs preferentially view, by systematically manipulating physical aspects of point-light displays representing both dogs' and humans' motion.

2. Materials and Methods

2.1. Subjects

Forty-eight dog-owner dyads were recruited through the database of volunteers at the Laboratory of Applied Ethology in the University of Padua. Twenty-eight dogs were pure-breeds (4 Australian Shepherds, 4 Border Collies, 3 Cocker Spaniels, 3 German Shepherds, 3 Golden Retrievers, 2 Weimaraners, 1 Basenji, 1 Czechoslovakian Wolfdog, 1 Dogue de Bordeaux, 1 English Setter, 1 Greyhound, 1 Staffordshire Bull Terrier, 1 Standard Poodle, 1 Vizsla, 1 Whippet) and 20 were mixed-breed dogs (4 small, ≤35 cm at the withers; 10 medium, >35 and <55 cm; 6 large, ≥55 cm). The sample consisted of 29 females and 19 males (mean age ± SD: 5.3 ± 2.6 years). The criteria for recruitment were that dogs had lived with their current owner for the last six months and that they were in good health condition. The study was conducted in accordance with relevant legislation about research involving animals, and, for the type of procedures involved, no formal ethical approval was required.

2.2. Stimuli

The stimuli consisted of white point-light displays representing walking humans or dogs on a black background (Figure 1). The point-light displays were created by video recording one male and one female from each species walking with lateral orientation from left-to-right or right-to-left at a constant speed for one complete cycle of their legs. Markers were placed on the following joints: atlas-occipital, shoulder, elbow, wrist, hip, knee, ankle and, for dogs only, the metatarsophalangeal and metacarpophalangeal joints. Videos were recorded at 120 frames per second. The videos were stabilised using Adobe After Effects CC 2017 (Version 14.2.1, Adobe Inc., San Jose, CA, USA), so they looked as if the person/dog was walking on a treadmill. The resultant movie clip was imported into Tracker [27], where the coordinates for each joint marker were recorded frame-by-frame. Using these coordinates, point-light animations were created using the BioMotion Toolbox [28] for Matlab (Mathworks Inc., Natick, MA, USA). The resulting animation was looped to create 15 s presentations with continuous motion. The BioMotion Toolbox was also used to create inverted and scrambled versions of the original point-light displays. In inverted versions the overall point-light displays were flipped upside-down, so that the spatial relationships between individual points were maintained, but the characteristics of their local motion were opposed to that of a biological entity with respect to gravity. In scrambled versions each individual dot composing the point-light display was randomly displaced to a different starting position compared to its original location, thereby disrupting the global coherence of the point-light display, whilst maintaining the characteristics of the local motion with respect to gravity. Inversion and scrambling could be combined to obtain inverted-scrambled point-light displays, so four different types of stimuli were created for both human and dog stimuli: upright coherent (UC), inverted coherent (IC), upright scrambled (US) and inverted scrambled (IS) (see Figure 1). Moreover, two versions of each stimulus were created, one where the animated figure appeared to be facing right and one facing left.

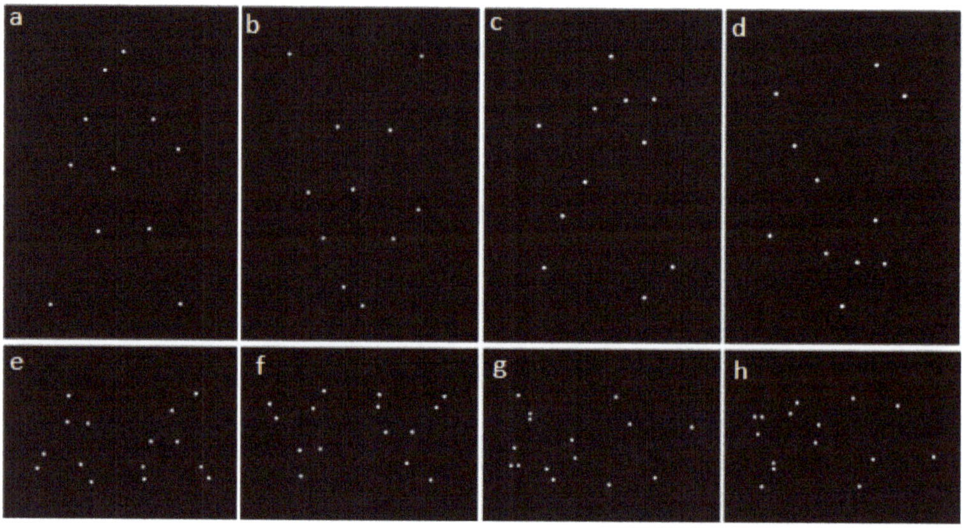

Figure 1. Screenshots exemplifying different types of stimuli used in the experiment: (**a**) human upright coherent, (**b**) human inverted coherent, (**c**) human upright scrambled, (**d**) human inverted scrambled, (**e**) dog upright coherent, (**f**) dog inverted coherent, (**g**) dog upright scrambled and (**h**) dog inverted scrambled.

2.3. Experimental Setting

The experiment was conducted in a quiet, dimly lit (approximately 4 cd/m^2) room (4.7 × 5.8 m) with a large plastic screen (2.4 × 3.4 m) at one end and a Toshiba TDP T100 projector (Toshiba corporation, Tokyo, Japan) mounted 2.15 m high on the wall opposite. Pairs of stimuli were projected onto the screen simultaneously. Human stimuli were sized approximately 130 × 80 cm and dog stimuli were sized approximately 65 × 100 cm. The distance between the centre of each point-light display was 1.80 m. During testing, dogs faced the screen at a distance of 1.65 m, either standing or sitting in between their owner's legs who was seated on a small stool behind them (Figure 2). Owners were instructed to gently hold the dog in place but look straight ahead so as not to influence the dog's behaviour. Trial presentation was controlled by an experimenter seated at the back of the room, using a MacBook Pro. A Canon XA20 (Canon, Tokyo, Japan) camcorder was mounted at floor level, 10 cm in front of the screen and facing the dog's head, to record dog's eye movements. Finally, two CCTV cameras were mounted on the ceiling, facing down towards the dog to record its head orientation.

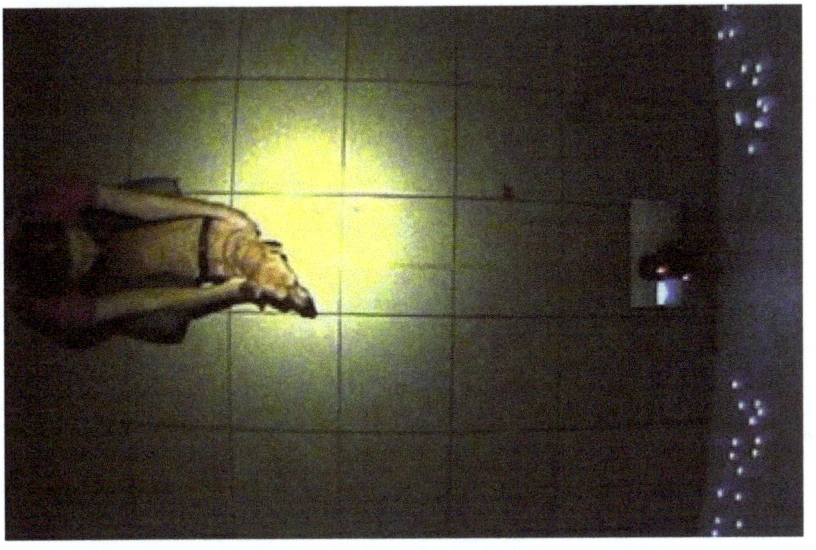

Figure 2. A video-still of the experimental setting, during a presentation.

2.4. Experimental Design

For each species, all possible combinations of manipulations were paired, therefore obtaining six different trial types per species, that displayed two different stimuli simultaneously (see Table 1). The two stimuli within each trial type were facing the same direction. To contain experimental subject habituation, each dog was only presented with three trial types per species (three dog point-light displays and three human point-light displays), totalling six trials per dog; in addition, the direction of movement of the figures was randomised and counterbalanced among the six trials. Overall, each trial type, for both dog and human point-light displays, was seen by 24 subjects. The presentation order of the six trials was pseudo-randomised, to ensure that each trial type was presented equally as often first, second, third, fourth, fifth or sixth, and that human and dog trials were presented in alternate fashion. Also, the side of presentation of the two stimuli, and the direction of movement of the figures, were counterbalanced across the dogs' sample.

Table 1. Combinations of stimuli presented in the six different trial types. Trials featuring these stimuli were created for both dog and human point-light displays. UC = upright coherent, IC = inverted coherent, US = upright scrambled, IS = inverted coherent.

Trial Type	Stimulus 1	Stimulus 2
1	UC	IC
2	UC	US
3	US	IC
4	US	IS
5	IS	UC
6	IS	IC

2.5. Test Procedure

Dogs were initially given ten minutes to become familiar with environment, including the experimenter. Before each trial, dogs were led into the testing room and positioned in front of the screen with their two left and two right paws either side of a central line marked on the floor. Each trial was started when dog was looking straight forward towards the presentation screen; and if the dog did not express the behaviour spontaneously, then their attention was captured by moving the projected computer mouse. At the start of the trial, the two point-light displays composing the trial-type were projected, and held on for 15 s, after which the stimuli disappeared, and a black screen appeared. Dogs were led out of the testing room at the end of each trial, and after a rest period of five minutes they were reintroduced for the start of a new trial.

2.6. Data Collection and Analysis

Using Observer XT software (version 12.5, Noldus, Groeningen, The Netherlands) a continuous sampling technique was used to collect data about dogs' visual orientation from the videos recorded during testing. Dogs' visual orientation was coded as "left" if the dog was looking at the point-light display to the dog's left, "right" if they were looking at the point-light display to the dog's right, and "elsewhere" if the dog was looking anywhere else in the room. If at any time it was not possible to tell where a dog was looking by the frontal video, then head orientation (videos from above) could be used but this was rarely needed. Inter-observer reliability was assessed using data collected by a second observer for dogs' visual orientation on 20% of videos and was revealed to be good (Pearson's $r = 0.85$). Data collected about the dogs' orientation were used to compute the total amount of time in which dogs looked at the stimuli as well the total amount of time the dogs looked at either stimulus. For the analysis, only data for dogs who looked at the display for a minimum of 5 s were included.

A generalised estimating equation (GEE) model was used to assess the influence of various physical characteristics of point-light displays on the amount of looking time dogs directed towards motion displays. In building the model, being scrambled (yes/no) and/or inverted (yes/no) were included as fixed factors, as were their first-order interactions. The dog's ID was included as a random factor, to account for the repeated sampling from each dog. Bonferroni-corrected post-hoc comparisons were performed when a significant effect was found for any of the factors included in the model.

As humans and dogs were never presented in the same display, the "species" factor was not included in the GEE model described above, and separate models were run on data collected from trials where dogs and where humans were presented. However, an additional analysis was performed on the total amount of attention paid by dogs to either stimulus to determine whether the presentation of dogs or humans had an overall effect in attracting dogs' attention. To this aim, a GEE model was run on total attention as dependent variable, the species displayed as a two-level factor, including the dog's ID as random factor.

All statistical analyses were conducted using SPSS (version 24, IBM, Armonk, NY, USA), with statistical significance level set at 0.05.

3. Results

An average of 15 dogs in each trial across all trial types (min: 10; max: 19) looked at the stimuli for more than 5.0 s. These dogs directed a minimum of 5.0 s towards both point-light displays, and a maximum of 15.0 s, with a mean ± SD of 9.6 ± 2.9 s, with no significant difference between displays showing dogs or humans (Wald $\chi^2 = 0.277$, $p = 0.599$). Of this looking time, dogs directed a minimum of 0.0 s towards each stimulus, and a maximum of 15.0 s (mean ± SD: 4.8 ± 4.2 s).

Effect of Stimulus Properties on Looking Time

Results of the GEE indicating the effect of factors influencing dogs' looking time towards human point-light displays are summarised in Table 2. With regards to human trials, there was no effect of inversion, scrambling or an interaction between the two at either time point. Results of the GEE indicating the effect of factors influencing dogs' looking time towards dog point-light displays are summarised in Table 3. No effect was found on dogs' looking time for scrambling and the interaction between scrambling and inversion at either time point. Conversely, a significant effect of inversion was found, with dogs preferentially looking at upright dog point-light displays (mean ± SE: 5.5 ± 0.4 s; 95% CI: (4.7, 6.2)) compared to inverted dog point-light displays (4.2 ± 0.3 s; 95% CI: (3.5, 4.8);

mean difference ± SE: 1.2 ± 0.6 s; 95% CI: (0.1, 2.4)). Mean ± SD looking time towards dog and human point-light displays with different manipulations are presented in Figure 3.

Table 2. Results of the Generalized Estimation Equation model on looking time to each stimulus during human trials. df = degrees of freedom.

Factor	Wald χ^2	df	p-Value
Inverted human	0.058	1	0.810
Scrambled human	0.400	1	0.841
Inverted × Scrambled human	0.023	1	0.880

Table 3. Results of the Generalized Estimation Equation model on looking time to each stimulus during dog trials. df = degrees of freedom.

Factor	Wald χ^2	df	p-Value
Inverted dog	4.198	1	0.040
Scrambled dog	0.347	1	0.556
Inverted × Scrambled dog	0.856	1	0.355

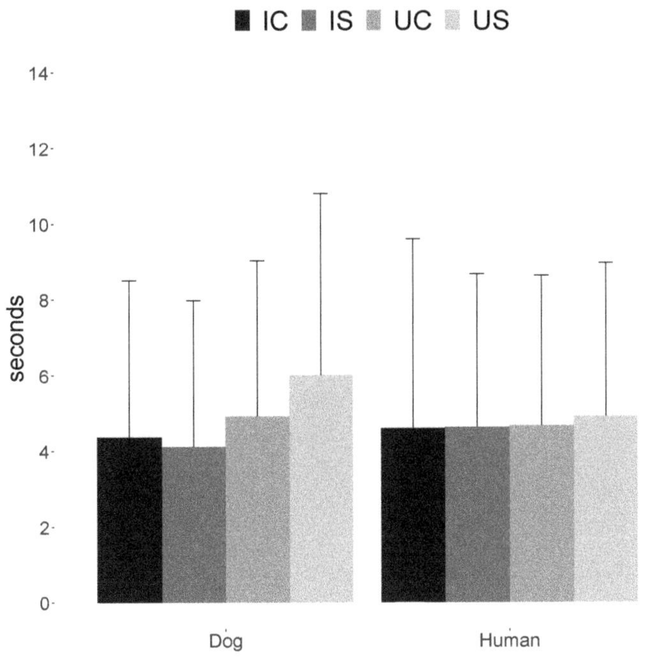

Figure 3. Mean ± SD amount of attention paid by dogs to upright-coherent (UC), upright-scrambled (US), inverted-coherent (IC) and inverted-scrambled (IS) light-point figures representing a walking dog or a walking human.

4. Discussion

The current study investigated which features of biological motion the dogs' directed more looking time towards, by presenting them with pairs of point-light displays of walking dogs or humans that contained aspects methodically manipulated (coherent/scrambled, upright/inverted). The results revealed that dogs directed significantly longer viewing times towards upright dog point-light displays, regardless of their global configuration. No bias in visual preference was observed when dogs were presented with any of the human point-light displays.

The finding that dogs significantly biased their looking time towards dog upright point-light displays corroborates with previous research which found that inverting point-light displays reduced the amount of looking time they attracted (e.g., chicks [20]; marmosets [23]) and impaired visual task performance (cats [11]; pigeons [14]). Traditionally, inversion was believed to impact an individual's ability to process stimuli holistically, a renowned effect observed in face processing [29,30]. However, in the case of biological motion stimuli, the effect of inversion is still present, even when viewing scrambled point-light displays [26], when holistic perception would not be possible. More relevant to the perception of biological motion is that inversion alters the kinematic properties of the moving dots, which no longer move in accordance with the laws of gravity. Thus, a detrimental effect of inversion on viewing times indicates that accordance with gravity is crucial to the detection of biological motion, as observed in chicks and humans [20,31]. This seems also to be the case for dogs in our experiment.

Nevertheless, accordance with gravity as such is not sufficient to explain why our dogs showed a bias in looking time for upright dog point-light displays, but not for upright human ones. A first explanation could be focused on the movement of specific dots. Particularly, human and dog stimuli differed in the amount of limb motion they contained. Not only was this due to the fact that humans are bipeds, but even more feet motion was present in dogs' point-light displays since two joint-dots were marked on every dog limb (ankle and metatarsophalangeal or metacarpophalangeal joints), whilst only one joint was marked on every human leg (ankle). Previous evidence showed that some point-lights provided more movement information than others, implying that there was something "special" about the motion of these dots. For example, Mather and co-authors [32] found that people selected the direction with an accuracy rate of 90% if the shoulder and hip dots, or elbow and knee dots were removed, but performed at near-chance levels if the wrist and ankle dots were removed. It could be argued that the arm movement of the human point-light display contains all the same biological movement that is contained in the legs, rising and falling under the influence of gravity. However, the kinematics of arms and limbs/legs movements are quite different. In accordance with this, Troje and Westhoff [26] also found that local feet motion was crucial for human participants to extract directional information from point-light displays, and based on this they suggested that the local motion contained in animals' feet was used as part of an evolutionary system for detecting animals within their visual environment. This idea is supported by Chang and Troje [33] who claim that it is the vertical acceleration pattern which feet motion contains that is essential to allow the visual system to identity an animal. Also, an electroencephalographic (EEG) study by Wang and co-authors [34] found that humans automatically responded to the characteristics of the local biological motion, but not of the global configuration.

A further aspect that needs to be considered when discussing the lack of effects of scrambling, is ambient luminance. An earlier study showed that humans have difficulty in discriminating scrambled from unscrambled biological motion figures at very low light levels [35]. The authors argued that such conditions only affect the perception of local motion to a limited extent but make it more difficult to assemble local signals into a global percept. Whether this is true, and at what light intensities this occurs, is a matter of debate. For instance, Burton and collaborators [36] report an impairment in the perception of (global) biological motion only in the scotopic range (e.g., when only rod photoceptors are active). Conversely, Billino and collaborators [37] report the greatest impairment in the mesopic range, attributing it to the conflicting, simultaneous activation of the rod and cone systems. Our experiment was conducted at light levels slightly above the threshold between the mesopic and photopic range of humans [38], which would predict no detrimental effect on our dogs' perceptual abilities. However, substantial differences exist between dogs' and humans' vision, including lower acuity at various luminance levels [39], and higher light sensitivity [40], which suggest that thresholds between the photopic, mesopic and scotopic range may also differ between the two species. In this sense, we cannot exclude that the relatively low ambient luminance contributed to the irrelevance of global configuration in driving dogs' attention to biological motion stimuli.

Another hypothesis that could explain the difference in looking time for upright dog point-light displays, but not for upright human ones might be related to different patterns of neural activity which underlie visual processing of conspecific and heterospecific movements. Previous studies in humans [41,42] and monkeys [43–46] indicate that visual sensitivity to actions depends on the observer's past motor experience with the action being observed. Using functional magnetic resonance imaging, Saygin and co-authors [47] found in humans that brain regions containing mirror-neurons are activated by viewing human point-light displays. Similar findings were also revealed during EEG studies with point-light displays showing different human actions [48] and emotions [49]. In line with these findings, Pinto and Shiffrar [50] found that people demonstrated greater visual sensitivity to coherent human motion than coherent horse motion perception in point-light displays. Thus far, mirror neurons have not been explicitly identified in dogs, but their presence can be assumed on the basis of dogs' performance during experiments investigating emotional contagion and mimicry (e.g., [51]). It is then possible that human motion is not able to attract attention in dogs because it does not activate their mirror neurons since they cannot physically preform the action themselves.

Our results support the hypothesis that the characteristic motion of the limbs of an animal in locomotion are crucial for biological motion detection [26], although detection in dogs may be selectively tuned to other quadrupeds. Whilst the evolutionary usefulness of such capacity is clear, allowing dogs to quickly identify other animals within their environment and act on that information accordingly, it also suggests that neither history of domestication, nor adult pet dogs' extensive exposure to humans, enhanced the salience of human bipedal motion in the same way. However, it remains unclear whether dogs recognised point-light displays as representing other dogs or at least a generic quadrupedal organism, or only perceived the characteristics of the local motion contained within the display. A study by Vallortigara and co-authors [19] suggests that chicks are not able to infer species information from movement, since they spontaneously approached point-light displays of predators as well as conspecifics. Further support for the idea that animals do not recognise the identity of point-light displays was provided by research in human infants [25] and chicks [20] which revealed no looking time bias towards upright coherent or upright scrambled/random point-light displays. The lack of any effect of scrambling in our experiment suggests that dogs may also not be able to recognise dogs in these types of stimuli.

At first glance, our results stand at odds with some previous findings. For instance, Kovacs and co-authors [16] suggested that dogs' preference for unmasked human biological motion was reduced by the administration of oxytocin; however, no main effect for biological motion preference was found in the placebo group dogs which falls in line with what we observed during human point-light trials. Ishikawa and co-authors [17] reported an overall looking time bias towards upright humans compared to inverted point-light displays, which clashes with the lack of inversion effect for humans in our study. However, this finding was observed when the human point-light displays were presented in the frontal orientation which the current experiment did not include. When constrained to dog point-light displays, Ishikawa and co-authors found no significant preferences for lateral upright compared to inverted dog point-light displays. A possible explanation for the lack of any preference found in their work could be the difference in the procedure. Unlike Ishikawa et al.'s experiment where the point-light displays were presented for 5 s, the current study measured dogs' looking time at 15 s time points allowing to capture more sustained looking time biases.

5. Conclusions

A systematic investigation of dogs' looking time allocation towards different features of point-light displays revealed that dogs did not show any bias towards features of human point-light displays, but preferentially viewed upright biological motion displays originated from walking dogs, compared to inverted ones and regardless of the global configuration. In line with previous research in animals, the finding champions the importance of limb motion in accordance with the laws of gravity for the detection of moving biological entities by dogs.

In spite of some discrepancies between the present findings and those of other studies in dogs, ours and previous research do converge on one remarkable aspect: that the extent of dog's attentional biases towards biological motion displays is rather limited. Considering that visual preference to biological motion seems well conserved across taxa, it seems unlikely that such weak bias merely reflects a scarce sensitivity of the species towards biological motion. Rather, other factors may have contributed to drive dogs' attention towards visual stimuli; for instance, both the novelty of a specific stimulus, as well as the familiarity of the subject with a certain category of stimuli, can interact in determining dogs' preferential looking at certain visual displays [52]. Thus, in ours, as well as in past research, these factors may have been competing to some extent with the attractiveness of figures depicting biological motion.

Author Contributions: Conceptualization, L.M. and P.M.; methodology, L.B.; formal analysis, P.M.; investigation, C.J.E. and M.L.; data curation, C.J.E.; writing—original draft preparation, C.J.E.; writing—review and editing, L.M. and P.M.

Funding: C.J.E. was supported by a post-doc grant from the University of Padua (Grant Nr. BIRD178748/17), M.L. was supported by a PhD grant from Fondazione Cariparo, and APC were funded by the PhD School in Veterinary Science, University of Padua.

Acknowledgments: We are very grateful to Carlo Poltronieri for his technical assistance, to Alessia Cocco, Stephanie Massaglia and Martina Violo for their help running the experiments, and to all the dogs' owners for volunteering their time.

Conflicts of Interest: The authors declare that they have no conflict of interest.

References

1. Johansson, G. Visual perception of biological motion and a model for its analysis. *Percept. Psychophys.* **1973**, *14*, 201–211. [CrossRef]
2. Barclay, C.D.; Cutting, J.E.; Kozlowski, L.T. Temporal and spatial factors in gait perception that influence gender recognition. *Percept. Psychophys.* **1978**, *23*, 145–152. [CrossRef]
3. Mather, G.; Murdoch, L. Gender discrimination in biological motion displays based on dynamic cues. *Proc. R. Soc. Lond. B Biol. Sci.* **1994**, *258*, 273–279. [CrossRef]
4. Dittrich, W.H.; Troscianko, T.; Lea, S.E.; Morgan, D. Perception of emotion from dynamic point-light displays represented in dance. *Perception* **1996**, *25*, 727–738. [CrossRef]
5. Cutting, J.E.; Kozlowski, L.T. Recognizing friends by their walk: Gait perception without familiarity cues. *Bull. Psychon. Soc.* **1977**, *9*, 353–356. [CrossRef]
6. Troje, N.F.; Westhoff, C.; Lavrov, M. Person identification from biological motion: Effects of structural and kinematic cues. *Percept. Psychophys.* **2005**, *67*, 667–675. [CrossRef]
7. Dittrich, W.H. Action categories and the perception of biological motion. *Perception* **1993**, *22*, 15–22. [CrossRef]
8. Blakemore, S.J.; Decety, J. From the perception of action to the understanding of intention. *Nat. Rev. Neurosci.* **2001**, *2*, 561. [CrossRef]
9. Parron, C.; Deruelle, C.; Fagot, J. Processing of biological motion point-light displays by baboons (*Papio papio*). *J. Exp. Psychol. Anim. Behav. Process.* **2007**, *33*, 381–391. [CrossRef]
10. Herman, L.M.; Morrel-Samuels, P.; Pack, A.A. Bottlenosed dolphin and human recognition of veridical and degraded video displays of an artificial gestural language. *J. Exp. Psychol. Gen.* **1990**, *119*, 215. [CrossRef]
11. Blake, R. Cats perceive biological motion. *Psychol. Sci.* **1993**, *4*, 54–57. [CrossRef]
12. Tomonaga, M. Visual search for biological motion patterns in chimpanzees (*Pan troglodytes*). *Psychologia* **2001**, *44*, 46–59.
13. Dittrich, W.H.; Lea, S.E.; Barrett, J.; Gurr, P.R. Categorization of natural movements by pigeons: Visual concept discrimination and biological motion. *J. Exp. Anal. Behav.* **1998**, *70*, 281–299. [CrossRef]
14. Troje, N.F.; Aust, U. What do you mean with "direction"? Local and global cues to biological motion perception in pigeons. *Vision Res.* **2013**, *79*, 47–55. [CrossRef]

15. MacKinnon, L.M.; Troje, N.F.; Dringenberg, H.C. Do rats (*Rattus norvegicus*) perceive biological motion? *Exp. Brain Res.* **2010**, *205*, 571–576. [CrossRef]
16. Kovács, K.; Kis, A.; Kanizsár, O.; Hernádi, A.; Gácsi, M.; Topál, J. The effect of oxytocin on biological motion perception in dogs (*Canis familiaris*). *Anim. Cogn.* **2016**, *19*, 513–522. [CrossRef]
17. Ishikawa, Y.; Mills, D.; Willmott, A.; Mullineaux, D.; Guo, K. Sociability modifies dogs' sensitivity to biological motion of different social relevance. *Anim. Cogn.* **2018**, *21*, 245. [CrossRef]
18. Regolin, L.; Tommasi, L.; Vallortigara, G. Visual perception of biological motion in newly hatched chicks as revealed by an imprinting procedure. *Anim. Cogn.* **2000**, *3*, 53–60. [CrossRef]
19. Vallortigara, G.; Regolin, L.; Marconato, F. Visually inexperienced chicks exhibit spontaneous preference for biological motion patterns. *PLoS Biol.* **2005**, *3*, 208. [CrossRef]
20. Vallortigara, G.; Regolin, L. Gravity bias in the interpretation of biological motion by inexperienced chicks. *Curr. Biol.* **2006**, *16*, 279–280. [CrossRef]
21. Miura, M.; Matsushima, T. Preference for biological motion in domestic chicks: Sex-dependent effect of early visual experience. *Anim. Cogn.* **2012**, *15*, 871–879. [CrossRef]
22. Nakayasu, T.; Watanabe, E. Biological motion stimuli are attractive to medaka fish. *Anim. Cogn.* **2014**, *17*, 559–575. [CrossRef]
23. Brown, J.; Kaplan, G.; Rogers, L.J.; Vallortigara, G. Perception of biological motion in common marmosets (*Callithrix jacchus*): By females only. *Anim. Cogn.* **2010**, *13*, 555–564. [CrossRef]
24. Atsumi, T.; Ide, M.; Wada, M. Spontaneous discriminative response to the biological motion displays involving a walking conspecific in mice. *Front. Behav. Neurosci.* **2018**, *12*, 263. [CrossRef]
25. Simion, F.; Regolin, L.; Bulf, H. A predisposition for biological motion in the newborn baby. *Proc. Natl. Acad. Sci.* **2008**, *105*, 809–813. [CrossRef]
26. Troje, N.F.; Westhoff, C. The inversion effect in biological motion perception: Evidence for a "life detector"? *Curr. Biol.* **2006**, *16*, 821–824. [CrossRef]
27. Brown, D. Version 4.11.0; Tracker Video Analysis and Modeling Tool. Available online: http://physlets.org/tracker/ (accessed on 9 November 2017).
28. Van Boxtel, J.J.A.; Lu, H. A biological motion toolbox for reading, displaying, and manipulating motion capture data in research settings. *J. Vis.* **2013**, *13*, 1–16. [CrossRef]
29. Valentine, T. Upside-down faces: A review of the effect of inversion upon face recognition. *Br. J. Psychol.* **1988**, *79*, 471–491. [CrossRef]
30. Farah, M.J.; Wilson, K.D.; Drain, H.M.; Tanaka, J.R. The inverted face inversion effect in prosopagnosia: Evidence for mandatory, face-specific perceptual mechanisms. *Vision Res.* **1995**, *35*, 2089–2093. [CrossRef]
31. Troje, N.F. Inverted gravity, not inverted shape impairs biological motion perception. *J. Vis.* **2004**, *4*, 227. [CrossRef]
32. Mather, G.; Radford, K.; West, S. Low-level visual processing of biological motion. *Proc. R. Soc. Lond. B Biol. Sci.* **1992**, *249*, 149–155. [CrossRef]
33. Chang, D.H.; Troje, N.F. Acceleration carries the local inversion effect in biological motion perception. *J. Vis.* **2009**, *9*, 19. [CrossRef]
34. Wang, L.; Yang, X.; Shi, J.; Jiang, Y. The feet have it: Local biological motion cues trigger reflexive attentional orienting in the brain. *NeuroImage* **2014**, *84*, 217–224. [CrossRef]
35. Grossman, E.D.; Blake, R. Perception of coherent motion, biological motion and form-from-motion under dim-light conditions. *Vision Res.* **1999**, *39*, 3721–3727. [CrossRef]
36. Burton, E.; Wattam-Bell, J.; Rubin, G.S.; Atkinson, J.; Braddick, O.; Nardini, M. Cortical processing of global form, motion and biological motion under low light levels. *Vision Res.* **2016**, *121*, 39–49. [CrossRef]
37. Billino, J.; Bremmer, F.; Gegenfurtner, K.R. Motion processing at low light levels: Differential effects on the perception of specific motion types. *J. Vis.* **2008**, *8*, 14. [CrossRef]
38. Lin, Y.; Chen, D.; Chen, W. The significance of mesopic visual performance and its use in developing a mesopic photometry system. *Build. Environ.* **2006**, *41*, 117–125. [CrossRef]
39. Lind, O.; Milton, I.; Andersson, E.; Jensen, P.; Roth, L.S. High visual acuity revealed in dogs. *PLoS ONE* **2017**, *12*, 1–12. [CrossRef]
40. Byosiere, S.E.; Chouinard, P.A.; Howell, T.J.; Bennett, P.C. What do dogs (*Canis familiaris*) see? A review of vision in dogs and implications for cognition research. *Psychon. Bull. Rev.* **2018**, *25*, 1798–1813. [CrossRef]

41. Jacobs, A.; Pinto, J.; Shiffrar, M. Experience, context, and the visual perception of human movement. *J. Exp. Psychol. Hum. Percept. Perform.* **2004**, *30*, 822. [CrossRef]
42. Casile, A.; Giese, M.A. Nonvisual motor training influences biological motion perception. *Curr. Biol.* **2006**, *16*, 69–74. [CrossRef]
43. Bonini, L. The extended mirror neuron network: Anatomy, origin, and functions. *Neuroscientist* **2016**, *23*, 56–67. [CrossRef]
44. Bruni, S.; Gerbella, M.; Bonini, L.; Borra, E.; Coudé, G.; Ferrari, P.F.; Fogassi, L.; Maranesi, M.; Rodà, F.; Simone, L.; et al. Cortical and subcortical connections of parietal and premotor nodes of the monkey hand mirror neuron network. *Brain Struct. Funct.* **2018**, *223*, 1713–1729. [CrossRef]
45. Giese, M.A.; Rizzolatti, G. Neural and computational mechanisms of action processing: Interaction between visual and motor representations. *Neuron* **2015**, *88*, 167–180. [CrossRef]
46. Simone, L.; Bimbi, M.; Rodà, F.; Fogassi, L.; Rozzi, S. Action observation activates neurons of the monkey ventrolateral prefrontal cortex. *Sci. Rep.* **2017**, *7*, 44378. [CrossRef]
47. Saygin, A.P.; Wilson, S.M.; Hagler, D.J.; Bates, E.; Sereno, M.I. Point-light biological motion perception activates human premotor cortex. *J. Neurosci.* **2004**, *24*, 6181–6188. [CrossRef]
48. Ulloa, E.R.; Pineda, J.A. Recognition of point-light biological motion: Mu rhythms and mirror neuron activity. *Behav. Brain Res.* **2007**, *183*, 188–194. [CrossRef]
49. Perry, A.; Troje, N.F.; Bentin, S. Exploring motor system contributions to the perception of social information: Evidence from EEG activity in the mu/alpha frequency range. *Soc. Neurosci.* **2010**, *5*, 272–284. [CrossRef]
50. Pinto, J.; Shiffrar, M. The visual perception of human and animal motion in point-light displays. *Soc. Neurosci.* **2009**, *4*, 332–346. [CrossRef]
51. Palagi, E.; Nicotra, V.; Cordoni, G. Rapid mimicry and emotional contagion in domestic dogs. *R. Soc. Open Sci.* **2015**, *2*, 150505. [CrossRef]
52. Racca, A.; Amadei, E.; Ligout, S.; Guo, K.; Meints, K.; Mills, D. Discrimination of human and dog faces and inversion responses in domestic dogs (*Canis familiaris*). *Anim. Cogn.* **2010**, *13*, 525–533. [CrossRef]

animals

MDPI

Article

Examining How Dog 'Acquisition' Affects Physical Activity and Psychosocial Well-Being: Findings from the BuddyStudy Pilot Trial

Katie Potter [1,*], Jessica E. Teng [2], Brittany Masteller [3], Caitlin Rajala [1] and Laura B. Balzer [4]

1 Department of Kinesiology, University of Massachusetts Amherst, Amherst, MA 01003, USA
2 Last Hope K9 Rescue, Boston, MA 02109, USA
3 Department of Exercise and Sports Studies, Smith College, Northampton, MA 01063, USA
4 Department of Epidemiology and Biostatistics, University of Massachusetts Amherst, MA 01002, USA
* Correspondence: katie.potter@umass.edu; Tel.: +1-413-545-6058

Received: 7 August 2019; Accepted: 4 September 2019; Published: 7 September 2019

Simple Summary: Dog owners are more physically active than non-dog owners, but the direction of the relationship between dog ownership and increased physical activity is unknown. In other words, it is unclear whether acquiring a dog causes a person to become more active, or whether more physically active people choose to acquire dogs. Given that regular physical activity is critical for the prevention and management of numerous chronic diseases, research supporting the hypothesis that dogs make people more active could inform programs and policies that encourage responsible dog ownership. In the BuddyStudy, we used dog fostering to mimic dog acquisition, and examined how taking a dog into one's home affected physical activity and psychosocial well-being. Nearly half of study participants saw large increases in physical activity and nearly three-quarters saw improvements in mood after fostering for six weeks. More than half met someone new in their neighborhood because of their foster dog. Most participants adopted their foster dog after the six-week foster period, and some maintained improvements in physical activity and well-being at 12 weeks. The results of this pilot study are promising and warrant a larger investigation.

Abstract: Dog owners are more physically active than non-dog owners, but evidence of a causal relationship between dog acquisition and increased physical activity is lacking. Such evidence could inform programs and policies that encourage responsible dog ownership. Randomized controlled trials are the 'gold standard' for determining causation, but they are prohibited in this area due to ethical concerns. In the BuddyStudy, we tested the feasibility of using dog fostering as a proxy for dog acquisition, which would allow ethical random assignment. In this single-arm trial, 11 participants fostered a rescue dog for six weeks. Physical activity and psychosocial data were collected at baseline, 6, and 12 weeks. At 6 weeks, mean change in steps/day was 1192.1 ± 2457.8. Mean changes on the Center for Epidemiologic Studies Depression Scale and the Perceived Stress Scale were -4.9 ± 8.7 and -0.8 ± 5.5, respectively. More than half of participants (55%) reported meeting someone new in their neighborhood because of their foster dog. Eight participants (73%) adopted their foster dog after the 6-week foster period; some maintained improvements in physical activity and well-being at 12 weeks. Given the demonstrated feasibility and preliminary findings of the BuddyStudy, a randomized trial of immediate versus delayed dog fostering is warranted.

Keywords: dog ownership; dog walking; physical activity; accelerometry; psychosocial well-being; prospective trial; animal-assisted intervention; dog rescue; foster dog; shelter dog

1. Introduction

Regular physical activity (PA) reduces the risk of cardiovascular disease, type II diabetes, depression, dementia, and some cancers [1]. For individuals living with one or more chronic conditions, regular PA is key to limiting disease progression, preventing co-morbid conditions, and improving physical function and quality of life [1]. The 2018 Physical Activity Guidelines for Americans recommend adults engage in 150 weekly minutes of moderate-intensity aerobic PA (equivalent in intensity to a brisk walk) to reap these health benefits [2]. The most recent national statistics suggest that fewer than one in two American adults meet this mark [3].

A growing international literature base has examined the relationship between dog ownership and PA levels. As discussed in two recent meta-analyses, several studies have demonstrated that dog owners, on average, are more active than non-dog owners [4], and that dog owners are more likely than non-dog owners to meet PA guidelines [5]. Importantly, multiple studies have examined this relationship in clinical populations, including patients with ischemic heart disease [6] and diabetes [7]. The primary limitation of the literature base is its cross-sectional nature; to date, only two studies have prospectively examined the relationship between dog ownership and PA [8,9]. While it is possible that dog acquisition leads to an increase in PA, it is also possible that more active individuals opt to become dog owners. A recent study reporting that dog owners are more likely to be home owners and have a higher annual household income than non-dog owners [10] may support the latter hypothesis, as socioeconomic status is a consistent correlate of leisure-time PA levels [11].

Determining the direction of the dog ownership–human health relationship has important implications. For example, if dog acquisition leads individuals to adopt a more active lifestyle, programming and policies that aim to improve public health might support responsible dog ownership (e.g., encourage pet-friendly lease agreements, provide financial support for veterinary care in low-income communities). Dog ownership could also become a prescriptive tool for physicians to facilitate patient PA, assuming risks are properly considered and mitigated [12]. If renters that can now have a dog in their home or patients that are 'prescribed' a dog choose to acquire a rescue dog, then these initiatives could simultaneously improve human health and dog welfare.

Rigorous prospective studies of the relationship between dog acquisition and PA are needed to inform public health policy and clinical practice. Studies must collect PA data on new owners before and after they acquire dogs, and compare it to data from a group of people that do not acquire dogs. Ideally, a randomized controlled study design would be used to ensure there are no baseline differences between groups that may affect the outcome. Although quasi-experimental (non-randomized) designs and sophisticated analyses [13–16] can help control for known confounding variables (variables that influence both acquisition of a dog and PA outcomes), there may be unknown confounding variables. To our knowledge, only one study used a randomized design to examine how taking a pet into the home affects pet owner health. In 2001, Allen et al. randomized 48 hypertensive individuals to a pet ownership plus ACE inhibitor condition or an ACE inhibitor only condition [17]. Individuals in the pet ownership group acquired a pet cat or dog at the time drug therapy began. Mental stress tests were conducted in participants' homes at baseline and 6 months. The researchers concluded that ACE inhibitor therapy alone reduced resting blood pressure, but that social support through pet ownership reduced the psychological response to mental stress. This study did not assess changes in PA.

Given the serious commitment involved in acquiring a pet, random assignment is no longer considered ethical. The purpose of the current, single-arm trial was to test the feasibility of using dog fostering as a proxy for dog acquisition, as dog fostering is a non-permanent commitment that allows for ethical random assignment. If deemed feasible, a randomized trial of immediate versus delayed dog fostering (or some attention control) would allow for rigorous examination of short-term changes in PA that occur upon taking a dog into one's home. If participants are given the option to permanently adopt their foster dogs after the study period, this approach may also allow for examination of long-term changes in PA. In the BuddyStudy, participants fostered a dog for six weeks and we collected PA, sedentary behavior, and psychosocial outcome data at three time-points to test the feasibility of assessment procedures and examine preliminary effectiveness.

2. Materials and Methods

2.1. Sample

BuddyStudy participants were non-dog owners who were willing and able to be the primary caregiver for a rescue dog for 6–8 weeks. Exclusion criteria included (1) <21 years old, (2) self-reported regular exercise over the past 6 months, (3) lack of reliable source of transportation, (4) presence of any conditions that limit ability to walk, (5) presence of uncontrolled hypertension or diabetes, (6) extensive upcoming travel plans, and (7) dog allergy. Inclusion/exclusion criteria aimed to identify a sample of inactive adults who could safely walk for exercise, and whose living arrangements allowed fostering. All participants were recruited through social media outlets affiliated with the University.

2.2. Community Partner

Dogs involved in this study were fostered through Last Hope K9 Rescue (LHK9), an all-volunteer, all-breed dog rescue organization based in Boston, MA. LHK9 is a foster-based rescue, meaning they rely on foster homes for their dogs and do not have a brick-and-mortar facility. All LHK9 dogs are evaluated and vetted prior to transport from their southern partners in Arkansas, and again prior to entering foster care in New England. Traditionally, LHK9 foster families foster a dog until the dog is adopted, usually between 3–4 weeks, and they are not allowed to adopt their foster dog if there is prior interest from other adopters. Procedures were modified for BuddyStudy participants, who fostered for an extended period of time (minimum six weeks) and were given the option to "foster-to-adopt" as part of the study.

2.3. Study Design

The BuddyStudy was a single-arm feasibility study. After fostering for six weeks, each participant could adopt the dog, transfer the dog to another foster family, or continue fostering until his/her foster dog was adopted into a permanent home. All participants provided written informed consent to participate in the study and signed a contract, which included a liability waiver, to foster with LHK9. This study was approved by the University's IRB and IACUC.

2.4. Procedures

2.4.1. Screening Procedures

Initial eligibility was determined via an online screening survey (Qualtrics). Preliminarily eligible individuals attended an orientation at the University where study procedures were explained and written informed consent obtained. Individuals were asked to complete an application to foster via the LHK9 website within 48 hours of attending orientation. LHK9 volunteers followed standard screening procedures for prospective foster home applications, including conducting home visits and calling personal references, current or recent veterinarians, and landlords (when applicable). Individuals deemed ineligible to foster were compensated $25.

2.4.2. Foster Procedures

Approved participants attended a virtual new foster orientation, and then were added to a private online group moderated by LHK9 volunteers. Each week, a foster coordinator would post a list of dogs (including photos, age, breed and any known background information) needing a foster home in New England. Young puppies (less than one year old), dogs with known medical or behavioral issues, or dogs who already had interested adopters waiting were not eligible for the BuddyStudy. After a participant matched with a foster dog, they picked up an 8-week supply of dog food, a slip lead, a six-foot non-retractable leash, dog toys, and a crate at the University. Each LHK9 foster dog was microchipped and fitted with a no-slip martingale collar with identification tag.

Per Massachusetts state law, all dogs coming into Massachusetts from out-of-state must be isolated in an approved facility for a 48-hour quarantine period before going to a foster home. LHK9 volunteers told participants what day their foster dogs were ready to be picked up at the quarantine facility, typically with 72 hours notice. If a participant was unable to pick up the dog on the pick-up date, they were asked to notify the research team for coverage. Each participant's 6-week study foster period began the day he/she picked up his/her foster dog.

In addition to the information provided during the virtual orientation, LHK9 provided all BuddyStudy participants with a Foster Information Packet with general dog care information, rescue protocols, and an extensive list of contacts and resources, including a 24/7 emergency line. Participants were asked to post all non-emergency questions in the private online group where a team of volunteers, including LHK9's medical and training coordinators, could respond and provide support. All preventative care (i.e., flea/tick, heartworm) for the duration of the foster period and any medical issues requiring veterinary care were coordinated and paid for by the rescue. Leash and collar safety were discussed, but no specific instructions regarding dog walking were provided.

2.5. Measures

Data were collected at baseline (pre-foster period), six-weeks (last week of foster period), and 12-weeks (post-foster period). Of note, some participants (adopters) had a dog in their home at 12 weeks and others (non-adopters) did not.

2.5.1. Feasibility

The primary purpose of the BuddyStudy was to determine the feasibility of using dog fostering as a proxy for dog ownership. Formal feasibility assessments focused on recruitment potential (number of completed applications, proportion of applicants deemed eligible, proportion enrolled), participant attrition (proportion dropping out prior to foster and proportion dropping out after getting the dog), data completeness, significant adverse events (number, type), and percentage of dogs adopted.

2.5.2. Device-Measured PA and Sedentary Behavior

The ActiGraph wGT3X-BT monitor (ActiGraph, Pensacola, FL, USA) was used to assess PA and sedentary behavior. The ActiGraph is a research-grade triaxial accelerometer deemed valid and reliable in free-living conditions [18,19]. Participants wore the device on an elastic band at their right hip during all waking hours (except when showering/swimming) for Seven consecutive days at all three time points. During each seven-day assessment period, participants logged all leisure-time PA, including dog walking specifically, to provide contextual information about their activity (as accelerometers only provide data on amount of PA, not type). ActiGraph data were processed using Actilife Version 6.13.3. (ActiGraph, Pensacola, FL, USA) to determine steps/day and PA minutes/day spent in each intensity category (sedentary, light, moderate, moderate-to-vigorous [MVPA]) based on the Freedson cut points [20]. A minimum of three weekdays and 1 weekend day with at least 10 hours of wear time was required for inclusion in analyses.

2.5.3. Self-Reported Dog Walking

Three questions from the Dogs And WalkinG Survey (DAWGS) were used to assess self-report dog walking behavior at six weeks among all participants and 12 weeks among adopters [21]. Questions included, "how many days do you walk your foster dog in a typical week?" (0–7 days); "how much time do you spend walking during your typical dog walk? (minutes); "on days you walk your dog, on average how many walks do you go on?" (1–5 or more). If participants reported a range for dog walking duration (e.g., 10–20 min), the lower value was used to calculate the average dog walking minutes/week.

2.5.4. Psychosocial Outcomes

Stress and depressive symptoms were assessed at all three time points. The 10-item Perceived Stress Scale (PSS) [22], which asks about thoughts and feelings during the last month, was used to evaluate changes in stress. This scale is widely used in behavioral health research and its psychometric properties have been established [23]. Scores can range from 0 to 40 with higher scores indicating more perceived stress. The PSS scale has no standard cut points; as a reference, an average score of 15.5 ± 7.4 was found in a large US sample (n = 968) in 2009 [24]. The 20-item Center for Epidemiologic Studies Depression Scale (CES-D) [25], which asks about feelings and behaviors over the past week, was used to measure changes in depressive symptoms; scores can range from 0 to 60 with scores ≥16 indicating risk for clinical depression. The CES-D has demonstrated high internal consistency, acceptable test–retest stability, concurrent validity, and construct validity when used in general American populations [25]. At 6 weeks, questions used in work by Wood and colleagues [26] were used to assess whether participants got to know people in their neighborhood since starting the study and, if so, whether the foster dog facilitated the interaction and whether the interaction developed into a friendship and/or provided a new source of social support (emotional support, informational support, appraisal support, or instrumental support). Participants also answered open-ended questions about the best part of fostering, the most challenging part of fostering, and the effect of fostering on quality of life.

2.6. Statistical Analyses

The purpose of this study was to test feasibility and therefore we did not perform inferential statistics. Quantitative data were summarized using means and standard deviations. Qualitative data from open-ended survey questions were coded by two coders (B.M. and C R.) to identify common themes seen throughout the responses. Each coder independently performed a content analysis to identify themes and categories to organize and refine the data. Following individual analyses, the coders compared their results and resolved any discrepancies prior to making conclusions. Direct quotes (de-identified) were extracted from the data to represent the general themes. Analyses of 12-week outcomes were restricted to participants who adopted their foster dog.

3. Results

3.1. Feasibility

One hundred and twenty-three individuals applied to participate in the BuddyStudy over the course of 6.5 weeks. About one-quarter of applicants (n = 28; 23%) were deemed preliminarily eligible and invited for study orientation. Eighteen individuals (15% of applicants) completed informed consent to enroll in the study. Twelve received the intervention (i.e., fostered a dog) and eleven completed the study. See Figure 1 for a detailed study flow chart. One participant dropped out of the study within days of picking up her dog, and the dog was transferred to a participant that had not yet matched with a dog. All other participants completed the 6-week foster period and completed 6-week and 12-week assessments. At six weeks, 11 of 11 participants provided valid ActiGraph data; at 12 weeks, 8 of 11 participants provided valid data. There were no significant adverse events. Of 11 rescue dogs in the study, 8 were adopted by their study foster family at the completion of the 6-week study foster period.

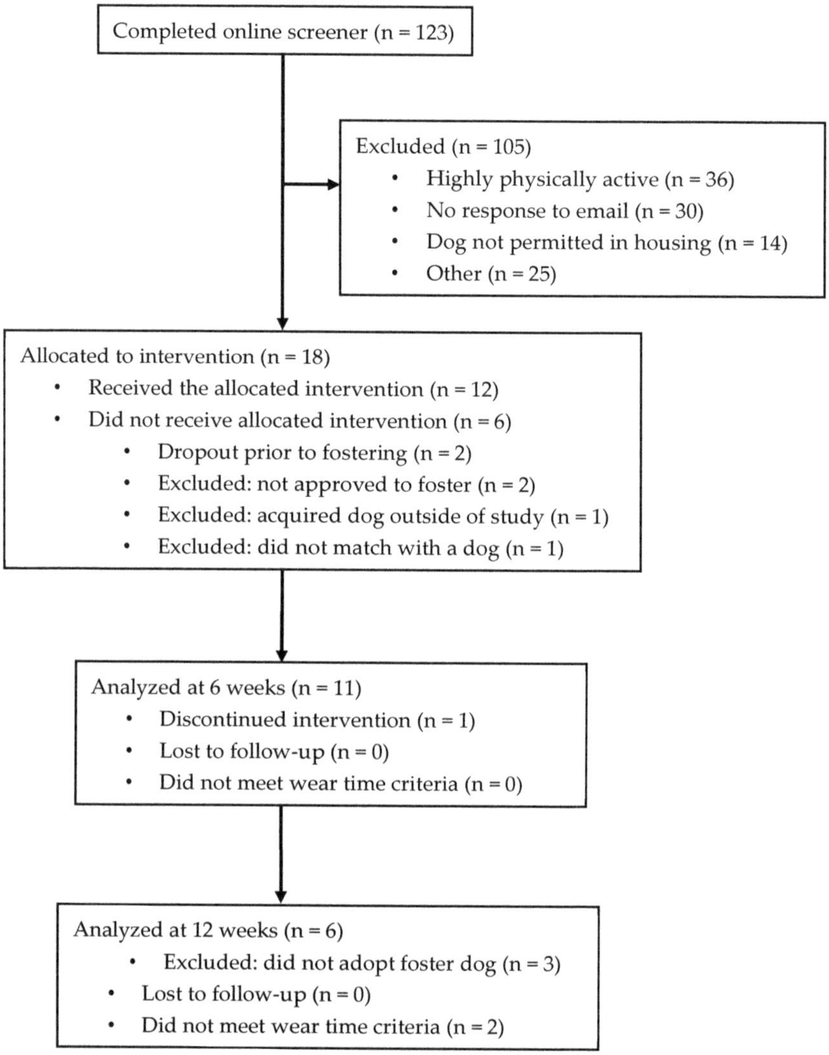

Figure 1. BuddyStudy flow chart.

3.2. Participant Characteristics

All participants who received the intervention (n = 12) were female (100%) and the majority were non-Hispanic white (83%). The average age was 37.8 ± 16.3 years (range 21–62 years). All had a college degree (67%) or were current college students (33%). The majority of participants reported living in a rural (42%) or suburban setting (50%); of seven participants with a yard, only one had a fenced area for their foster dog. Participants averaged 6932.7 ± 2418.9 steps/day, 33.6 ± 19.7 MVPA minutes/day, and 572.4 ± 65.3 sedentary minutes/day at baseline. The average score on the stress measure (PSS) was 15.0 ± 6.9 and the average score on the depressive symptom measure (CES-D) was 13.9 ± 12.5 at baseline.

3.3. *PA and Sedentary Behavior*

Average steps/day and MVPA minutes/day as measured by the ActiGraph, as well as self-reported dog walking data, are presented by time point in Table 1. At 6 weeks, participants reported dog walking 6.5 ± 0.9 days/week (range 5–7) and increased steps/day by 1192.1 ± 2457.8 and MVPA minutes/day by 12.7 ± 20.9 from baseline. The majority reported 10–15 min walk durations (n = 4) or walks ≥30 min in duration (n = 5). Most participants (n = 10) reported taking more than 1 walk/day on days that they walked their foster dog. Nearly half of the sample (n = 5; 45%) increased their steps by >2000 steps/day and their MVPA by >20 min/day. Individual changes in PA are presented in Figure 2.

Table 1. BuddyStudy physical activity (PA) and psychosocial outcomes by time point.

Outcome	Baseline (*n* = 11) [1]	Week 6 (*n* = 11)	Week 12 (*n* = 8) [2]
Device-measured PA			
Daily steps	6976.6 ± 2532.0	8168.7 ± 3827.8	7528.1 ± 4356.9 [3]
MVPA minutes per day	35.3 ± 19.7	48.0 ± 33.6	37.8 ± 32.2 [3]
Sedentary minutes per day	574.0 ± 65.3	524.3 ± 68.3	512.3 ± 94.3 [3]
Self-reported dog walking			
Days w/ at least 1 walk	-	6.5 ± 0.9	6.8 ± 0.5
Minutes per typical walk	-	23.0 ± 16.8	18.1 ± 17.3
Minutes of dog walking per week	-	340.3 ± 243.7	242.5 ± 196.1
Depressive symptoms (CES-D; scale range 0–60)	14.4 ± 13.0	9.5 ± 10.0	7.9 ± 8.9
Perceived stress (PSS; score range 0–40)	15.6 ± 6.9	14.7 ± 5.8	10.0 ± 7.4
Social facilitation			
n (%) met person in neighborhood through foster dog	-	6 (55%)	-
n (%) consider person met through foster dog to be a friend	-	1 (9%)	-
n (%) received social support from person met through foster dog	-	5 (45%)	-

Results reported as mean ± standard deviation unless otherwise noted. [1] Only includes baseline data from participants who completed the study. [2] Adopters, only. [3] n = 6 (n = 2 adopters had invalid ActiGraph data at 12 weeks). Abbreviations: MVPA = moderate-to-vigorous physical activity; CES-D = Center for Epidemiologic Studies-Depression scale; PSS = Perceived Stress Scale.

At 12 weeks, participants who still had a dog in their home (n = 8) reported walking 6.8 ± 0.5 days/week (range 6–7). The majority reported that typical dog walks were ≥30 min in duration (n = 3) or that walks were sometimes as short as 5 min (n = 4). Most still reported walking more than once per day (n = 6) on days that they walked their dog. Adopters who provided valid ActiGraph data at this time point (n = 6) maintained an increase of 552.7 ± 3557.1 steps/day and 8.8 ± 18.8 MVPA minutes/day from baseline.

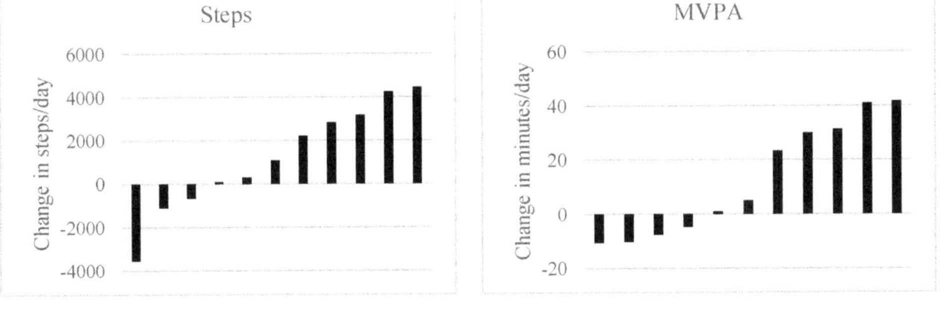

Figure 2. *Cont.*

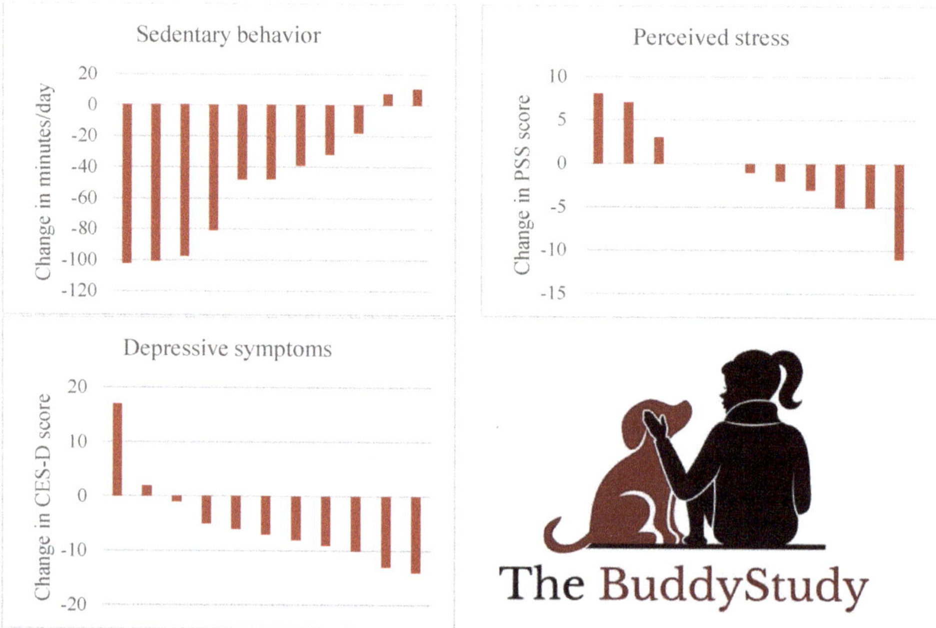

Figure 2. Individual participant changes from baseline to 6 weeks in physical activity, sedentary behavior, and psychosocial outcomes in the BuddyStudy (n = 11). Black bars indicate positive change is better; red bars indicate negative change is better. Physical activity and sedentary behavior were measured via ActiGraph accelerometer. Abbreviations: MVPA = moderate-to-vigorous physical activity; PSS = Perceived Stress Scale; CES-D = Center for Epidemiologic Studies- Depression scale.

Sedentary behavior data are also presented in Table 1. At 6 weeks, participants decreased sedentary minutes/day by 49.8 ± 41.1. Individual changes in sedentary behavior at 6 weeks are presented in Figure 2. At 12 weeks, participants who still had a dog in the home and provided valid ActiGraph data (n = 6) were sedentary for 56.4 ± 59.6 fewer minutes/day, on average, than baseline.

3.4. Psychosocial Outcomes

Average scores on the stress (PSS) and depressive symptom (CES-D) measures by time point are also presented in Table 1. At 6 weeks, participants reported a reduction of 0.8 ± 5.5 on the PSS and 4.9 ± 8.7 on the CES-D. Additionally, more than half of participants (6/11; 55%) reported meeting someone new in their neighborhood because of their foster dog. While only 1 participant considered someone they met through their foster dog to be a friend, 5 of 11 reported receiving some form of social support through an acquaintance they met through their dog. Individual changes in perceived stress and depressive symptoms are presented in Figure 2. At 12 weeks, participants who still had a dog in their home (n = 8) scored 4.8 ± 7.8 points lower on the stress measure (PSS) and 7.2 ± 12.1 points lower on the depressive symptom measure (CES-D), on average, than at baseline.

Responses to open-ended survey questions at 6 weeks are summarized in Table 2. When asked to describe the best part of fostering a dog, common themes included feelings of fun, happiness, joy, companionship, and family bonding. The most commonly reported challenge of fostering was the stress and responsibility involved with taking care of a dog. Participants reported that fostering a dog either improved their quality of life or that the impact was mixed; no participants reported that fostering had a predominately negative impact on their quality of life.

Table 2. Open-ended participant responses at 6 weeks in the BuddyStudy.

Question/Prompt	N Endorsing [1]	Exemplary Quotes
"What has been the best thing about fostering a dog?"		
Fun/happiness/joy	5	*"My pup is a continuous source of happiness in my life - coming home is much more enjoyable than it ever has been. To know she is always at home waiting for me gives me a sense of joy and comfort."*
Companionship	3	*"The continuous love and kisses and snuggles! He is a constant companion that always cheers me up."*
Family bonding	3	*"The best thing about fostering X has been our family working together to care for him."*
"What has been the most challenging thing about fostering a dog?"		
Stress/responsibility	4	*"In a busy household, I already feel like I have a pile of people to care for, be available for, keep healthy, feed, etc. Adding another one can sometimes feel overwhelming (especially when she wasn't feeling well) … "* *"The learning curve of understanding my foster dogs needs and how to juggle that with my own needs has been tricky at times."*
"Please write a brief reflection on how fostering a dog has affected your quality of life."		
Improved quality of life	5	*"It's improved my quality of life. He's brought a lot of joy into my life, and I think I'm less stressed since he's around."* *"Fostering has provided me with a happier and healthier lifestyle. Coming home to a pup that is excited to see you is the best feeling. Fostering has also encouraged me to go on many more walks and seek a more active lifestyle."*
Mixed impact on quality of life	3	*"It has been overwhelming, exciting, stressful, relaxing, wonderful and difficult."* *"It has been very mixed. I have been happy to have his company, and gratified to help him learn, but it has been very difficult to do anything but take care of him for the last six weeks … "*

[1] Only themes endorsed by at least three participants are included in the table.

4. Discussion

Randomized controlled trials provide the highest level of evidence of treatment effectiveness in clinical research. The purpose of the BuddyStudy was to pilot a novel approach that would allow researchers to investigate the PA and health benefits of 'getting' a dog using a randomized controlled design. The BuddyStudy used dog fostering as a proxy for dog acquisition, as fostering is a non-permanent commitment and thereby allows for ethical random assignment. The pilot demonstrated that the design is feasible for implementation on a larger scale, based on the high degree of community interest in the project, high retention among participants who fostered a dog, and absence of significant adverse events. Further, most participants permanently adopted their foster dog, which may allow for the examination of long-term changes in PA with dog acquisition.

The BuddyStudy pilot also demonstrated the preliminary effectiveness of taking a dog into one's home for increasing PA. At the end of the six-week foster period, around half of participants increased their steps by >2000 steps/day and MVPA by >20 minutes/day. These are clinically meaningful increases, as previous research has demonstrated that increases in the range of 1000–2000 steps/day have been associated with reduced risk of type II diabetes [27], cardiovascular disease [28,29], and all-cause mortality [30–32] and the 2018 Physical Activity Guidelines for Americans [2] recommend adults accumulate 150 MVPA minutes/week to improve health and prevent chronic disease. Based on this BuddyStudy data, we anticipate that a sample size of 136 (68/arm) would be needed to detect, with at least 80% power, a difference of 1192 steps/day (change from baseline) between randomized arms in a full-scale trial. Future studies may wish to match self-reported dog walking data with accelerometer data by timestamp [33] to determine the number of steps and MVPA minutes accumulated specifically during dog walking. It should be noted that some participants decreased their daily steps and MVPA minutes from baseline to six weeks. This may have been due to the onset of cold weather, as baseline assessments were completed in September/October and six-week assessments were completed in December in New England.

Many BuddyStudy participants also decreased sedentary time by >45 min/day. There is growing scientific interest in the health risks of too much sitting [34], and some countries now include sedentary behavior guidelines along with PA guidelines. For example, Australia's government recommends that adults minimize the amount of time spent in prolonged sitting and break up long periods of sitting as often as possible [35]. Our findings are in line with three recent studies that have examined dog ownership in relation to sedentary behavior. In a large epidemiological investigation, Garcia et al. found that dog ownership was associated with a lower likelihood of being sedentary for ≥8 hours/day among postmenopausal women [36]. In two separate studies with older adults, dog ownership was associated with an average of 21 fewer minutes of sedentary time/day as measured by ActiGraph accelerometers [37] and with fewer sitting events as measured by active PAL monitors [38].

The BuddyStudy pilot also demonstrated the preliminary effectiveness of taking a dog into one's home for improving psychosocial well-being. At the end of the six-week foster period, most participants who fostered a dog reported decreases in depressive symptoms. When asked to reflect on how fostering a dog affected their quality of life, multiple participants mentioned increased joy, fun, and companionship, which may explain improvements in mood. Alternatively, few participants reported reductions in stress at six weeks. This likely reflects the significant responsibility and time commitment involved in fostering a rescue dog, which many participants acknowledged when asked about the most challenging thing about the foster experience. Finally, many participants reported meeting someone new in their neighborhood because of their foster dog. The phenomenon of dogs as social facilitators has been previously demonstrated [26], and may be one of the most important health benefits of dog ownership given the powerful influence of social relationships, or lack thereof, on human health and longevity [39,40].

To date, no randomized trials have examined the influence of dog 'acquisition' on dog owner PA. Two non-randomized trials have collected self-report PA data on individuals before and after they acquired a dog, and compared it to data from individuals who did not acquire a dog; both found increases in self-reported recreational walking among new dog owners [8,9]. A large randomized trial of immediate versus delayed fostering that employs device-based PA assessment would build on these studies in two key ways: by better controlling for confounding variables that might explain observed increases in PA (with randomization) and by improving the precision of PA assessment (with accelerometry and a larger sample). A larger sample would also allow for the examination of previously demonstrated correlates of dog walking (e.g., owner income, sense of obligation to walk the dog [41]) as moderators and mediators of change in PA.

A number of lessons were learned from the BuddyStudy pilot and should be considered prior to scale-up. First, given the extensive foster screening process and significant commitment that fostering requires, researchers should anticipate substantial attrition between the informed consent and dog matching steps. Second, running the trial in small waves (n = 5–8) may help to not overwhelm the rescue organization and also account for seasonality, which is likely to impact dog-related PA. Third, future trials should be scheduled to avoid major holidays. In the BuddyStudy pilot, we had to board a foster dog for a week after Thanksgiving coverage plans fell through last minute. Finally, and more generally, the pilot taught us that, while feasible, this study is logistically challenging and will require a sizable research team and budget to be properly conducted. In addition to traditional costs of conducting a randomized trial, the budget should cover costs relating to dog transportation, quarantine boarding, preventive medical care for all study dogs, dog foster supplies (e.g., crates, food, leashes), and reimbursement for short-term boarding, training, and medical care, like dewormer medication and vet visits (all of which are typically covered by rescue organizations).

Finally, while the BuddyStudy study design ultimately uses dog fostering to examine how 'acquiring' a dog influences PA and health, it simultaneously raises awareness about the need for foster homes to reduce shelter euthanasia rates and recruits new volunteers for the cause. An estimated 6–8 million cats and dogs enter US animal shelters each year and 3 million are euthanized [42]. Across the US, shelters are filled with dogs that need to be cared for and exercised by volunteers. Foster families are also needed to make room in the shelters for more surrendered and abandoned dogs. Foster-based dog rescue organizations, like LHK9 Rescue, transport dogs from shelters in overpopulated parts of the country to areas that tend to have fewer dogs for people to adopt locally. The more foster families these organizations have, the more dogs they can pull in to rescue. All 11 rescue dogs involved in the BuddyStudy pilot were permanently adopted, eight by their study foster parents.

The primary limitations of this study are the small sample size and lack of control group, which limit our ability to draw conclusions about effectiveness. The study sample was also highly homogenous (100% women; 83% non-Hispanic white). Future studies, including multi-site trials, could examine effectiveness in a more diverse population and across different geographic regions. It should be noted that the vast majority of animal welfare volunteers are women, and therefore this approach may ultimately appeal more to women [43]. Importantly, women are less active and therefore in greater need of PA intervention than men [3]. Finally, our two primary psychosocial variables were assessed over different time frames (past week for depressive symptoms, past month for perceived stress) and future studies should consider using measures with more similar time frames. Strengths of this study include the innovative approach taken to address a critical gap in the literature, which required a novel University-Dog Rescue partnership. The use of accelerometry to measure PA is also a strength, as the literature on dogs and PA is heavily reliant on self-report PA measures prone to recall and social desirability biases.

5. Conclusions

Given the demonstrated feasibility and preliminary findings of the BuddyStudy pilot, a randomized trial of immediate versus delayed dog fostering is warranted and will provide the most rigorous evidence to date of the effects of dog acquisition on human PA and health. Given the major disease and economic burden caused by physical inactivity [44], as well as the ongoing pet homelessness epidemic [42], this line of research has significant societal implications. Demonstration of a positive causal relationship between dog acquisition and increased PA could lead to the testing and implementation of evidence-based public health policies and programs that encourage responsible dog ownership.

Author Contributions: Conceptualization, K.P.; methodology, K.P.; validation, K.P.; formal analysis, K.P., B.M. and C.R.; investigation, K.P., B.M., C.R.; resources, K.P. and J.E.T.; data curation, K.P.; writing—original draft preparation, K.P.; writing—review and editing, J.E.T., B.M., C.R., and L.B.B.; visualization, K.P.; supervision, K.P. and J.E.T.; project administration, K.P. and J.E.T.

Funding: This research received no external funding.

Acknowledgments: The authors thank Last Hope K9 Rescue and all of its volunteer coordinators and directors for the significant time, energy, and resources dedicated to this research project and to responsible rescue; we want to also thank LHK9 foster families in both Arkansas and New England who give their homes and hearts to Last Hope rescue dogs. The authors thank Nick Diehl and Grace McCarthy for their assistance with foster screening procedures. Potter thanks Rena Wing for her support and assistance in conceptualizing the study design.

Conflicts of Interest: The authors declare no conflict of interest.

References

1. *2018 Physical Activity Guidelines Advisory Committee 2018 Physical Activity Guidelines Advisory Committee Scientific Report*; U.S. Department of Health and Human Services: Washington, DC, USA, 2018; p. 779.
2. Powell, K.E.; King, A.C.; Buchner, D.M.; Campbell, W.W.; DiPietro, L.; Erickson, K.I.; Hillman, C.H.; Jakicic, J.M.; Janz, K.F.; Katzmarzyk, P.T. The Scientific Foundation for the Physical Activity Guidelines for Americans, 2nd Edition. *J. Phys. Act. Health* **2018**. [CrossRef] [PubMed]
3. Blackwell, D.L.; Clarke, T.C. State Variation in Meeting the 2008 Federal Guidelines for Both Aerobic and Muscle-strengthening Activities Through Leisure-time Physical Activity Among Adults Aged 18–64: United States, 2010–2015. *Natl. Health Stat. Rep.* **2018**, *112*, 1–22.
4. Christian, H.E.; Westgarth, C.; Bauman, A.; Richards, E.A.; Rhodes, R.E.; Evenson, K.R.; Mayer, J.A.; Thorpe, R.J. Dog ownership and physical activity: A review of the evidence. *J. Phys. Act. Health* **2013**, *10*, 750–759. [CrossRef] [PubMed]
5. Soares, J.; Epping, J.N.; Owens, C.J.; Brown, D.R.; Lankford, T.J.; Simoes, E.J.; Caspersen, C.J. Odds of Getting Adequate Physical Activity by Dog Walking. *J. Phys. Act. Health* **2015**, *12* (Suppl. 1), S102–S109. [CrossRef] [PubMed]
6. Dunn, S.L.; Sit, M.; DeVon, H.A.; Makidon, D.; Tintle, N.L. Dog Ownership and Dog Walking: The Relationship with Exercise, Depression, and Hopelessness in Patients with Ischemic Heart Disease. *J. Cardiovasc. Nurs.* **2018**, *33*, E7–E14. [CrossRef] [PubMed]
7. Riske, J.; Janert, M.; Kahle-Stephan, M.; Nauck, M.A. Owning a Dog as a Determinant of Physical Activity and Metabolic Control in Patients with Type 1 and Type 2 Diabetes Mellitus. *Exp. Clin. Endocrinol. Diabetes* **2019**. [CrossRef] [PubMed]
8. Cutt, H.E.; Knuiman, M.W.; Giles-Corti, B. Does getting a dog increase recreational walking? *Int. J. Behav. Nutr. Phys. Act.* **2008**, *5*, 17. [CrossRef]
9. Serpell, J. Beneficial effects of pet ownership on some aspects of human health and behaviour. *J. R. Soc. Med.* **1991**, *84*, 717–720. [CrossRef]
10. Saunders, J.; Parast, L.; Babey, S.H.; Miles, J.V. Exploring the differences between pet and non-pet owners: Implications for human-animal interaction research and policy. *PLoS ONE* **2017**, *12*, e0179494. [CrossRef]
11. O'Donoghue, G.; Kennedy, A.; Puggina, A.; Aleksovska, K.; Buck, C.; Burns, C.; Cardon, G.; Carlin, A.; Ciarapica, D.; Colotto, M.; et al. Socio-economic determinants of physical activity across the life course: A "DEterminants of DIet and Physical ACtivity" (DEDIPAC) umbrella literature review. *PLoS ONE* **2018**, *13*. [CrossRef]
12. Hodgson, K.; Barton, L.; Darling, M.; Antao, V.; Kim, F.A.; Monavvari, A. Pets' Impact on Your Patients' Health: Leveraging Benefits and Mitigating Risk. *J. Am. Board Fam. Med.* **2015**, *28*, 526–534. [CrossRef] [PubMed]
13. van der Laan, M.J.; Rose, S. *Targeted Learning: Causal Inference for Observational and Experimental Data*; Springer Series in Statistics; Springer-Verlag: New York, NY, USA, 2011; ISBN 978-1-4419-9781-4.

14. Taubman, S.L.; Robins, J.M.; Mittleman, M.A.; Hernán, M.A. Intervening on risk factors for coronary heart disease: An application of the parametric g-formula. *Int. J. Epidemiol.* **2009**, *38*, 1599–1611. [CrossRef] [PubMed]
15. Hernán, M.A.; Lanoy, E.; Costagliola, D.; Robins, J.M. Comparison of dynamic treatment regimes via inverse probability weighting. *Basic Clin. Pharmacol. Toxicol.* **2006**, *98*, 237–242. [CrossRef] [PubMed]
16. Robins, J. A new approach to causal inference in mortality studies with a sustained exposure period—application to control of the healthy worker survivor effect. *Appl. Math. Model* **1986**, *7*, 1393–1512. [CrossRef]
17. Allen, K.; Shykoff, B.E.; Izzo, J.L., Jr. Pet Ownership, but Not ACE Inhibitor Therapy, Blunts Home Blood Pressure Responses to Mental Stress. *Hypertension* **2001**, *38*, 815–820. [CrossRef]
18. Lyden, K.; Keadle, S.K.; Staudenmayer, J.; Freedson, P.S. A method to estimate free-living active and sedentary behavior from an accelerometer. *Med. Sci. Sports Exerc.* **2014**, *46*, 386–397. [CrossRef] [PubMed]
19. Aadland, E.; Ylvisåker, E. Reliability of the Actigraph GT3X+ Accelerometer in Adults under Free-Living Conditions. *PLoS ONE* **2015**, *10*, e0134606. [CrossRef]
20. Freedson, P.S.; Melanson, E.; Sirard, J. Calibration of the Computer Science and Applications, Inc. accelerometer. *Med. Sci. Sports Exerc.* **1998**, *30*, 777–781. [CrossRef]
21. Richards, E.A.; McDonough, M.H.; Edwards, N.E.; Lyle, R.M.; Troped, P.J. Development and psychometric testing of the Dogs and WalkinG Survey (DAWGS). *Res. Q. Exerc. Sport* **2013**, *84*, 492–502. [CrossRef]
22. Cohen, S.; Kamarck, T.; Mermelstein, R. A global measure of perceived stress. *JHSB* **1983**, *24*, 385–396.
23. Lee, E.-H. Review of the psychometric evidence of the perceived stress scale. *Asian Nurs. Res.* **2012**, *6*, 121–127. [CrossRef]
24. Cohen, S.; Janicki-Deverts, D. Who's stressed? Distributions of psychological stress in the United States in probability samples from 1983, 2006, and 2009. *J. Appl. Soc. Psychol.* **2012**, *42*, 1320–1334. [CrossRef]
25. Radloff, L.S. The CES-D Scale: A Self-Report Depression Scale for Research in the General Population. *Appl. Psychol. Meas.* **1977**, *1*, 385–401. [CrossRef]
26. Wood, L.; Martin, K.; Christian, H.; Nathan, A.; Lauritsen, C.; Houghton, S.; Kawachi, I.; McCune, S. The Pet Factor—Companion Animals as a Conduit for Getting to Know People, Friendship Formation and Social Support. *PLoS ONE* **2015**, *10*, e0122085. [CrossRef] [PubMed]
27. Kraus, W.E.; Yates, T.; Tuomilehto, J.; Sun, J.-L.; Thomas, L.; McMurray, J.J.V.; Bethel, M.A.; Holman, R.R. Relationship between baseline physical activity assessed by pedometer count and new-onset diabetes in the NAVIGATOR trial. *BMJ Open Diabetes Res. Care* **2018**, *6*, e000523. [CrossRef] [PubMed]
28. Yates, T.; Haffner, S.M.; Schulte, P.J.; Thomas, L.; Huffman, K.M.; Bales, C.W.; Califf, R.M.; Holman, R.R.; McMurray, J.J.V.; Bethel, M.A.; et al. Association between change in daily ambulatory activity and cardiovascular events in people with impaired glucose tolerance (NAVIGATOR trial): A cohort analysis. *Lancet* **2014**, *383*, 1059–1066. [CrossRef]
29. Jefferis, B.J.; Parsons, T.J.; Sartini, C.; Ash, S.; Lennon, L.T.; Papacosta, O.; Morris, R.W.; Wannamethee, S.G.; Lee, I.-M.; Whincup, P.H. Does total volume of physical activity matter more than pattern for onset of CVD? A prospective cohort study of older British men. *Int. J. Cardiol.* **2019**, *278*, 267–272. [CrossRef]
30. Dwyer, T.; Pezic, A.; Sun, C.; Cochrane, J.; Venn, A.; Srikanth, V.; Jones, G.; Shook, R.P.; Shook, R.; Sui, X.; et al. Objectively Measured Daily Steps and Subsequent Long Term All-Cause Mortality: The Tasped Prospective Cohort Study. *PLoS ONE* **2015**, *10*, e0141274.
31. Jefferis, B.J.; Parsons, T.J.; Sartini, C.; Ash, S.; Lennon, L.T.; Papacosta, O.; Morris, R.W.; Wannamethee, S.G.; Lee, I.-M.; Whincup, P.H. Objectively measured physical activity, sedentary behaviour and all-cause mortality in older men: Does volume of activity matter more than pattern of accumulation? *Br. J. Sports Med.* **2018**. [CrossRef]
32. Lee, I.-M.; Shiroma, E.J.; Kamada, M.; Bassett, D.R.; Matthews, C.E.; Buring, J.E. Association of Step Volume and Intensity With All-Cause Mortality in Older Women. *JAMA Intern Med.* **2019**. [CrossRef]
33. Richards, E.A.; Troped, P.J.; Lim, E. Assessing the Intensity of Dog Walking and Impact on Overall Physical Activity: A Pilot Study Using Accelerometry. *Open J. Prev. Med.* **2014**, *2014*. [CrossRef]
34. Young, D.R.; Hivert, M.-F.; Alhassan, S.; Camhi, S.M.; Ferguson, J.F.; Katzmarzyk, P.T.; Lewis, C.E.; Owen, N.; Perry, C.K.; Siddique, J.; et al. Sedentary Behavior and Cardiovascular Morbidity and Mortality: A Science Advisory from the American Heart Association. *Circulation* **2016**, *134*, e262–e279. [CrossRef] [PubMed]

35. Australia's Physical Activity and Sedentary Behaviour Guidelines and the Australian 24-Hour Movement Guidelines. Available online: https://www1.health.gov.au/internet/main/publishing.nsf/Content/health-pubhlth-strateg-phys-act-guidelines (accessed on 6 September 2019).
36. Garcia, D.O.; Wertheim, B.C.; Manson, J.E.; Chlebowski, R.T.; Volpe, S.L.; Howard, B.V.; Stefanick, M.L.; Thomson, C.A. Relationships between dog ownership and physical activity in postmenopausal women. *Prev. Med.* **2015**, *70*, 33–38. [CrossRef] [PubMed]
37. White, M.N.; King, A.C.; Sallis, J.F.; Frank, L.D.; Saelens, B.E.; Conway, T.L.; Cain, K.L.; Kerr, J. Caregiving, Transport-Related, and Demographic Correlates of Sedentary Behavior in Older Adults: The Senior Neighborhood Quality of Life Study. *JAH* **2016**, *28*, 812–833. [CrossRef] [PubMed]
38. Dall, P.M.; Ellis, S.L.H.; Ellis, B.M.; Grant, P.M.; Colyer, A.; Gee, N.R.; Granat, M.H.; Mills, D.S. The influence of dog ownership on objective measures of free-living physical activity and sedentary behaviour in community-dwelling older adults: A longitudinal case-controlled study. *BMC Public Health* **2017**, *17*, 496. [CrossRef] [PubMed]
39. Holt-Lunstad, J.; Smith, T.B.; Layton, J.B. Social relationships and mortality risk: A meta-analytic review. *PLoS Med.* **2010**, *7*, e1000316. [CrossRef] [PubMed]
40. Holt-Lunstad, J.; Smith, T.B.; Baker, M.; Harris, T.; Stephenson, D. Loneliness and social isolation as risk factors for mortality: A meta-analytic review. *Perspect. Psychol. Sci.* **2015**, *10*, 227–237. [CrossRef]
41. Westgarth, C.; Christley, R.M.; Christian, H.E. How might we increase physical activity through dog walking?: A comprehensive review of dog walking correlates. *Int. J. Behav. Nutr. Phys. Act.* **2014**, *11*, 83. [CrossRef]
42. Pets by the Numbers. Available online: https://www.animalsheltering.org/page/pets-by-the-numbers (accessed on 13 January 2019).
43. Neumann, S.L. Animal Welfare Volunteers: Who Are They and Why Do They Do What They Do? *Anthrozoös* **2010**, *23*, 351–364. [CrossRef]
44. Ding, D.; Lawson, K.D.; Kolbe-Alexander, T.L.; Finkelstein, E.A.; Katzmarzyk, P.T.; van Mechelen, W.; Pratt, M. Lancet Physical Activity Series 2 Executive Committee The economic burden of physical inactivity: A global analysis of major non-communicable diseases. *Lancet* **2016**, *388*, 1311–1324. [CrossRef]

animals

MDPI

Article

Investigating the Role of Prolactin as a Potential Biomarkerof Stress in Castrated Male Domestic Dogs

Jara Gutiérrez [1,*], Angelo Gazzano [1], Federica Pirrone [2], Claudio Sighieri [1] and Chiara Mariti [1]

1 Department of Veterinary Sciences, University of Pisa, 56124 Pisa, Italy
2 Department of Veterinary Sciences, University of Milano, 26900 Milan, Italy
* Correspondence: jara.gutierrez@vet.unipi.it

Received: 2 August 2019; Accepted: 10 September 2019; Published: 12 September 2019

Simple Summary: Although cortisol is usually considered the main reference for the assessment of stress, in some animal species it has been shown that prolactin can be used as a biomarker of both acute and chronic stress. Behavioural parameters can also be used to assess the state of welfare and stress. This study was aimed at evaluating the possible relationship between serum prolactin, serum cortisol and behavioural signs of stress in domestic dogs. To reduce the possible influence of some factors, the study was performed on a homogeneous sample formed by 40 castrated male Spanish Greyhound dogs housed in a dog shelter. The weak negative correlation found between serum cortisol and prolactin values agrees with results obtained in other studies, indicating that prolactin response might be an alternative to cortisol response.

Abstract: Prolactin has been recently regarded as a potential biomarker of both acute and chronic stress in several species. Since only few studies until now have focussed on domestic dogs, this study was aimed at evaluating whether prolactin, cortisol and stress behaviour correlated with each other in sheltered dogs. Both cortisol and prolactin analysis were performed in serum samples through a hormone-specific ELISA kit. For each dog, a stress score was calculated by summing the number of occurrences of stress-related behaviours. The presence/absence of fear during the time spent in the collection room was also scored for each individual. Results revealed a weak negative correlation between cortisol and prolactin levels. Neither of the hormones was correlated with the stress score, nor did their values seem to be influenced by showing fear in the collection room. The weak negative correlation found between cortisol and prolactin values agrees with results obtained in other studies, indicating that prolactin response might be an alternative to cortisol response. This, together with the high serum prolactin levels compared to those reported by other authors for healthy domestic dogs, may indicate that prolactin might be a good biomarker of chronic stress, and although further studies are needed to better understand the potential role of prolactin in the evaluation of canine welfare.

Keywords: behaviour; cortisol; dog; male; prolactin; shelter; Spanish Greyhound; stress

1. Introduction

Although prolactin hormone is known for its function in the stimulation of the growth of the mammary gland and the lactation processes, it has more than 300 different biological activities, homeostatic roles and physiological functions in the organism, e.g., electrolyte balance, luteal function, regulation of the immune system, osmoregulation, angiogenesis, and maintenance of the inter-oestrous interval [1]. In fact, the important genetic role of the prolactin receptor (PRL-R) in energy balance and metabolic adaptation has been recently evidenced in rodents, proving that prolactin has essential roles for the metabolism of glucose, insulin and lipids, as well as in promoting a positive energy balance [2]. Furthermore, prolactin has been found to be an index of acute stress in some non-human animal species, such as rats [3–5], domestic ruminants [6], donkeys [7], cattle [8], and sheep [9]; and different specific functions of prolactin during the stress response have also been demonstrated [2], such as mediation in the epidermal adaptation to environmental stress in fish [10].

It has been recently found that prolactin signalling in tuberoinfundibular dopamine (TIDA) neurons was reduced in mice exposed to acute stress, with a consequent potential decline in their inhibitory influence on prolactin secretion [11]. This suggests that prolactin secreted in response to acute stress may activate prolactin receptors in certain tissues involved in physiological adaptation to stress [11]. In humans, prolactin is stimulated by suckling, perception of visual, acoustic and olfactory stimuli, as well as by stress [12], including psychological and psychosocial stresses [13], and experimental stress-related conditions, such as hypoglycaemia [14], surgery [14], parachute jumping in military recruits [15], and compulsory swimming in non-swimmers [16]. In fact, cortisol, prolactin and day-to-day changes in anxiety indexes measured by questionnaires were found to be significantly correlated [17].

However, the scientific literature reports divergent results with both unchanged [18,19] and decreased prolactin levels [20–22] in response to stress, questioning the belief that stress stimulates prolactin secretion [23]. In addition, while cortisol levels were found to correlate with anxiety levels the day before a surgical intervention, no significant correlation between prolactin and anxiety was found [24].

As for the domestic dog (*Canis familiaris*), to our knowledge the potential link between stress and prolactin levels has been scarcely investigated. One study assessing surgery-related stress in dogs found that cortisol increased and prolactin decreased in the post-surgery period compared to basal levels [25], the latter being in contrast with increased post-operative prolactin response found in humans after surgery [26]. The authors suggested that this contrast may be showing that different species can exert differences in the activation of prolactin feedback regulatory systems [25].

A few studies have reported that dogs with generalized anxiety had hyperprolactinemia, but dogs with phobias or mild anxiety did not [27,28]. Prolactin blood levels were positively correlated with the score obtained on an individual scale for quantifying the presence of anxiety-related behaviours through a physical and behavioural evaluation [28]. In fact, dogs suffering from different kinds of generalized anxiety showed a significant decrease in dopamine blood levels [29], which can justify measuring prolactin, since its secretion is mainly controlled by dopaminergic neurones [28]. However, the involvement of prolactin in emotional responses in dogs seems to be more complex, as its circulating concentrations have been shown to increase also during positive interaction with humans [30].

Cortisol is commonly regarded as the "stress hormone"; cortisol and its metabolites have in fact been quantified in various sample matrices such as blood, saliva, urine and faeces [31], and more recently in hair [32–34], as an endocrine response to stress. Cortisol levels have been shown to increase in adverse conditions such as isolation, restriction of movement, regrouping or transport [35]. In fact, cortisol is secreted following the activation of the hypothalamic–pituitary–adrenal (HPA) axis, one of the major stress response systems [36]. Nonetheless, increased HPA activity is not stress-specific, since it can also be caused by metabolic processes, positive affective states, maternal behaviour and physical activity [37–40]. For this reason, some authors suggest that an increase in cortisol levels should be regarded as an indicator of arousal [37] rather than of stress.

The reliability of cortisol in the assessment of stress has further been questioned due to many factors that can affect the interpretation of cortisol levels such as individual variability in the response to a stressful exposure [41,42], a high inter-individual variability in baseline cortisol levels between dogs [43], distress for blood sampling [44], and possible variations due to the circadian rhythm [44], the latter being not clearly identified in domestic dogs: in fact, no circadian rhythm for cortisol has been reported for laboratory [45] and working dogs [46,47] for intervals of 24–28 h. All these factors suggest caution when comparing cortisol levels in groups of different individual dogs, and when assessing long-term cortisol secretion from fluids such as blood, urine and saliva [44]. The evaluation of cortisol levels can be further complicated by changes with respect to the duration of stress. Whereas in the presence of an acute stressor serum, cortisol rapidly increases and then returns to the basal levels, in a chronic stress situation, a prolonged exposure to the stressor can lead to the suppression and deregulation of the HPA axis [48], which generally does not mean a normalization of circulating cortisol levels, and thus, they are a very useful measure of chronic stress [49].In recent years, hair cortisol analysis has also been supported as a reliable reflection of long-term cortisol secretion [50,51].

Cortisol and prolactin seem to both be involved in the stress response and their levels are likely to be somehow associated. For instance, in vitro experiments showed higher basal levels of corticosterone in hyperprolactinemic rats than in normal and hypoprolactinemic ones; and prolactin was found to exert a stimulating effect on ACTH-induced corticosterone secretion in acute restraint stress [52]. Nevertheless, the response of both prolactin and cortisol hormones to a stressful situation may be divergent, and some authors suggest that prolactin release can act as an alternative form of the cortisol response to stress [22].

In dogs, the evaluation of stress is often a combination of physiological and behavioural parameters, the latter being usually considered a reliable indicator [53,54]. For instance, after the induction of chronic stress through a model of social and spatial restriction, dogs living in small indoor kennels showed significantly lower postures [55], as well as an increase in cortisol levels [56] compared to those living in enriched spacious outdoors housing in groups. When exposed to short-kennelling environment, dogs were generally more active [36,57], and their cortisol:creatinine ratios (C/Cr) were significantly higher compared to when they were in home environment [36].

The aim of this opportunistic study was to evaluate whether there is a correlation between serum cortisol and prolactin concentrations and between them and behavioural indicators of stress and fear in domestic dogs. To reduce the impact of possible affecting factors such as sex, neutering state, breed and housing condition, this study was carried out on a homogeneous sample of castrated male Spanish Greyhound dogs housed for more than 155 days in a dog shelter. This breed has a complex social context in the south of Spain, where this study was performed, with reports of abuse and neglect being frequent for them. Therefore, they could have experienced previous common adverse experiences prior to being rescued, which could reinforce the homogeneity of the sample.

The hypothesis of this study was that, as previously suggested by other authors, prolactin could be used as a biomarker of stress in domestic dogs and, for this reason, that it was correlated with other measures related to stress, such as serum cortisol, the presence of fear or the frequency of stress-related behaviours in dogs. Prolactin secretion from the pituitary gland is used as a marker for lactotropic axis activation [58], it has been found to be an index of acute stress in some species, and have specific functions during the stress response [2]. However, due to the controversial findings in the literature, we did not make any predictions about the possible increase or decrease of prolactin levels in a stressful condition.

This study includes some elements of novelty such as being, to our knowledge, the first including multiple measures of stress, both hormonal and behavioural (prolactin, cortisol and behaviours related to stress and fear), investigating their possible correlation, and involving such a homogeneous sample of domestic dogs, with all belonging to one single breed, being castrated males, and living in the same environment.

2. Materials and Methods

The procedure was communicated to the Ethics Committee of the University of Pisa, Italy and it received a favourable opinion with Decision N.09/2018.

2.1. Subjects and Place

Samples were collected between 3rd September and 26th November 2018 from 40 sheltered castrated adult male Spanish Greyhound dogs (mean age ± standard deviation = 46.5 ±20.8 months; min. = 19; max. = 112 months). Dogs were housed at the Fundación Benjamin Mehnert (Seville, Spain).

Dogs that entered the shelter were either wandering alone, probably after having been abandoned, occasionally hit by a car, or coming directly from hunter discards or reports of abuse and mistreatment.

All of them were hosted in the shelter for a period longer than 155 days (270.9 ± 115.6 days), and had a healthy condition, regularly checked by a veterinarian.

The shelter included a covered space of 2000 m^2 divided into 2 rows and 3 main corridors, with a total of 60 boxes, each of which was provided with 1-3 plastic beds (88 × 58 × 28 cm), depending on the number of dogs sharing the box. Apart from the feeder and the water bucket, there were no other objects such as toys in the boxes. Dogs went out to the outer courtyards in small groups twice a day.

Dogs were usually living in a group of 3–4 dogs (92.5% of cases), all Spanish Greyhounds, sharing the same box for days or weeks; in several cases (n = 3), five dogs were housed together. The shelter staff organized dogs in groups that ensured the greatest possible social stability, thus partially counterbalancing the risk of stress due to overcrowding or conflicts.

2.2. Video Analysis

A 10-min video was recorded of each dog using a video camera (JVC GZ-MG130E, Yokohama, Japan) located about one meter away from the box fence (the maximum possible distance to cover the whole box) and at a stable height of 100 cm. Videos were recorded between 8:30 a.m. and 3:00 p.m. To facilitate the recognition of each dog in the videos, their most peculiar physical characteristics were registered, and a picture was taken.

Videos were analysed using BORIS (Behavioral Observation Research Interactive Software, University of Turin, Turin, Italy) [59]. A list of 18 possible signs of stress in dogs was created [54,60–63], and for each of them the number of displays (observations) was recorded. A stress score was calculated for each dog, by summing up the total number of times each of the following stress-related behaviour was displayed: yawning, shaking, paw lifting, tongue out, eliminating, growling, turn head, tuck tail, cowering, trembling, circling, pacing, hiding, panting, salivation, howling, whining and behaviour against the fence (any behaviour in which the dog physically contacted the fence, such as jumping on the fence or standing up against it).

Videos were analysed by an observer having experience with video analysis of dog behaviour. To measure observational accuracy, a second observer, also having previous experience in behavioural analysis of videos, analysed 3 videos using the same software (BORIS), and inter-observer agreement was calculated.

2.3. Presence of Fear in the Collection Room

After recording the video, 2–3 dogs were contemporaneously led on the leash to a room within the shelter facility, from here called collection room. This movement required around two minutes. In the collection room, dogs were restrained using a leash tied to a table.

The presence of fear around the time of blood sampling was qualitatively scored (yes/no), as a result of a global consideration of the dog behaviour during the time spent in the collection room (20.6 ± 7.0 min). Fear was scored with a "yes" for those dogs expressing at least one of the behaviours related to fear [64], such as the tail tucked, trembling, cowering, and hiding behind objects or in a corner, avoidance of eye contact and approach, and freezing.

2.4. Serum Samples Collection and Storing

Blood (3–4 mL) was drawn from the cephalic vein in the collection room by a veterinarian, part of the shelter staff and familiar to all the dogs, avoiding stress for the dog as much as possible. Only the veterinarian was involved in both restraining the dog during the collection and blood withdrawing.

Based on previous literature about the times at which prolactin and cortisol reach their maximum values and decreased in blood after a stressful situation in humans [65,66], samples were collected about 10 min (ranging from 5 to 15 min) after entering the collection room. When the procedure (identification, blood sampling, and clinical examination) was completed for all the dogs in room, they were moved together to their box.

Tubes containing the blood were kept upright refrigerated (4 °C) in a padded box. After 50–90 min, blood was centrifuged at 3000 rpm for 20 min (using Nahita 2615 Auxilab SL, Beriain, Spain). Serum samples were stored in Eppendorf tubes kept vertically in the freezer (−18 °C) and maintained frozen until being analysed.

2.5. ELISA Kit

Prolactin and cortisol concentrations from canine serum were measured using an EIA (enzyme immunoassay) kit (Demeditec Diagnostics®, Kiel, Germany for prolactin; Diametra®, Segrate, Italy for cortisol), according to the manufacturer's instructions.

2.6. Statistical Analysis

A Shapiro-Wilk test ($p < 0.05$) was applied to the three parameters (prolactin, cortisol, and stress score) to investigate the normality of data. None of them was found to have a normal distribution; therefore, further analyses were carried out using non-parametric statistics.

A comparison between dogs showing fear and dogs not showing fear in the collection room was made for each parameter, through a Mann-Whitney U-test ($p < 0.05$).

A possible correlation between serum prolactin levels (ng/mL), serum cortisol levels (ng/mL) and stress score was investigated using Spearman's rank correlation coefficient ($p < 0.05$).

3. Results

Regarding the videos, the percentage of inter-observer agreement was excellent (92.1%).

Results of stress-related behaviours are reported in Table 1, showing the number and relative percentage of dogs displaying each of the 18 analysed behaviours, and the number of occurrences (times the behaviour was displayed). Tongue out behaviour occurred with the highest frequency (total frequency = 101) and tuck tail with the lowest (total frequency = 1). Four behaviours (growling, hiding, salivation and trembling) were not displayed for any of the dogs (Table 1).

Table 1. Number of dogs and relative percentage (%, calculated on the total sample of 40 dogs) displaying each of the 18 stress-related behaviours, and number of occurrences (times the behaviour was displayed).

Behaviour	Number and % of Dogs Displaying the Behaviour	Number of Occurrences	Behaviour	Number and % of Dogs Displaying the Behaviour	Number of Occurrences
Tongue out	20; 50.0%	101	Circling	3; 7.5%	3
Panting	4; 10.0%	38	Eliminating	3; 7.5%	3
Yawning	15; 37.5%	34	Howling	2; 5.0%	3
Paw lifting	1; 2.5%	21	Turn head	2; 5.0%	2
Behaviour against the fence	3; 7.5%	12	Tuck tail	1; 2.5%	1
Shaking	9; 22.5%	10	Growling	0; 0.0%	0
Pacing	3; 7.5%	7	Hiding	0; 0.0%	0
Cowering	4; 10.0%	6	Salivation	0; 0.0%	0
Whining	4; 10.0%	5	Trembling	0; 0.0%	0

Thirteen dogs showed fear in the collection room and 27 did not. Results of the three parameters for individual dogs are reported in Figure 1.

Prolactin values varied between 2.51 and 43.75 ng/mL, with a mean ± standard deviation of 10.82 ± 9.80 ng/mL. No difference in prolactin concentrations was observed when comparing dogs with and without fear in the collection room ($u = 153.00$; $p = 0.516$; see Figure 2).

Cortisol values ranged 4.43–85.14 ng/mL (25.97 ± 20.33 ng/mL). No difference was found in cortisol levels between fearful and non-fearful dogs in the collection room ($u = 160.00$; $p = 0.654$; see Figure 2).

Stress score values varied between 0 and 27 (6.15 ± 6.77). Stress scores did not differ between dogs with and without fear in the collection room ($u = 143.00$; $p = 0.344$; see Figure 2).

Serum prolactin levels and serum cortisol levels showed a weak negative correlation (Spearman Rho = −0.319; $p = 0.045$). The stress score did not show a correlation with either prolactin (Spearman Rho = 0.017; $p = 0.918$) or with cortisol levels (Spearman Rho = −0.155; $p = 0.342$).

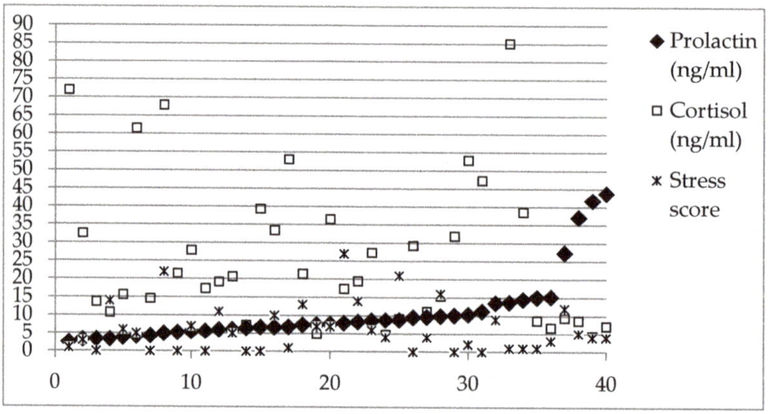

Figure 1. Results of serum prolactin, serum cortisol and stress score. Values for prolactin in serum (ng/mL), cortisol in serum (ng/mL) and stress score for each dog (n = 40) are shown. Data are organized according to increasing value of prolactin.

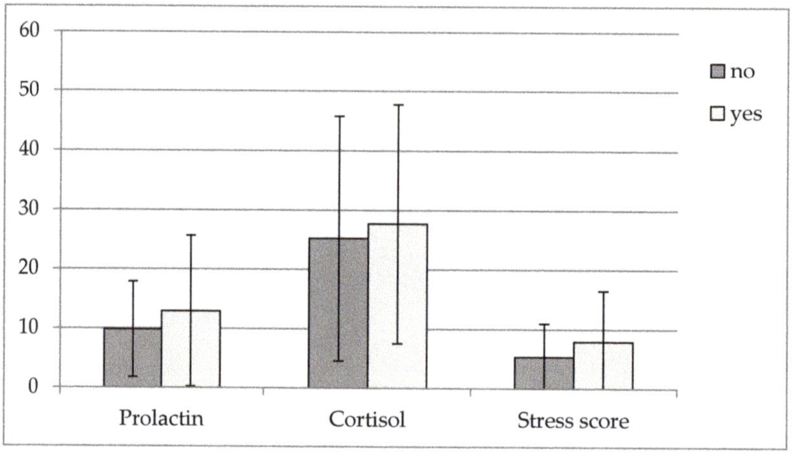

Figure 2. Results of serum prolactin, cortisol and stress score for dogs showing and not showing fearful in the collection room. Values for prolactin in serum (ng/mL), cortisol in serum (ng/mL) and stress score in dogs evaluated as showing (yes; n = 13) or not showing (no; n = 27) fearful behaviour in the collection room.

4. Discussion

Scientific literature about prolactin normal ranges in dogs is relatively scarce. It is well documented, considering its role in lactation, that prolactin increases in pregnant and above all in lactating bitches, as well as in pseudopregnant bitches [67,68], with values ranging from 0 to around 40 ng/mL at 10 weeks after oestrus [67,69]. As for healthy male dogs, studies carried out on intact males have found different but similar ranges [70–72], which can be summarised as a normal range in intact male ranging from non-detectable to 6 ng/mL [71]. Some factors such as breed seem to be responsible for differences in prolactin secretion in intact male dogs [71,72], while age seems not to be influential [71]. The results of the current study refer to castrated males, to dogs housed in a shelter, and to a canine breed not previously involved in studies on prolactin, therefore their interpretation should be performed cautiously. However, data reported in Figure 1, as well as the mean value, show that most of the involved dogs had prolactin concentrations higher than the threshold indicated by Corrada et al. [71], and some dogs had values closer or higher than those reported by Pageat et al. [28] for anxious dogs treated with fluoxetine or selegiline (around 13 ng/mL).

Results on serum cortisol concentrations provided by this study showed a high individual variability, in accordance with Bennett and Hayssen [43]. The values obtained here are included within the wide range (5–100 ng/mL) for baseline cortisol measured by an ELISA kit in client-owned dogs' serum [73]. However, other studies have reported a narrower range, varying according to the physiological state of dogs and the used analytical methods [74–77], making it difficult to compare and frame the values found in the current study. Generally speaking, the mean serum cortisol value obtained in this study was higher than the expected for healthy domestic dogs. The high cortisol values might be due to many factors, including a potential chronic stress condition lived by dogs in the shelter, with a consequent deregulation of HPA axis [48,56]. This will be discussed later.

Looking at the relationships between the investigated parameters, serum prolactin values were not correlated with stress score and they were weakly negatively correlated with serum cortisol levels. Several authors have reported that stressful situations can lead to higher circulating prolactin levels in various species [3–9,11–16], suggesting that prolactin might be a good biomarker of acute stress. Results of the current study seem not to support this hypothesis. However, other studies found that stress-linked cortisol changes did not affect prolactin levels [19,78,79] or have even reported a negative correlation between cortisol and prolactin levels [21,22,30,80,81]. Sobrinho et al. [22] suggested that this negative correlation might be due to the fact that prolactin release acts as an alternative form, rather than an extension of, the more common cortisol response, thus each hormone could be released in response to specific emotions [22].In fact, neural pathways responsible for cortisol and prolactin responses to stress are different in rodents [82], who often release prolactin in response to acute psychological stress [83]. Likewise, in sheltered dogs, whereas serum prolactin levels decreased in response to stress, no changes were observed in serum cortisol concentrations [58].

Prolactin has a direct influence on oxytocin secretion, and both hormones modulate the neuroendocrine acute stress response during maternal behaviour, having an anxiolytic effect in pregnant and lactating rats [84,85]. However, prolactin response to stress does not seem to be limited to females: in a study by Siracusa et al., there was no significant interaction between gender and change in serum prolactin concentrations among time points for dogs exposed to a synthetic, dog-appeasing pheromone both before and after surgery [58].

It must be noticed that the way in which individuals respond to stress and fear may depend on their individual history. Dogs of the current study have experienced situations of abuse, neglect and/or abandonment prior to entering the shelter, which can be all considered stressful situations. It can be hypothesized that their individual history may significantly have affected their way to respond to an acute stress situation. In fact, dopamine release has been shown to increase in response to acute stress in rats if they had previous exposure to chronic stress [86]. This was in agreement with the decrease found in prolactin concentrations in sheltered dogs with assumed chronic stress that were exposed to acute stress [25].

In the current study, prolactin concentrations did not correlate neither to stress score nor to fear behaviour in the collection room. This result, together with the shortage of publications on prolactin response to stress in dogs [58], strongly suggests the need for future research, focused on acute stressors that might provide additional information on the possible use of prolactin as a biomarker of acute stress in domestic dogs. It must be noted that most dogs did not show fearful behaviours in the collection room. This could not be a consequence of their familiarity with this place, since visiting veterinary facilities occurred only for veterinary care, and surgical procedures. However, the presence of both familiar people (shelter staff) and familiar dogs (box mates) likely decreased the stress of novelty and blood withdraw in the collection room.

A possible explanation for cortisol results of the current study is that some dogs belonging to the sample could be chronically stressed and their cortisol levels mirroring such condition: in fact, sheltered dogs are frequently exposed to stressful conditions [87], and those of the current sample had experienced abandonment, abuse and/or neglect.

In other terms, the absence of correlation found between cortisol levels, stress score and fear behaviour in the collection room, as previously reported for the relationship between cortisol and prolactin levels, might be explained by the fact that cortisol values were affected by factors other than those analysed in this study. For instance, the long permanence in the shelter, lived by dogs in our sample, might have led to chronic stress, leading to the suppression and deregulation of the HPA axis [48]. In fact, several studies have reported cortisol variations depending on how long dogs have been in the shelter [88–92]. This point, together with other weaknesses previously reported for cortisol, make even more relevant the need for alternative easily measurable biomarkers of stress.

It could be argued that, if prolactin levels remain constantly elevated during a long period of stress, then this hormone could be a good biomarker of chronic stress in the dog. If prolactin levels in long-stay sheltered dogs were elevated, while cortisol levels varied according to the welfare state of the dog, e.g., decreasing along with the permanence time in the shelter [90–92], then a negative correlation between prolactin and cortisol would be expected, as has been partially reported in the current study. It is advisable that further research investigate whether prolactin is a good candidate for being a biomarker of chronic stress in dogs. For this, it must be taken into account that, at the moment, the use of biological matrixes other than blood for measuring prolactin is not advisable in dogs [93].

This study presents some weaknesses and limitations. One limitation is that this study was focused on assessing the possible association between prolactin and other measures of stress, but no analysis was performed to investigate the possible impact of factors such as the number of dogs per box, the variability on previous individual experiences, etc. In addition, shared experiences such as living in the same cage might have an impact and lead to a cage effect, which was not investigated in the current research. Future studies, involving a wider sample, should take these factors into account. Another limitation is that, although on one hand the homogeneity of the sample may have reduced variability in the data, the specific characteristics of dogs (breed, sex, sexual status, previous experience, and management) require caution in generalizing results to the whole canine population, in particular to pet dogs, whose previous experiences might lead them to respond differently to stress. A potential weakness of this study is related to the lack of knowledge about the timing of prolactin release and decline after a stressful event, especially in the canine species. Future research should better investigate this point.

5. Conclusions

The study of the serum levels of cortisol and prolactin, and stress behaviour in dogs has not resulted in any positive correlation between variables, and the display of fearful behaviour at the moment of blood sampling seems to not have influenced the assessed parameters. Cortisol levels may depend not only on the observational factors scored in this study (stress score and fearful behaviour), but also on the possible presence of a chronic stress. This might explain the weak negative correlation found with prolactin values. This weak negative correlation, together with the high serum prolactin levels compared to those reported by other authors for healthy domestic dogs, may indicate that prolactin could be a good biomarker of chronic stress.

The evaluation of welfare in kennelled animals is essential to ensure and improve their quality of life, and the importance of a multiparametric approach has been stressed [48]. The study of behavioural and physiological parameters other than cortisol might be very helpful to increase the robustness of these studies and consequently provide a useful tool to facilitate adoptions, apply proper environmental enrichment improvements, or target dogs for behavioural or clinical intervention.

Author Contributions: First Author (J.G.) was responsible for data curation, investigation, and writing (original draft + review & editing). Second author (A.G.) was responsible for conceptualization, supervision, and writing (review & editing). Third author (F.P.) was responsible for methodology, interpretation of data and writing (review & editing). Forth author (C.S.) was responsible for writing (review & editing). Last author (C.M.) was responsible for conceptualization, methodology, interpretation of data, supervision, and writing (review & editing).

Funding: This research received no external funding.

Acknowledgments: We appreciate the collaboration of the dog shelter *Fundación Benjamin Mehnert* (Seville, Spain).

Conflicts of Interest: The authors declare no conflict of interest.

References

1. Bole-Feysot, C.; Goffin, V.; Edery, M.; Binart, N.; Kelly, P.A. Prolactin (PRL) and its receptor: Actions, signal transduction pathways and phenotypes observed in PRL receptor knockout mice. *Endocr. Rev.* **1998**, *19*, 225–268. [CrossRef] [PubMed]
2. Bernard, V.; Young, J.; Binart, N. Prolactin—A pleiotropic factor in health and disease. *Nat. Rev. Endocrinol.* **2019**, *15*, 356–365. [CrossRef] [PubMed]
3. Euker, J.S.; Meites, J.; Riegle, G.D. Effects of acute stress on serum LH and prolactin in intact, castrate and dexamethasone-treated male rats. *Endocrinology* **1975**, *96*, 85–92. [CrossRef] [PubMed]
4. Döhler, K.-D.; Gärtner, K.; von zur Mühlen, A.; Döhler, U. Activation of anterior pituitary. thyroid and adrenal gland in rats after disturbance stress. *Eur. J. Endocrinol.* **1977**, *86*, 489–497. [CrossRef] [PubMed]
5. Van Vugt, D.A.; Bruni, J.F.; Meites, J. Naloxone inhibition of stress-induced increase in prolactin secretion. *Life Sci.* **1978**, *22*, 85–89. [CrossRef]
6. Alamer, M. The role of prolactin in thermoregulation and water balance during heat stress in domestic ruminants. *Asian J. Anim. Vet. Adv.* **2011**, *6*, 1153–1169. [CrossRef]
7. Kataria, N.; Kataria, A. Ambience-associated variation in serum biomarkers of oxidative stress in donkeys of arid tracts in India. *Egypt. J. Biol.* **2013**, *15*, 44–47. [CrossRef]
8. Yayou, K.I.; Ito, S.; Yamamoto, N.; Kitagawa, S.; Okamura, H. Relationships of stress responses with plasma oxytocin and prolactin in heifer calves. *Physiol. Behav.* **2010**, *99*, 362–369. [CrossRef]
9. Kataria, N.; Kataria, A.K. Use of prolactin as an indicator of stress in Marwari sheep from arid tracts in India. *Comp. Clin. Path.* **2011**, *20*, 333–336. [CrossRef]
10. Manzon, L.A. The role of prolactin in fish osmoregulation: A review. *Gen. Comp. Endocrinol.* **2002**, *125*, 291–310. [CrossRef]
11. Kirk, S.E.; Xie, T.Y.; Steyn, F.J.; Grattan, D.R.; Bunn, S.J. Restraint stress increases prolactin-mediated phosphorylation of signal transducer and activator of transcription 5 in the hypothalamus and adrenal cortex in the male mouse. *J. Neuroendocrinol.* **2017**, *29*, 1–9. [CrossRef]
12. Rojas Vega, S.; Hollmann, W.; Strüder, H.K. Influences of Exercise and Training on the Circulating Concentration of Prolactin in Humans. *J. Neuroendocrinol.* **2012**, *24*, 395–402. [CrossRef]
13. Lennartsson, A.-K.; Jonsdottir, I.H. Prolactin in response to acute psychosocial stress in healthy men and women. *Psychoneuroendocrinology* **2011**, *36*, 1530–1539. [CrossRef] [PubMed]
14. Noel, G.L.; Suh, H.K.; Stone, J.G.; Frantz, A.G. Human prolactin and growth hormone release during surgery and other conditions of stress. *J. Clin. Endocrinol. Metab.* **1972**, *35*, 840–851. [CrossRef] [PubMed]
15. Noel, G.L.; Dimond, R.C.; Earll, J.M.; Frantz, A.G. Prolactin, thyrotropin, and growth hormone release during stress associated with parachute jumping. *Aviat. Sp. Environ. Med.* **1976**, *47*, 543–547.
16. Vaernes, R.; Ursin, H.; Darragh, A.; Lambe, R. Endocrine response patterns and psychological correlates. *J. Psychosom. Res.* **1982**, *26*, 123–131. [CrossRef]
17. Jeffcoate, W.J.; Lincoln, N.B.; Selby, C.; Herbert, M. Correlation between anxiety and serum prolactin in humans. *J. Psychosom. Res.* **1986**, *30*, 217–222. [CrossRef]

18. Baumgartner, A.; Gräf, K.J.; Kürten, I. Prolactin in patients with major depressive disorder and in healthy subjects. III. Investigation of basal and Post-TRH prolactin during different forms of acute and chronic psychological stress. *Biol. Psychiatry* **1988**, *24*, 286–298. [CrossRef]
19. Modell, E.; Goldstein, D.; Reyes, F. Endocrine and behavioral responses to psychological stress in hyderandrogenic women. *Int. J. Gynecol. Obstet.* **1991**, *34*, 90. [CrossRef]
20. Arnetz, B.B.; Fjellner, B.; Eneroth, P.; Kallner, A. Stress and psoriasis: Psychoendocrine and metabolic reactions in psoriatic patients during standardized stressor exposure. *Psychosom. Med.* **1985**, *47*, 528–541. [CrossRef]
21. Gerra, G.; Zaimovic, A.; Mascetti, G.G.; Gardini, S.; Zambelli, U.; Timpano, M.; Raggi, M.A.; Brambilla, F. Neuroendocrine responses to experimentally-induced psychological stress in healthy humans. *Psychoneuroendocrinology* **2001**, *26*, 91–107. [CrossRef]
22. Sobrinho, L.G.; Simões, M.; Barbosa, L.; Raposo, J.F.; Pratas, S.; Fernandes, P.L.; Santos, M.A. Cortisol, prolactin, growth hormone and neurovegetative responses to emotions elicited during an hypnoidal state. *Psychoneuroendocrinology* **2003**, *28*, 1–17. [CrossRef]
23. Biondi, M.; Picardi, A. Psychological stress and neuroendocrine function in humans: The last two decades of research. *Psychother. Psychosom.* **1999**, *68*, 114–150. [CrossRef] [PubMed]
24. Brooks, J.E.; Herbert, M.; Walder, C.P.; Selby, C.; Jeffcoate, W.J. Prolactin and stress: Some endocrine correlates of pre-operative anxiety. *Clin. Endocrinol. (Oxf.)* **1986**, *24*, 653–656. [CrossRef] [PubMed]
25. Siracusa, C.; Manteca, X.; Cerón, J.; Martínez-Subiela, S.; Cuenca, R.; Lavin, S.; Garcia, F.; Pastor, J. Perioperative stress response in dogs undergoing elective surgery: Variations in behavioural, neuroendocrine, immune and acute phase responses. *Anim. Welf.* **2008**, *17*, 259–273.
26. Marrocco-Trischitta, M.M.; Tiezzi, A.; Svampa, M.G.; Bandiera, G.; Camilli, S.; Stillo, F.; Petasecca, P.; Sampogna, F.; Abeni, D.; Guerrini, P. Perioperative stress response to carotid endarterectomy: The impact of anesthetic modality. *J. Vasc. Surg.* **2004**, *39*, 1295–1304. [CrossRef]
27. Pageat, P.; Gaultier, E. Current research in canine and feline pheromones. *Vet. Clin. North Am.-Small Anim. Pract.* **2003**, *33*, 187–211. [CrossRef]
28. Pageat, P.; Lafont, C.; Falewée, C.; Bonnafous, L.; Gaultier, E.; Silliart, B. An evaluation of serum prolactin in anxious dogs and response to treatment with selegiline or fluoxetine. *Appl. Anim. Behav. Sci.* **2007**, *105*, 342–350. [CrossRef]
29. Berteselli, G.V.; Servida, F.; Dall'Ara, P.; Verga, M.; Piola, E.; Puricelli, M.; Palestrini, C. Evaluation of immunological, stress and behavioural parameters in dogs (Canis familiaris) with anxiety related disorders. In *Current Issues and Research in Veterinary Behavioral Medicine*; Mills, D., Levine, E., Landsberg, G., Horwitz, D., Duxbury, M., Meyer, K., Huntley, L.R., Reich, M., Willard, J., Eds.; Purdue University Press: West Lafayette, IN, USA, 2005; pp. 18–22. ISBN 1557534098.
30. Odendaal, J.S.J.; Meintjes, R.A. Neurophysiological correlates of affiliative behaviour between humans and dogs. *Vet. J.* **2003**, *165*, 296–301. [CrossRef]
31. Cook, N.J. Review: Minimally invasive sampling media and the measurement of corticosteroids as biomarkers of stress in animals. *Can. J. Anim. Sci.* **2012**, *92*, 227–259. [CrossRef]
32. Van Uum, S.H.M.; Sauvé, B.; Fraser, L.A.; Morley-Forster, P.; Paul, T.L.; Koren, G. Elevated content of cortisol in hair of patients with severe chronic pain: A novel biomarker for stress. *Stress* **2008**, *11*, 483–488. [CrossRef] [PubMed]
33. Kirschbaum, C.; Tietze, A.; Skoluda, N.; Dettenborn, L. Hair as a retrospective calendar of cortisol production-Increased cortisol incorporation into hair in the third trimester of pregnancy. *Psychoneuroendocrinology* **2009**, *34*, 32–37. [CrossRef] [PubMed]
34. Heimbürge, S.; Kanitz, E.; Otten, W. The use of hair cortisol for the assessment of stress in animals. *Gen. Comp. Endocrinol.* **2019**, *270*, 10–17. [CrossRef] [PubMed]
35. Miranda-de la Lama, G.C.; Rivero, L.; Chacón, G.; Garcia-Belenguer, S.; Villarroel, M.; Maria, G.A. Effect of the pre-slaughter logistic chain on some indicators of welfare in lambs. *Livest. Sci.* **2010**, *128*, 52–59. [CrossRef]
36. Part, C.E.; Kiddie, J.L.; Hayes, W.A.; Mills, D.S.; Neville, R.F.; Morton, D.B.; Collins, L.M. Physiological, physical and behavioural changes in dogs (Canis familiaris) when kennelled: Testing the validity of stress parameters. *Physiol. Behav.* **2014**, *133*, 260–271. [CrossRef] [PubMed]
37. Mason, G.; Mendl, M. Why is there no simple way of measuring animal welfare? *Anim. Welf.* **1993**, *2*, 301–319.

38. Mormède, P.; Andanson, S.; Aupérin, B.; Beerda, B.; Guémené, D.; Malmkvist, J.; Manteca, X.; Manteuffel, G.; Prunet, P.; van Reenen, C.G.; et al. Exploration of the hypothalamic-pituitary-adrenal function as a tool to evaluate animal welfare. *Physiol. Behav.* **2007**, *92*, 317–339. [CrossRef] [PubMed]
39. Angle, C.T.; Wakshlag, J.J.; Gillette, R.L.; Stokol, T.; Geske, S.; Adkins, T.O.; Gregor, C. Hematologic, serum biochemical, and cortisol changes associated with anticipation of exercise and short duration high-intensity exercise in sled dogs. *Vet. Clin. Pathol.* **2009**, *38*, 370–374. [CrossRef] [PubMed]
40. Ralph, C.R.; Tilbrook, A.J. Invited Review: The usefulness of measuring glucocorticoids for assessing animal welfare. *J. Anim. Sci.* **2016**, *94*, 457–470. [CrossRef]
41. Sapolsky, R.M.; Romero, L.M.; Munck, A.U. How do glucocorticoids influence stress responses? Integrating permissive, suppressive, stimulatory, and preparative actions. *Endocr. Rev.* **2000**, *21*, 55–89.
42. Trevisi, E.; Bertoni, G. Some physiological and biochemical methods for acute and chronic stress evaluationin dairy cows. *Ital. J. Anim. Sci.* **2009**, *8*, 265–286. [CrossRef]
43. Bennett, A.; Hayssen, V. Measuring cortisol in hair and saliva from dogs: Coat color and pigment differences. *Domest. Anim. Endocrinol.* **2010**, *39*, 171–180. [CrossRef] [PubMed]
44. Lee, D.Y.; Kim, E.; Choi, M.H. Technical and clinical aspects of cortisol as a biochemical marker of chronic stress. *BMB Rep.* **2015**, *48*, 209. [CrossRef] [PubMed]
45. Takahashi, Y.; Ebihara, S.; Nakamura, Y.; Takahashi, K. A model of human sleep-related growth hormone secretion in dogs: Effects of 3, 6, and 12 hours of forced wakefulness on plasma growth hormone, cortisol, and sleep stages. *Endocrinology* **1981**, *109*, 262–272. [CrossRef] [PubMed]
46. Kemppainen, R.J.; Sartin, J.L. Evidence for episodic but not circadian activity in plasma concentrations of adrenocorticotrophin, cortisol and thyroxine in dogs. *J. Endocrinol.* **1984**, *103*, 219–226. [CrossRef] [PubMed]
47. Kolezvská, J.; Brunclík, V.; Svoboda, M. Circadian Rhythm of Cortisol Secretion in Dogs of Different Daily Activities. *Acta Vet. Brno.* **2003**, *72*, 599–605. [CrossRef]
48. Protopopova, A. Effects of sheltering on physiology, immune function, behavior, and the welfare of dogs. *Physiol. Behav.* **2016**, *159*, 95–103. [CrossRef] [PubMed]
49. Beerda, B.; Schilder, M.B.H.; Van Hooff, J.A.R.A.M.; De Vries, H.W.; Mol, J.A. Behavioural and hormonal indicators of enduring environmental stress in dogs. *Anim. Welf.* **2000**, *9*, 49–62.
50. Stalder, T.; Kirschbaum, C.; Alexander, N.; Bornstein, S.R.; Gao, W.; Miller, R.; Stark, S.; Bosch, J.A.; Fischer, J.E. Cortisol in hair and the metabolic syndrome. *J. Clin. Endocrinol. Metab.* **2013**, *98*, 2573–2580. [CrossRef]
51. Russell, E.; Koren, G.; Rieder, M.; Van Uum, S. Hair cortisol as a biological marker of chronic stress: Current status, future directions and unanswered questions. *Psychoneuroendocrinology* **2012**, *37*, 589–601. [CrossRef]
52. Jaroenporn, S.; Nagaoka, K.; Kasahara, C.; Ohta, R.; Watanabe, G.; Taya, K. Physiological roles of prolactin in the adrenocortical response to acute restraint stress. *Endocr. J.* **2007**, *54*, 703–711. [CrossRef] [PubMed]
53. Beerda, B.; Schilder, M.B.H.; Van Hooff, J.A.R.A.M.; De Vries, H.W.; Mol, J.A. Behavioural, saliva cortisol and heart rate responses to different types of stimuli in dogs. *Appl. Anim. Behav. Sci.* **1998**, *58*, 365–381. [CrossRef]
54. Mariti, C.; Gazzano, A.; Moore, J.L.; Baragli, P.; Chelli, L.; Sighieri, C. Perception of dogs' stress by their owners. *J. Vet. Behav. Clin. Appl. Res.* **2012**, *7*, 213–219. [CrossRef]
55. Beerda, B.; Schilder, M.B.H.; Van Hooff, J.A.R.A.M.; De Vries, H.W.; Mol, J.A. Chronic stress in dogs subjected to social and spatial restriction. I. Behavioral responses. *Physiol. Behav.* **1999**, *66*, 233–242. [CrossRef]
56. Beerda, B.; Schilder, M.B.H.; Bernadina, W.; Van Hooff, J.A.R.A.M.; De Vries, H.W.; Mol, J.A. Chronic stress in dogs subjected to social and spatial restriction. II. Hormonal and immunological responses. *Physiol. Behav.* **1999**, *66*, 243–254. [CrossRef]
57. Cozzi, A.; Mariti, C.; Ogi, A.; Sighieri, C.; Gazzano, A. Behavioral modification in sheltered dogs. *Dog Behav.* **2016**, *2*, 1–12.
58. Siracusa, C.; Manteca, X.; Cuenca, R.; Del Mar Alcalá, M.; Alba, A.; Lavín, S.; Pastor, J. Effect of a synthetic appeasing pheromone on behavioral, neuroendocrine, immune, and acute-phase perioperative stress responses in dogs. *J. Am. Vet. Med. Assoc.* **2010**, *237*, 673–681. [CrossRef] [PubMed]
59. Friard, O.; Gamba, M. BORIS: A free, versatile open-source event-logging software for video/audio coding and live observations. *Methods Ecol. Evol.* **2016**, *7*, 1325–1330. [CrossRef]
60. Gazzano, A.; Migoni, S.; Guardini, G.; Bowen, J.; Fatjò, J.; Mariti, C. Stress in aggressive dogs towards people: Behavioral analysis during consultation. *Dog Behav.* **2015**, *1*, 6–13.

61. Stephen, J.M.; Ledger, R.A. An audit of behavioral indicators of poor welfare in kenneled dogs in the United Kingdom. *J. Appl. Anim. Welf. Sci.* **2005**, *8*, 79–96. [CrossRef] [PubMed]
62. Dreschel, N.A.; Granger, D.A. Physiological and behavioral reactivity to stress in thunderstorm-phobic dogs and their caregivers. *Appl. Anim. Behav. Sci.* **2005**, *95*, 153–168. [CrossRef]
63. Rooney, N.; Gaines, S.; Hiby, E. A practitioner's guide to working dog welfare. *J. Vet. Behav. Clin. Appl. Res.* **2009**, *4*, 127–134. [CrossRef]
64. Polgár, Z.; Blackwell, E.J.; Rooney, N.J. Assessing the welfare of kennelled dogs—A review of animal-based measures. *Appl. Anim. Behav. Sci.* **2019**, *213*, 1–13. [CrossRef]
65. Meyerhoff, J.L.; Oleshansky, M.A.; Mougey, E.H. Psychologic stress increases plasma levels of prolactin, cortisol, and POMC-derived peptides in man. *Psychosom. Med.* **1988**, *50*, 295–303. [CrossRef] [PubMed]
66. Richter, S.D.; Schürmeyer, T.H.; Schedlowski, M.; Hädicke, A.; Tewes, U.; Schmidt, R.E.; Wagner, T.O.F. Time kinetics of the endocrine response to acute psychological stress. *J. Clin. Endocrinol. Metab.* **1996**, *81*, 1956–1960. [PubMed]
67. Jöchle, W. Prolactin in canine and feline reproduction. *Reprod. Domest. Anim.* **1997**, *32*, 183–193. [CrossRef]
68. Gobello, C.; Colombani, M.; Scaglia, H.; De La Sota, R.L.; Goya, R.G. Heterogeneity of circulating prolactin in the bitch. *Reprod. Nutr. Dev.* **2001**, *41*, 505–511. [CrossRef]
69. Graf, K.J.; El Etreby, M.F. Endocrinology of reproduction in the female beagle dog and its significance in mammary gland tumorigenesis. *Acta Endocrinol. Suppl.* **1979**, *222*, 1–34.
70. Graef, K.J.; Friedreich, E.; Matthes, S.; Hasan, S.H. Homologous radioimmunoassay for canine prolactin and its application in various physiological states. *J. Endocrinol.* **1977**, *75*, 93–103. [CrossRef]
71. Corrada, Y.; Rimoldi, I.; Arreseigor, S.; Marecco, G.; Gobello, C. Prolactin reference range and pulsatility in male dogs. *Theriogenology* **2006**, *66*, 1599–1602. [CrossRef]
72. Urhausen, C.; Seefeldt, A.; Eschricht, F.; Koch, A.; Hoppen, H.; Beyerbach, M.; Möhrke, C.; Dieleman, S.; Günzel-Apel, A. Concentrations of Prolactin, LH, Testosterone, TSH and Thyroxine in Normospermic Dogs of Different Breeds. *Reprod. Domest. Anim.* **2009**, *44*, 279–282. [CrossRef]
73. Lane, M.B.; Flatland, B.; Olin, S.J.; Fecteau, K.A.; Rick, M.; Giori, L. Analytic performance evaluation of a veterinary-specific ELISA for measurement of serum cortisol concentrations of dogs. *J. Am. Vet. Med. Assoc.* **2018**, *253*, 1580–1588. [CrossRef] [PubMed]
74. Cafazzo, S.; Maragliano, L.; Bonanni, R.; Scholl, F.; Guarducci, M.; Scarcella, R.; Di Paolo, M.; Pontier, D.; Lai, O.; Carlevaro, F.; et al. Behavioural and physiological indicators of shelter dogs' welfare: Reflections on the no-kill policy on free-ranging dogs in Italy revisited on the basis of 15years of implementation. *Physiol. Behav.* **2014**, *133*, 223–229. [CrossRef] [PubMed]
75. Ishioka, K.; Hatai, H.; Komabayashi, K.; Soliman, M.M.; Shibata, H.; Honjoh, T.; Kimura, K.; Saito, M. Diurnal variations of serum leptin in dogs: Effects of fasting and re-feeding. *Vet. J.* **2005**, *169*, 85–90. [CrossRef] [PubMed]
76. Chen, C.L.; Gelatt, K.N.; Gum, G.G. Serum hydrocortisone (cortisol) values in glaucomatous and normotensive Beagles. *Am. J. Vet. Res.* **1980**, *41*, 1516–1518.
77. Garnier, F.; Benoit, E.; Virat, M.; Ochoa, R.; Delatour, P. Adrenal cortical response in clinically normal dogs before and after adaptation to a housing environment. *Lab. Anim.* **1990**, *24*, 40–43. [CrossRef] [PubMed]
78. Avella, M.; Schreck, C.B.; Prunet, P. Plasma prolactin and cortisol concentrations of stressed coho salmon, Oncorhynchus kisutch, in fresh water or salt water. *Gen. Comp. Endocrinol.* **1991**, *81*, 21–27. [CrossRef]
79. Kant, G.J.; Leu, J.R.; Anderson, S.M.; Mougey, E.H. Effects of chronic stress on plasma corticosterone, ACTH and prolactin. *Physiol. Behav.* **1987**, *40*, 775–779. [CrossRef]
80. Cook, C.J. Oxytocin and prolactin suppress cortisol responses to acute stress in both lactating and non-lactating sheep. *J. Dairy Res.* **1997**, *64*, 327–339. [CrossRef]
81. Bazan, J.F.; Bole-Feysot, C.; Goffin, V.; Edery, M.; Binart, N.; Kelly, P.A.; Carter, D.A.; Lightman, S.L.; Andrews, Z.B.; Kokay, I.C.; et al. Oxytocin responses to stress in lactating and hyperprolactinaemic rats. *Endocrinology* **2001**, *46*, 532–537.
82. Van De Kar, L.D.; Blair, M.L. Forebrain pathways mediating stress-induced hormone secretion. *Front. Neuroendocrinol.* **1999**, *20*, 1–48. [CrossRef] [PubMed]
83. Ratner, A.; Yelvington, D.B.; Rosenthal, M. Prolactin and corticosterone response to repeated footshock stress in male rats. *Psychoneuroendocrinology* **1989**, *14*, 393–396. [CrossRef]

84. Neumann, I.D.; Torner, L.; Wigger, A. Brain oxytocin: Differential inhibition of neuroendocrine stress responses and anxiety-related behaviour in virgin, pregnant and lactating rats. *Neuroscience* **1999**, *95*, 565–575. [CrossRef]
85. Grattan, D.R. Behavioural significance of prolactin signalling in the central nervous system during pregnancy and lactation. *Reproduction* **2002**, *123*, 497–506. [CrossRef] [PubMed]
86. Cuadra, G.; Zurita, A.; Lacerra, C.; Molina, V. Chronic stress sensitizes frontal cortex dopamine release in response to a subsequent novel stressor: Reversal by naloxone. *Brain Res. Bull.* **1999**, *95*, 567–575. [CrossRef]
87. Coppola, C.L.; Grandin, T.; Enns, R.M. Human interaction and cortisol: Can human contact reduce stress for shelter dogs? *Physiol. Behav.* **2006**, *87*, 537–541. [CrossRef] [PubMed]
88. Hennessy, M.B. Hypothalamic-pituitary-adrenal responses to brief social separation. *Neurosci. Biobehav. Rev.* **1997**, *21*, 11–29. [CrossRef]
89. Hennessy, M.B.; Voith, V.L.; Mazzei, S.J.; Buttram, J.; Miller, D.D.; Linden, F. Behavior and cortisol levels of dogs in a public animal shelter, and an exploration of the ability of these measures to predict problem behavior after adoption. *Appl. Anim. Behav. Sci.* **2001**, *73*, 217–233. [CrossRef]
90. Stephen, J.M.; Ledger, R.A. A longitudinal evaluation of urinary cortisol in kennelled dogs, Canis familiaris. *Physiol. Behav.* **2006**, *87*, 911–916. [CrossRef]
91. Rooney, N.J.; Gaines, S.A.; Bradshaw, J.W.S. Behavioural and glucocorticoid responses of dogs (Canis familiaris) to kennelling: Investigating mitigation of stress by prior habituation. *Physiol. Behav.* **2007**, *92*, 847–854. [CrossRef]
92. Righi, C.; Menchetti, L.; Orlandi, R.; Moscati, L.; Mancini, S.; Diverio, S. Welfare Assessment in Shelter Dogs by Using Physiological and Immunological Parameters. *Animals* **2019**, *9*, 340.
93. Gutiérrez, J.; Gazzano, A.; Torracca, B.; Meucci, V.; Mariti, C. Determination of prolactin in canine saliva: Is it possible to use a commercial ELISA kit? *Animals* **2019**, *9*, 418. [CrossRef] [PubMed]

animals

MDPI

Article

Socioeconomic Influences on Reports of Canine Welfare Concerns to the Royal Society for the Prevention of Cruelty to Animals (RSPCA) in Queensland, Australia

Hao Yu Shih [1,*], Mandy B. A. Paterson [1,2] and Clive J. C. Phillips [1]

1 Centre for Animal Welfare and Ethics, University of Queensland, White House Building (8134), Gatton Campus, Gatton, QLD 4343, Australia; mpaterson@rspcaqld.org.au (M.B.A.P.); c.phillips@uq.edu.au (C.J.C.P.)

2 Royal Society for the Prevention of Cruelty to Animals Queensland, Brisbane, QLD 4076, Australia

* Correspondence: haoyu.shih@uqconnect.edu.au

Received: 28 June 2019; Accepted: 18 September 2019; Published: 23 September 2019

Simple Summary: The role of the socioeconomic status of dog owners in canine welfare concerns is not fully understood. We conducted a retrospective study of 107,597 canine welfare complaints attended by the Royal Society for the Prevention of Cruelty to Animals (RSPCA) Queensland from 2008 to 2018. We explored the relationship between the owner's socioeconomic status and reported (rather than confirmed) complaints about welfare of dogs. The socioeconomic status of the owner was estimated from the postcode of where the alleged welfare issue occurred, using government statistics for Socio-Economic Indexes of different regions of Australia. Reported complaints were correlated with socioeconomic scores. There was a lower median socioeconomic score in our study group compared to the entire Queensland state, indicating that alleged canine welfare concerns were more likely to be reported in areas with inhabitants of low socioeconomic status. The status was also low if the complaint was about a crossbred rather than a purebred dog. Among the purebred dogs, complaints involving working dogs, terriers, and utility breeds were associated with the lowest socioeconomic scores. The following complaints were associated with low socioeconomic status: cruelty, insufficient food and/or water, a dog not being exercised, a dog being confined/tethered, failure to provide shelter or treatment, overcrowding, a dog being in poor condition or living in poor conditions. Increased status was observed in alleged cases of a dog being left in a hot car unattended.

Abstract: Human–dog relationships are an important contributor to the welfare of dogs, but little is known about the importance of socioeconomic status of the dogs' owners. We conducted a retrospective study of canine welfare complaints, using Australian government statistics on the socioeconomic status of the inhabitants at the location of the alleged welfare issue. The socioeconomic score of inhabitants at the relevant postcode was assumed to be that of the plaintiff. Our dataset included 107,597 complaints that had been received by RSPCA Queensland between July 2008 and June 2018, each with the following information: the number of dogs involved, dog(s) age, breed(s), suburb, postcode, date received, and complaint code(s) (describing the type of complaint). The median index score for relative social advantage of the locations where the alleged welfare concern occurred was less than the median score for the population of Queensland, suggesting that welfare concerns in dogs were more commonly reported in areas with inhabitants of low socioeconomic status. It was also less if the dog being reported was not of a recognised breed, compared to dogs of recognised breeds. Dogs reported to be in the gundog breed group were in the most socioeconomically advantaged postcodes, followed by toy, hound, non-sporting, working dog, terrier, and utility breed groups. Reports of alleged cruelty, insufficient food and/or water, a dog being not exercised or being confined/tethered, failure to provide shelter or treatment, overcrowding, a dog being in poor condition or living in poor conditions were most likely to be made in relation to dogs in low socioeconomic postcodes. Reports of dogs being left in a hot vehicle unattended were more likely to be made in relation to dogs in high socioeconomic postcodes. It is concluded that both canine welfare complaints and dogs in specific breed groups appear to be related to the owner's socioeconomic status. This study may be used to improve public awareness and to tailor educational campaigns toward different populations.

Keywords: dog; animal welfare; animal cruelty; RSPCA; shelter; socioeconomic

1. Introduction

Dogs (*Canis familiaris*) share intimate relationships with humans as they are one of the most popular pets in our society [1]. According to the 2016 statistics, 3,555,000 Australian households (38.5%) own at least one dog, accounting for 4,759,700 dogs in Australia [1]. In the household, owners may develop a strong attachment to their dogs and vice versa, which creates a mutually beneficial relationship [2,3]. For instance, dog ownership apparently reduces owners' physiological and psychological stresses [3,4], as well as providing a safe haven for their dogs [5]. However, the human–dog relationship does not always provide mutual benefits and sometimes may even break down [6].

Animal cruelty refers to any morally or legally unacceptable behaviour which causes animals to experience physiological, psychological, and/or behavioural discomfort [7,8]. Animal welfare concerns are reported in different forms and in different species, and dogs are one of the most commonly reported victims [9]. Concerns may involve violence [10,11], injuries caused by their involvement in the sporting industry [12,13], abandonment [14,15], inappropriate surgical procedures [16], and neglecting care of the dog [9,17]. Not only do these issues compromise the welfare of dogs, some of these welfare concerns (e.g., intentional abuse and dog fighting) have been recognised as sentinels of other social issues [6,9], in particular domestic violence [18–20] and sexual assaults [21]. Some have also been reported to be a precursor of antisocial behaviours among young people [7,22,23].

To address these cruelty issues, animal protection legislation has been enacted worldwide [8,9,24]. In Queensland, Australia, animals are protected by *Animal Care and Protection Act 2001* (ACPA) [8]. This state-based legislation empowers the State to appoint inspectors, some of whom are employed by the Royal Society for the Prevention of Cruelty to Animals, Queensland (RSPCA Qld), who investigate potential breaches of the Act and enforce compliance with the Act [8]. There are two main offences under the ACPA: failure to fulfil 'duty of care' responsibilities and cruelty. There are a number of other specified offences. The Act recognises that a person who has charge of an animal owes that animal a duty of care. Failure to provide such care is the basis of the "breach of duty of care" offence. This offence covers such actions as not providing sufficient food, water, exercise, veterinary care, and suitable living conditions. It is not only the owner that has a duty of care towards an animal; anyone who is even temporarily in charge of an animal has this responsibility. The second major offence is "animal cruelty" and according to Section 18 of the ACPA cruelty describes any action that causes unjustifiable and unnecessary physical and mental discomfort to animals, inappropriate confinement or transport, unreasonable injuries and inhumane death [8]. A cruel act can be committed by anyone towards an animal, whether it is their own animal, another domestic animal or even a wild animal [8]. It is important to note, that under the ACPA, it is not necessary for a person to have the intention of being cruel for the offence to be proven in Queensland. If an action carried out by a person causes pain and suffering and the action was intentional, the person may be charged with cruelty. The intention to carry out the action must be proved but not the intention to be cruel. If a lack of action deprives an animal of its fundamental needs, then the person who has a duty of care towards the animal may be charged with a breach of their duty of care or cruelty depending on the circumstances. Intention may be considered during sentencing however [8]. Other offences under the Act include unreasonable abandonment or release, the carrying out of prohibited surgical procedures (e.g., tail docking, ear cropping, debarking, etc.); being involved in, or having items used for, a prohibited event, such as dog or cock fighting; and allowing an animal to injure or kill another animal [8].

Potential cases are reported to RSPCA Qld through various means. RSPCA Qld has a "Cruelty Complaints" telephone number manned 24 h a day, seven days a week and complaints also come in through emails. These complaints are primarily made by members of the public but a few are also made by veterinarians and veterinary nurses, council officers, and other government and non-government employees visiting a location as part of their duties. Animals surrendered to the RSPCA or that come in as strays may be investigated by RSPCA Qld inspectors if cruelty or neglect is suspected. All cases reported by sources mentioned above were the focus of this study.

Apart from aiding law enforcement, identification of risk factors associated with animal welfare concerns would be beneficial in the development of awareness and in education of the general public. Risk factors include dogs' breed or breed group [12,15,25–27], dogs' ages [15,28], behaviours [15,26], owners' characteristics [15], and the complainant's socioeconomic status [6,20,22,29]. Among all these risk factors, the role of household socioeconomic level in crimes and animal welfare concerns is not yet well understood [6,20,22]. It has been proposed that socioeconomic status is negatively correlated with the frequency of crimes, domestic violence, and neglecting and abusing animals [6,22,29–31]. However, a tautologous relationship has also been postulated, because a dog owner's socioeconomic status may affect his or her ability to provide for all aspects of good welfare.

Therefore, this study aimed to investigate the relationships between socioeconomic status, dog breeds, and different types of dog welfare complaints. The socioeconomic status was quantified using a socioeconomic score derived from the postcode of where the alleged welfare issue occurred and government statistics for Socio-Economic Indexes of different postcode regions of Australia. We hypothesized that owners from relatively poor socioeconomic postcode regions would be more likely to be the subject of complaints about an absence of key resources for dogs, such as insufficient food, water, living space or veterinary care, lack of shelter, and poor living conditions. We also hypothesized that complaints concerning owners from poor socioeconomic regions would relate more to dog fighting, because of a known potential association with low socioeconomic status [32]. Finally, some research suggests that low socioeconomic status people are less likely to travel with their dogs, and therefore it is possible that low socioeconomic status is less likely to relate to dogs being left in a hot vehicle [33]. This is the third report in a series relating to the analysis of RSPCA Qld canine welfare complaint data [see also 27,28].

2. Materials and Methods

From July 2008 to June 2018, RSPCA Qld received 129,036 canine welfare complaints. Some involving more than one dog were recorded as multiple complaints sharing the same case number, while others were recorded as one complaint with multiple animals. To avoid sample bias due to multiple entries, we only retained the first complaint of case numbers with multiple entries, discarding 21,439 entries as a result. There remained 107,597 canine welfare complaints for this retrospective study. The data analysis was originally undertaken on the entire dataset and then repeated with the reduced number. Finding the complaint distribution and demographics to be similar, we opted for the reduced dataset to avoid pseudoreplication. Animal welfare complaints that fell within the geographical zone of responsibility of RSPCA Qld (determined by a Memorandum of Understanding between RSPCA Qld and Biosecurity Queensland, the Government Department tasked with the administration of ACPA) were investigated by RSPCA Qld inspectors. All other complaints were referred by RSPCA Qld to Biosecurity Queensland to be investigated by their inspectors. All complaints received by RSPCA Qld were included in this study, regardless of which authority investigated them.

All complaints were recorded in ShelterBuddy® (RSPCA, Queensland, Australia), the RSPCA Qld database. The following information was requested from the reporter of each incident at the time of taking the complaint: the number of dogs involved and their age, breed(s) (if known), the "complaint code(s)", suburb, postcode and date. All cases were investigated either by RSPCA Qld inspectors (n = 100,432) or Biosecurity Qld inspectors (n = 7165). It is recognized that some of the calls, once investigated, were found not to relate to a breach of the ACPA or to a genuine welfare concern. However, the outcome data for complaints were not analysed in this research, which focused solely on the complaint calls received by RSPCA Qld.

Dogs were classified according to two broad age ranges, dog and puppy, based on reporters' interpretation. It was important to recognise that the information recorded from the complainant may be inaccurate or inaccurately interpreted, e.g., a small dog is commonly referred to as a puppy in Queensland. Records regarding breed and the number of dogs involved were based on either complainants' initial reports or comments from trained inspectors, again recognising inaccuracies with identification of the breed. The "complaint code" was selected by the staff member receiving the call or email from a drop-down menu of 18 possible complaints (Appendix A) [28]. Multiple "complaint codes" were able to be selected for each case according to the description of what was alleged to have happened to the dog(s), and each was treated as a separate code for analysis.

2.1. Socioeconomic Scores

Australia is spatially divided into regions by postcodes; our dataset included the postcode of the location of the dog being reported, which was taken as a proxy measure for the socioeconomic status of the owner. We also reviewed the Australian government's Socio-Economic Indexes for Areas (SEIFA) developed in 2011 [34], which rank postcode regions in Australia by the socioeconomic level of inhabitants. Four indices are assigned to each area to describe the local socioeconomic status: (1) the Index of Relative Socio-Economic Disadvantage, (2) the Index of Relative Socio-Economic Advantage and Disadvantage, (3) the Index of Economic Resources, and (4) the Index of Education and Occupation. Each index is ranked by decile, percentile, and score. Among these four indices, the Index of Relative Socio-Economic Advantage and Disadvantage (IRSAD) correlates well with the other three and is suitable for comparing the entire range of areas, and was therefore considered the most appropriate descriptor of the socioeconomic level of inhabitants of each postcode [34]. The IRSAD score is a weighted combination of selected indicators of advantage and disadvantage: household income, availability of internet connection, educational level, occupation, employment rate, property, mortgage, and health. Nationwide, the score is standardized with a mean of 1000 and a standard deviation of 100, with a mean, median and standard deviation in Queensland of 999, 1014 and 65 respectively. An area receives a score of 1000 if all of the above indicators are equal to the national average; the score for an area increases or decreases if the indicators are greater or less than the national average, respectively [34]. The index is positively associated with socio-economic advantage and negatively associated with socio-economic disadvantage, thus a region with a high IRSAD score is more likely to have people with high socioeconomic levels dwelling within it [34].

2.2. Dog Breeds

Any breed in our dataset that was recognized by the Australian National Kennel Council (ANKC) [35], New Zealand Kennel Club (NZKC) [36], American Kennel Club (AMKC) [37], or United Kennel Club (UKC) [38] was considered a recognized breed (RB) (see Appendix B list of recognized breeds). Any other reported breed in our data was considered an unrecognised breed (UB), including all crossbred dogs without any identified breed. If more than one dominant breed was listed, the first mentioned was used. For instance, Great Dane × Bull Arab was categorized as Great Dane (Appendix B).

To achieve a secondary representation of breed recognition, RB breeds were amalgamated into the following seven breed groups based on the breed inclusion categories of the ANKC: toys, terriers, gundogs, hounds, working dogs, utility, and non-sporting. Breeds not listed by the ANKC but recognised by the NZKC, AMKC, or UKC were categorized into one of the seven groups based on the description of each kennel club. Some breeds (e.g., Australian Koolie and Bull Arab) were listed by the council registration, it being an obligation of all dog owners in Queensland to register their dogs with the local council. As a result, they were on the breed list (Appendix B), however they were not recognised as breeds by any major kennel club worldwide. Therefore, these breeds were categorized as unrecognized breeds, UB. If the breed description was left blank, the dogs' breed was considered unknown (n = 15,576/107,597), and these complaints were excluded from any data analysis related to breed factors. Our previous study suggested that, compared to specific breeds, breed groups and the dichotomization into RB and UB provided better agreements between the public and the trained RSPCA inspectors, and were therefore used for genotype identification in a public reporting system [27]. Therefore, this study used breed groups and RB/UB dichotomization rather than specific breeds for statistical analyses.

2.3. Statistical Analysis

Data was analysed using the statistical package Minitab® 17.3.1. (Minitab, LLC., State College, PA, USA). The distributions of IRSAD scores of both our entire study group and the RB/UB differentiation were not normal. Box–Cox and John transformations were used, but the assumption of normal distribution of residuals still could not be met. Therefore, a one-sample sign test was used to compare the median IRSAD scores between owners of dogs involved in an alleged incident in our dataset and the entire Queensland population as recorded by SEIFA. A Mann–Whitney test was conducted to compare the IRSAD scores of postcodes where there had been reports of RB and those of UB. As for the IRSAD scores of owners of dogs of the different breed groups, normality was assessed by visual inspection of residual plots. All complaints were independent but the assumption of equal variance could not be met based on the Bartlett's test ($p = 0.018$). Consequently, Welch's ANOVA followed by the Games-Howell pairwise comparisons were used to compare IRSAD scores between dogs of the different breed groups.

To determine factors influencing complaint codes, the IRSAD score, dog's age (dog or puppy), and breeds (UB or RB) were entered into eleven binary logistic regression models as fixed factors, using non-linear logit models with an alpha value for variables to enter the model of 0.15 [39]. Complaint codes were entered into the model as outcomes. Each case was independent from each other, and little multicollinearity was observed for all independent variables, with the variance inflation factors being less than 5. The linearity of the independent variable (the IRSAD score for the postcode of the dog) and the log-odds was assessed by plotting the independent variable and the log-odds fitted with a linear regression line. The assumption of linearity was considered met when the P value of a straight-line regression was less than 0.05. Eight complaint codes—Abandonment, Baiting/poisoning, Causing a captive animal to be injured/killed by dog, Dog fighting or other prohibited offence, Emergency relief, Keeping or using an animal for blooding/coursing a dog, Prohibition order breached, and Tail docking or other surgical procedure—did not fulfill the linearity assumption. Therefore, to fulfill the linearity assumption, Causing a captive animal to be injured/killed by dog, Dog fighting or other prohibited offence, Keeping or using an animal for blooding/coursing a dog, and Knowingly allowing an animal to kill/injure another were combined and categorized as a new code—Dog fighting. Baiting/poisoning, Cruelty, Prohibition order breached, and Tail docking or other surgical procedure were combined and categorized as a new code—Cruel act. Abandonment did not meet the linearity assumption statistically, but the graph was linear by observation, so the code was still used to construct a stepwise forward binary logistic model. Finally, the IRSAD score was removed from the regression model of Emergency relief because it did not meet the linearity assumption and the code was rarely cited in the past decade (0.01%, n = 8) [28]. Eleven stepwise forward binary logistic regression models were constructed to examine how different fixed factors (IRSAD score, dog's age and breeds) correlated with different outcomes (9 complaint codes and 2 combined complaint codes). Separate models were constructed for each code with the same input variable. In this paper, we focused on the relationship of IRSAD scores with complaint codes and breed factors. The relationship of other variables with the complaint codes have been reported separately [27,28].

3. Results

3.1. Descriptive Statistics

The median IRSAD score of owners being reported (median = 975) was significantly lower ($p < 0.001$) than that recorded for the population of Queensland (median = 1,014), and the Q1 and Q3 values were less (Table 1). Owners of reported UB dogs had significantly lower IRSAD scores (median = 970) than those reported owning RB dogs (median = 981) (p <0.001) (Table 2). Mean IRSAD scores for the postcodes of the different breed groups reported were significantly different (p <0.001) (Table 3), with gundogs (994 ± 63.0) being the highest, followed by toy (986 ± 64.3) and hound (984 ± 63.7), then non-sporting (980 ± 64.6) and working dogs (977 ± 63.0), terriers (977 ± 62.7) and utility dogs (976 ± 62.2).

Table 1. Descriptive analysis of Index of Relative Socio-Economic Advantage and Disadvantage (IRSAD) scores of the postcode of dogs in our study and those determined for the population of Queensland.

	Study Group	Queensland	*p* Value (One-Sample Sign Test)
Median	975	1014	<0.001
Q1	935	967	
Q3	1021	1039	

Q1: the first quartile; Q3: the third quartile.

Table 2. Index of Relative Socio-Economic Advantage and Disadvantage (IRSAD) scores of the postcode of owners of dogs reported to be recognised breeds (RB) and unrecognised breeds (UB).

	UB (n = 35, 080)	RB (n = 56, 663)	*p* Value (Mann-Whitney Test)
Median	970	981	<0.001
Q1	934	935	
Q3	1014	1024	

Q1: the first quartile; Q3: the third quartile.

Table 3. Total numbers and Index of Relative Socio-Economic Advantage and Disadvantage (IRSAD) scores of the postcode of reported dogs in each breed group ($p < 0.001$).

Breed Group	N	Mean ± (SD)	Grouping *				
Gundogs	4394	994 ± (63.0)	A				
Toys	5203	986 ± (64.3)		B			
Hounds	3152	984 ± (63.7)		B	C		
Non Sporting	5056	980 ± (64.6)			C	D	
Working Dogs	14,049	977 ± (63.0)				D	E
Terrier	15,979	977 ± (62.7)					E
Utility	8830	976 ± (62.2)					E

* Means not sharing a letter are significantly different ($p < 0.05$) by Games-Howell pairwise comparisons. SD: standard deviation.

3.2. Complaint Codes

The IRSAD scores of postcodes of dogs who were or were not reported for each complaint code significantly differed for the nine regression models (Table 4). The odds ratio (OR) was defined as a one unit increase in the IRSAD score leading to a corresponding x-fold decrease or increase in the odds of the cited event. The following codes were associated with dogs from postcodes with low IRSAD scores: cruel act (OR = 0.9994, $p < 0.001$), insufficient food and/or water (OR = 0.9981 $p < 0.001$), no exercise/confined/tethered (OR = 0.9979, $p < 0.001$), no shelter (OR = 0.9990, p <0.001), no treatment (OR = 0.9986, $p < 0.001$), overcrowding (OR = 0.9982, $p = 0.002$), poor dog condition (OR = 0.9974, $p < 0.001$) and poor living conditions (OR = 0.9996, $p = 0.002$). A single code was associated with dogs in postcodes with a high IRSAD score, hot animal in a car (OR = 1.0067, $p < 0.001$).

Table 4. Median Index of Relative Socio-Economic Advantage and Disadvantage (IRSAD) scores of the postcode of reported dogs for whom each complaint code was used, and the odds ratio and the 95% confidence interval (CI) of the IRSAD score in the logistic regression model of complaint codes. Outcomes of these models were different complaint codes and the fixed factor was the IRSAD score of the owners' postcode.

Complaint Code	IRSAD Score of the Owner's Postcode When This Code Cited	IRSAD Score of the Owner's Postcode When This Code Not Cited	IRSAD Odds Ratio and (CI)	p Value
Abandonment	976	975	- [a]	- [a]
Dog fighting *	969	975	0.9991 (0.9979, 1.0003)	0.143
Cruel act +	972	975	0.9994 (0.9991, 0.9997)	<0.001
Emergency relief	960	975	- [b]	- [b]
Hot animal in car	999	975	1.0067 (1.0063, 1.0071)	<0.001
Insufficient food and/or water	971	979	0.9981 (0.9978, 0.9983)	<0.001
No exercise/confined/tethered	970	979	0.9979 (0.9977, 0.9982)	<0.001
No shelter	973	977	0.9990 (0.9987, 0.9994)	<0.001
No treatment	972	977	0.9986 (0.9983, 0.9988)	<0.001
Overcrowding	968	977	0.9982 (0.9970, 0.9993)	0.002
Poor dog condition	969	979	0.9974 (0.9972, 0.9976)	<0.001
Poor living condition	975	977	0.9996 (0.9993, 0.9998)	0.002

[a] Not selected by the binary logistic regression model (alpha value to enter > 0.15). [b] The IRSAD did not meet the assumption of being linear to the logit so was not included in the binary logistic regression model. * Dog fighting is a combined code including Causing captive animal to be injured/killed by dog, Dog fighting or other prohibited offence, Keeping or using an animal for blooding/coursing a dog, Knowingly allowing an animal to kill/injure another. + Cruel act is a combined code including Baiting/poisoning, cruelty, Prohibition order breached, Tail docking or other surgical procedure.

4. Discussion

4.1. Socioeconomic Status of Dog Owners

A better understanding of relationships between the socioeconomic status of dog owners and specific welfare issues could help to elucidate the reasons for the issues, as well as helping to target specific sectors of the population for education about dog welfare. Compared to the median IRSAD score across the entire Queensland state, our study group had significantly lower IRSAD scores, indicating that the reported dogs were in postcodes with inhabitants at a lower socioeconomic level. It is possible that dog owners generally come from postcodes with inhabitants at low socioeconomic status, but because not all welfare issues were more frequently reported in lower socioeconomic regions and one was more frequent in higher socioeconomic regions, we consider that there may be an important relationship between welfare issues and socioeconomic status.

The score differences contributed by the variables appear small, and the odds ratios for IRSAD scores in the regression model were close to one. This is probably because the distribution of IRSAD scores across different regions in Queensland is narrow, with an interquartile range of 71.1 and a standard deviation of just 64.8, compared with a range of 558. Also, our dataset only covers the coastal area of Queensland which is relatively homogenous in terms of the socioeconomic level of inhabitants. The difference between our median score and that in Queensland, 38.3, represents about 59% of one SD in that region. Since about 68% of values lie within 1 SD of the mean, it can be seen that the differences in IRSAD scores in the variables tested in this study are meaningful and reflect the range in values from a significant proportion of the total Queensland population, about 40% (68% × 59%). Such a small difference is important because it should enable us to predict differences in commonly reported dog welfare concerns across populations and regions in a large and relatively homogenous area. The majority of previous studies have focused on socioeconomics and animal abuse [20,22,29]. The relationship between socioeconomic level and other welfare concerns in relation to canines [6] has received little research attention. This study may bridge this gap by determining some of the factors that relate to specific welfare concerns.

A key feature of the IRSAD score is that it might positively relate to the financial circumstances of the owner, notwithstanding the previously mentioned concern that the owner may not be represented by the status of the entire postcode, since it includes the % of people whose annual household income is < AUS $20,799 and whose rent is less than AUS $166/week [34]. This could indicate some constraints on the part of the owner in providing for the welfare of his or her dog, such as provision of adequate food, shelter, or other resources. This is discussed later in relation to individual complaint codes. The IRSAD score is also positively associated with the inhabitants' educational level, a key component being whether the members of the household progressed past year 11 in school, assuming they were over 15 years of age. This could affect whether the owner has sufficient knowledge to care for his or her dog, providing suitable nutrition, for example. Important detractors in the IRSAD include unemployment and the percentage of employed people classified as "labourers" living in that postcode [34]. This could relate to whether the owner has sufficient financial resources to care for his or her dog. However, these results should be interpreted with caution because the IRSAD score relates to the entire postcode, which may include substantial variation within and between regions. Clearly the circumstance of the owner or owners may be different to that of other inhabitants in the postcode in question.

Broad correlations between socioeconomic status and human behaviour towards animals have been noted previously [6,40,41]. People with high socioeconomic levels are more likely to advocate for animal welfare and to volunteer for animals [40], whereas those with lower socioeconomic backgrounds are more likely to be involved in animal neglect and abuse [6]. Our findings are in agreement with these previous conclusions.

It is also important to remember that in this study we analyzed only reported data, i.e., alleged welfare issues, which may not all reflect actual cases. It is known that complainants will report a neighbour out of spite, misread a situation, or report actions which may not actually represent a breach of the ACPA. The negative correlation between the socioeconomic status of the dog's owner and the risk of most reported welfare issues may also be explained by the fact that individuals may be more prepared or likely to report welfare issues in low socioeconomic postcodes. Since household income, home ownership, and full-time employment were reported to negatively correlate with dog ownership, it is less likely that the higher number of reports in low socioeconomic regions is because people from low socioeconomic regions own more dogs [42]. However, the same study reported living in rural locations was associated with higher odds of owning a dog [42], and we found a tendency for low socioeconomic regions to overlap non-urban regions [43]. Therefore, the socioeconomic level and dwelling in rural regions may both increase the possibility of being reported; yet we could not validate these hypotheses as the IRSAD score is a generalization of postcode regions ignoring the within postcode variation. However, inconclusive results were reported in another study, that found that ones' personal and household income levels were not associated with the propensity of being reported for animal cruelty [44].

4.2. Breed

Not only can socioeconomic level of dog owners be related to animal welfare concerns, it can also be linked with breed factors. Reported cases involving UB dogs were potentially related to owners being socioeconomically disadvantaged compared with cases involving RB dogs. Similar results of breeds' predisposition to welfare concerns and socioeconomic levels were observed when we examined the RB dogs. Reported cases involving utility breeds, terriers, and working dogs were associated with the three least socioeconomically advantaged groups. Utility breeds, terriers, and working dogs were also more commonly reported for canine welfare concerns [27]. Although this study involved only reported but not confirmed cases, these results potentially support previous studies suggesting that people with lower income have higher risks of mistreating animals [20,22], which might be further associated with specific dog breeds, and linked to some breed-specific complaints [27]. Nevertheless, it is important to note that the differences in socioeconomic scores among different breed groups were statistically significant but small, and there may be some inaccuracies and biases when breeds were reported. Therefore, over-interpretation should be avoided.

4.3. Complaint Types

There were clear correlations between the socioeconomic level and complaint type being reported. For example, allegedly committing cruel acts was associated with lower socioeconomic level.

Previous research focusing on crime and animal cruelty has found negative relationships between the socioeconomic level and tendencies to commit crime, including being cruel to dogs and cats [22,45]. However, a study comparing animal cruelty between rural and urban regions found that rural residents mostly targeted cats rather than dogs [10], and another study investigating community demographics of animal cruelty reports found no differences among urban, town, and rural residents in the likelihood of being reported for animal cruelty [43].

Except for canine abuse, most complaints for which the owner was from a low socioeconomic background were neglect related. The aforementioned explanation is that people with lower socioeconomic backgrounds may lack the ability (e.g., money, space, or transportation) to manage animal care and welfare [6]. Moreover, this finding is in line with a previous study that people of lower socioeconomic status tended to moralize transgressions that did not cause obvious harm to animals [46,47]. Less affluent people may therefore be more likely to view an animal welfare compromise that is not overtly cruel as a moral but not a legal issue [6,46,47]. Consequently, there is a risk that people with lower socioeconomic backgrounds may tend to neglect the fundamental needs of their dogs, including failing to provide appropriate nutrients, adequate living conditions, and medical treatments, which increases the chance of them being reported. In this respect, the first of our hypotheses was supported, that complainants from relatively poor socioeconomic postcode regions would be more likely to complain about an absence of key resources for dogs. This may relate to the factors included in the IRSAD score that are relevant to educational level. The IRSAD score of different postcodes across Australia is highly correlated with the Index of Education and Occupation score, with a Spearman's rank correlation of 0.85 [34]. This indicates a potential relationship between the low levels of education and neglect-related canine welfare concerns, either because people who are less educated do not consider deprivation as an act of neglect, or they are simply lacking in knowledge about the welfare and care of dogs [6,29].

Among these neglect-related complaints, the finding associated with insufficient (medical) treatments seems contradictory to our previous study [27]. This study reveals that insufficient veterinary treatments were less commonly reported in regions of lower socioeconomic status, which is supported by the fact that household income limits owners' access or willingness to provide veterinary care [48,49]. In addition, UB dogs were more commonly reported in lower socioeconomic regions. Consequently, it would be expected that UB dogs would be the subject of more complaints about poor veterinary care. However, according to our previous study, RB but not UB dogs were more likely to be reported with insufficient veterinary treatments [27]. The potential predisposition of RB dogs to a complaint about lack of veterinary care may be influenced not only by owners' socioeconomic status, but also by other factors. These factors may include morality [10], attitudes to the welfare of breeding dogs [50], human–animal bond [49], and registration rate [51], which can outweigh or confound the effects of socioeconomic level. For instance, people who are affluent but have less moral conviction may prefer to purchase an RB dog from a breeder rather than adopting an UB dog from a shelter, and the owners with less moral conviction may be less likely to bring their sick dogs to a veterinary clinic [49]. Besides, the potentially low registration rate of UB dogs [27] may encourage owners to abandon their dogs when medical care is required, leading to a reduced but inaccurate prevalence of UB dogs being reported for lack of treatment. However, these hypotheses cannot be confirmed in this study as they were all reported not confirmed cases.

Another inconclusive finding was that the difference in socioeconomic levels between reported cases citing and not citing abandonment was small. Dog ownership is positively correlated to household income [30,42,52]. However, it has been suggested that people with lower household income are less likely to relinquish their dogs, for financial reasons [14]. Therefore, it has been argued that other variables, including problematic behaviours [15,53,54], nature of the dog [15,22,53], human factors [53,55], and the human–dog bond [14,56] may also play important roles in determining the benefits of dog ownership. This finding again supports the previous assumption that other factors also critically influence some complaint types.

Similarly, we hypothesized that complaints about 'blood sports', such as 'Knowingly allowing an animal to kill / injure another' and 'Dog fighting or other prohibited offence' would be associated with lower socioeconomic backgrounds. However, the IRSAD score did not significantly differ between reported cases cited and not cited with this code. Previous research exploring financial aspects of dog-fighting in the UK has pointed out that this kind of 'blood sporting' was more popular among working-class men, as a way of life and an alternative expression of masculinity [32]. Nevertheless, a small proportion of middle-class people might also be involved as business owners or as a hobby [32], and thus increase the average socioeconomic scores. In addition, only a small number of cases were reported involving those alleged complaints, with even fewer being confirmed [28]; thereby, it might not be enough to test for statistically significant differences with validity.

Although most complaints were related to socioeconomic disadvantaged people, one complaint type was reported more commonly among more socioeconomically advantaged people. Those with relatively higher socioeconomic levels were more likely to be reported leaving their dogs unattended in a hot car. This finding partially supports the previous study that people with higher socioeconomic level, mainly living in urban areas, are more likely to own cars and take their dogs for outdoor activities [41], and thus have a greater chance of leaving their dogs alone in a car. In high socioeconomic regions and in high density urban regions it would be more likely that owners are reported when they left their dogs in a vehicle in a busy public area.

Although this study reports unconfirmed dog welfare complaints, the results reflect the major welfare concerns in dog populations in higher or lower socioeconomic background. Considering the relevance of different complaint reasons and different socioeconomic levels, intervention strategies for the prevention of animal neglect or cruelty should be directed differently in high and low socioeconomic regions. Studies of confirmed welfare issues are required. Interventions are recommended to be taken in lower socioeconomic areas to explore whether the high number of reports is driven by actual offenses or by higher public awareness. These results can also be used to increase public awareness and promote public education. For instance, councils of relatively higher socioeconomic regions are recommended to place more emphasis on enforcing that people do not leave dogs in a hot vehicle unattended. Councils of relatively lower socioeconomic regions can highlight information for owners on the basic needs of dogs (e.g., the amount of water and food consumption). Another important implication of this study is that it provides information regarding the correlation of socioeconomic backgrounds and the preference for dog types, which could help develop more tailored educational programs that target different populations.

4.4. Limitations and Need for Future Research

Several limitations were identified in this study. First, the dataset consisted of cases reported but not confirmed, so results only reflect a likelihood of correlations between socioeconomics and different types of welfare concerns in dogs. Besides, the socioeconomic data was acquired by linking to the postcodes where an alleged welfare concern occurred (used as a proxy for the owners' postcode), which appeared to be a generalization and might be an ecological fallacy. Therefore, the results should be cautiously interpreted. Future study could try to obtain a more direct measure of socioeconomic status, for example, the household income of each individual. Second, total numbers of residents in each post code are not accurately known. Therefore, we cannot calculate the exact prevalence of each welfare issue in different areas. If the prevalence in certain regions is particularly high, then sampling bias may occur. Third, breed recognition was based on comments made by complainants or trained inspectors, which may not be accurate. Finally, the data was obtained from populations within Queensland, and thus wider geographical generalization should be made cautiously.

5. Conclusions

This dataset was analyzed based on reported but not confirmed cases of canine welfare concerns, so the results reflect the tendency rather than fact. Results correlate the socioeconomic level with different dog breeds. The relationships between socioeconomic levels and different complaint types are also identified. Reported dogs of unrecognizable breeds came from postcodes with lower socioeconomic status to those reporting dogs of recognizable breeds. Among RB dogs, reports concerning utility breeds, terriers, and working dogs were more common than dogs reported in socioeconomically disadvantaged areas, but it is not clear to what extent these breed groups are more prevalent in these areas. People living in lower socioeconomic regions were more likely to be reported to be involved in canine welfare concerns, especially neglect-related complaints, and abusing dogs. In contrast, people living in higher socioeconomic areas were more alleged to leave their dogs unattended in a hot vehicle. This study provides detailed information which may help in the development of tailored strategies for different populations to combat welfare concerns in dogs. However, the differences of socioeconomic level were relatively small so the results should be interpreted cautiously. Finally, more risk factors and their roles in different complaint types should also be identified in order to give a better picture of canine welfare concerns.

Author Contributions: Conceptualization, C.J.C.P., H.Y.S. and M.B.A.P.; methodology, H.Y.S., C.J.C.P. and M.B.A.P.; software, H.Y.S. and C.J.C.P.; validation, H.Y.S., C.J.C.P. and M.B.A.P.; formal analysis, H.Y.S.; investigation, H.Y.S.; resources, M.B.A.P. and C.J.C.P.; data curation, C.J.C.P., H.Y.S. and M.B.A.P.; writing—original draft preparation, H.Y.S.; writing—review and editing, C.J.C.P., M.B.A.P. and H.Y.S.; visualisation, H.Y.S.; supervision, C.J.C.P. and M.B.A.P.; project administration, C.J.C.P. and M.B.A.P.

Funding: This research received no external funding.

Acknowledgments: We thank RSPCA, Qld for providing the database of canine welfare complaints, and RPSCA inspectors for the consultation.

Conflicts of Interest: Mandy Paterson is employed as the principal scientist by RSPCA, Qld, but none of the authors received any interest or financial support from people or organizations that inappropriately influenced this study.

Appendix A

Table A1. Description of each complaint code, alleging a welfare issue.

Complaint Code	Description
Abandonment	An animal was abandoned/left by the owner either at their abode or somewhere else such as in the bush.
Baiting/Poisoning	An animal was poisoned or planned to be poisoned.
Causing captive animal to be injured/killed by dog	A person let a captive animal be injured/killed by a dog.
Cruelty	A person was reported to have abused an animal.
Dog fighting or other prohibited offence	A person was reported as allowing dogs to fight or conducting other specifically prohibited acts.
Emergency relief	Emergency relief is required for an animal left unattended because its owner experienced an emergency (e.g., flood or being hit by a car).
Hot animal in car	An animal was left unattended in a car during hot weather.
Insufficient food and/or water	An animal has insufficient food and/or water.
Keeping or using animal for blooding/coursing a dog	A person used a live bait for blooding/coursing a dog.
Knowingly allowing an animal to kill/injure another	A person allows one animal to kill/injuring another one, and does nothing to stop them.
No exercise/confined/tethered	An animal is confined or tethered and not given a suitable amount of exercise.
No shelter	An animal is not provided with suitable shelter provisions.
No treatment	An animal did not receive appropriate medical treatment when needed.
Overcrowding	The number of animals was too high for the living space provided.
Poor dog condition	The general condition of an animal is poor. (e.g., messy/matted coat, pussy eyes, etc.)
Poor living condition	The living environment of the animal is poor.
Prohibition order breached	An owner violated a prohibition order [a].
Tail docking or other surgical procedure	Tail docking or other surgical procedure (e.g., declaw removal, etc.) was conducted on an animal.
Unknown	Unknown

[a] Prohibition order-A prohibition order is given by the court when a person convicted of an animal welfare offense must not possess any or specific animal for a prescribed period of time [8].

Appendix B

Table A2. Breed list.

Breed Reported	Associated Listed Breed	Breed Group
Affenpinscher	Affenpinscher	Toys
Afghan hound	Afghan Hound	Hounds
Airedale terrier	Airedale Terrier	Terrier
Akita	Akita	Utility
Alaskan husky	Siberian Husky	Utility
Alaskan malamute	Alaskan Malamute	Utility
American bulldog	American Bulldog	Non sporting
American foxhound	Foxhound	Hounds
American pit bull terrier	Pit Bull Terrier	Terrier

Table A2. *Cont.*

Breed Reported	Associated Listed Breed	Breed Group
American Staffordshire terrier	Staffordshire Terrier	Terrier
American water spaniel	American Water Spaniel	Gundogs
Anatolian shepherd dog	Anatolian Shepherd Dog	Utility
Australian bandog	Cross Breed	UB
Australian bulldog	Australian Bulldog	Non sporting
Australian bulldog cross	Australian Bulldog	Non sporting
Australian cattle dog	Australian Cattle Dog	Working dogs
Australian koolie	Coolie/Koolie	UB
Australian sheepdog	Australian Sheepdog	Working dogs
Australian shepherd	Australian Shepherd	Working dogs
Australian silky terrier	Australian Silky Terrier	Toys
Australian stumpy tail cattle dog	Australian Stumpy Tail Cattle Dog	Working dogs
Australian terrier	Australian Terrier	Terrier
Bandogge mastiff	Cross Breed	UB
Basenji	Basenji	Hounds
Basset fauve de bretagne	Basset Fauve De Bretagne	Hounds
Basset hound	Basset Hound	Hounds
Beagle	Beagle	Hounds
Bearded collie	Bearded Collie	Working dogs
Bedlington terrier	Bedlington Terrier	Terrier
Belgian shepherd	Belgian Shepherd	Working dogs
Belgian shepherd-Groenendael	Belgian Shepherd	Working dogs
Belgian shepherd-Laekenois	Belgian Shepherd	Working dogs
Belgian shepherd-Malinois	Belgian Shepherd	Working dogs
Belgian shepherd-Tervueren	Belgian Shepherd	Working dogs
Bernese mountain dog	Bernese Mountain Dog	Utility
Bichon fries	Bichon Frise	Toys
Bloodhound	Bloodhound	Hounds
Bluetick coohound	Bluetick Coohound	Hounds
Border collie	Border Collie	Working dogs
Border collie × Labrador	Border Collie	Working dogs
Border collie, miniature	Border Collie	Working dogs
Border terrier	Border Terrier	Terrier
Borzoi	Borzoi	Hounds
Boston terrier	Boston Terrier	Non sporting
Bouvier des flandres	Bouvier Des Flandres	Working dogs
Boxer	Boxer	Utility
Boxer cross	Boxer	Utility
Boxer × bullmastif	Boxer	Utility
Boxer × American Staffordshire terrier	Boxer	Utility
Bracco Italiano	Bracco Italiano	Gundogs
Briard	Briard	Working dogs
British bulldog	British Bulldog	Non sporting
Brittany	Brittany	Gundogs
Bull Arab	Bull Arab	UB
Bull Arab × greyhound	Bull Arab	UB
Bull terrier	Bull terrier	Terrier
Bull terrier cross	Bull terrier	Terrier
Bull Terrier, Miniature	Bull terrier	Terrier
Bulldog	British bulldog	Non sporting
Bulldog cross	British bulldog	Non sporting
Bullmastiff	Bullmastiff	Utility
Bullmastiff cross	Bullmastiff	Utility
Bullmastiff × wolfhound × Great dane	Bullmastiff	Utility
Cane corso (Italian mastiff)	Cane corso	Utility
Canaan dog	Canaan dog	Non sporting
Cairn terrier	Cairn terrier	Terrier
Cattle dog	Australian cattle dog	Working dogs

Table A2. *Cont.*

Breed Reported	Associated Listed Breed	Breed Group
Cattle dog cross	Australian cattle dog	Working dogs
Cavalier King Charles spaniel	Cavalier King Charles spaniel	Toys
Central Asian shepherd dog	Central Asian shepherd dog	Utility
Cesky terrier	Cesky terrier	Terrier
Chesapeake bay retriever	Chesapeake bay retriever	Gundogs
Chihuahua	Chihuahua	Toys
Chihuahua cross	Chihuahua	Toys
Chihuahua × Jack Russell	Chihuahua	Toys
Long hair chihuahua	Chihuahua	Toys
Chinese crested dog	Chinese crested dog	Toys
Chinese crested dog—powder puff	Chinese crested dog	Toys
Chow chow	Chow chow	Non sporting
Clumber spaniel	Clumber spaniel	Gundogs
Cocker spaniel	Cocker spaniel	Gundogs
Cocker spaniel, American	Cocker spaniel	Gundogs
Cocker spaniel, English	Cocker spaniel	Gundogs
Collie	Collie	Working dogs
Collie rough	Collie	Working dogs
Collie smooth	Collie	Working dogs
Corgi	Corgi	Working dogs
Corgi, Cardigan Welsh	Corgi	Working dogs
Corgi, Pembroke Welsh	Corgi	Working dogs
Corgi × fox Terrier	Corgi	Working dogs
Coton de tulear	Coton de tulear	Toys
Cross breed	Cross breed	UB
Curly coated retriever	Curly coated retriever	Gundogs
Dachshund	Dachshund	Hounds
Dachshund, long-haired	Dachshund	Hounds
Dachshund, miniature	Dachshund	Hounds
Dalmatian	Dalmatian	Non sporting
Dalmatian cross	Dalmatian	Non sporting
Dandie dinmont terrier	Dandie dinmont terrier	Terrier
Deerhound	Deerhound	Hounds
Dingo	Cross breed	UB
Dingo cross	Cross breed	UB
Dobermann	Dobermann	Utility
Dogue de bordeaux	Dogue de bordeaux	Utility
Dunker	Dunker	Hounds
Dutch shepherd	Dutch shepherd	Working dogs
English foxhound	Foxhound	Hounds
English pointer	English pointer	Gundogs
English mastiff	English mastiff	Utility
English setter	English setter	Gundogs
English springer spaniel	Springer spaniel	Gundogs
English toy terrier	English toy terrier	Toys
Field spaniel	Field spaniel	Gundogs
Finnish lapphund	Finnish lapphund	Working dogs
Flat coated retriever	Flat coated retriever	Gundogs
Formosan mountain dog (Taiwan Dog)	Formosan mountain dog	Utility
Fox Terrier	Fox terrier	Terrier
Fox terrier, smooth	Fox terrier	Terrier
Foxhound	Foxhound	Hounds
French bulldog	French bulldog	Non sporting
German coolie	Coolie/koolie	UB
German hunting terrier	German hunting terrier	Terrier
German pinscher	German pinscher	Utility
German shepherd	German shepherd	Working dogs

Table A2. *Cont.*

Breed Reported	Associated Listed Breed	Breed Group
German shepherd cross	German shepherd	Working dogs
German shorthaired pointer	German shorthaired/wirehaired pointer	Gundogs
German spitz	Spitz	Non sporting
German wirehaired pointer	German shorthaired/wirehaired pointer	Gundogs
Glen of Imaal terrier	Glen of Imaal terrier	Terrier
Golden retriever	Golden retriever	Gundogs
Gordon setter	Gordon setter	Gundogs
Great dane	Great dane	Non sporting
Great dane × bull Arab	Great dane	Non sporting
Great dane × bullmastiff	Great dane	Non sporting
Great pyrenees	Great pyrenees	Working dogs
Greater Swiss mountain dog	Greater Swiss mountain dog	Working dogs
Greyhound	Greyhound	Hounds
Griffon bruxellois	Griffon bruxellois	Toys
Harrier	Harrier	Hounds
Havanese	Havanese	Toys
Hungarian vizsla	Hungarian vizsla	Gundogs
Husky	Siberian husky	Utility
Husky cross	Siberian husky	Utility
Ibizan hound	Ibizan hound	Hounds
Irish red & white setter	Irish setter	Gundogs
Irish setter	Irish setter	Gundogs
Irish terrier	Irish terrier	Terrier
Irish water spaniel	Irish water spaniel	Gundogs
Irish wolfhound	Irish wolfhound	Hounds
Italian greyhound	Italian greyhound	Toys
Italian spinone	Italian spinone	Gundogs
Jack Russell terrier	Jack Russell terrier	Terrier
Japanese chin	Japanese chin	Toys
Japanese spitz	Spitz	Non sporting
Kangal shepherd dog	Kangal shepherd dog	Utility
Keeshond	Keeshond	Non sporting
Kelpie	Kelpie	Working dogs
Kelpie cross	Kelpie	Working dogs
Kelpie × staffordshire terrier	Kelpie	Working dogs
Kelpie × border collie	Kelpie	Working dogs
Kelpie × cattle dog	Kelpie	Working dogs
Kelpie × labrador	Kelpie	Working dogs
Kelpie × dingo	Kelpie	Working dogs
Kerry blue terrier	Kerry blue terrier	Terrier
King Charles spaniel	King Charles spaniel	Toys
Kuvasz	Kuvasz	Working dogs
Labrador retriever	Labrador retriever	Gundogs
Labrador retriever cross	Labrador retriever	Gundogs
Labradoodle	Labrador retriever	Gundogs
Lagotto Romagnolo	Lagotto Romagnolo	Gundogs
Lakeland terrier	Lakeland terrier	Terrier
Large Munsterlander	Large Munsterlander	Gundogs
Leonberger	Leonberger	Utility
Large terrier cross	Terrier	Terrier
Lancashire heeler	Lancashire heeler	Working dogs
Lhasa apso	Lhasa apso	Non sporting
Louisiana Catahoula leopard dog	Louisiana Catahoula leopard dog	Working dogs
Löwchen	Löwchen	Toys
Lurcher	Cross breed	UB

Table A2. *Cont.*

Breed Reported	Associated Listed Breed	Breed Group
Maltese	Maltese	Toys
Maltese cross	Maltese	Toys
Manchester terrier	Manchester terrier	Terrier
Maremma sheepdog	Maremma sheepdog	Working dogs
Mastiff	Mastiff	Utility
Mastiff cross	Mastiff	Utility
Mastiff × bull Arab	Mastiff	Utility
Medium terrier	Terrier	Terrier
Medium terrier cross	Terrier	Terrier
Miniature fox terrier	Fox Terrier	Terrier
Miniature pinscher	Miniature pinscher	Toys
Neapolitan mastiff	Neapolitan mastiff	Utility
New Zealand huntaway	New Zealand huntaway	Working dogs
Newfoundland	Newfoundland	Utility
Norfolk terrier	Norfolk terrier	Terrier
North Queensland bullhound	Cross breed	UB
Norwegian elkhound	Norwegian elkhound	Hounds
Norwich terrier	Norwich terrier	Terrier
Nova Scotia duck tolling retriever	Nova Scotia duck tolling retriever	Gundogs
Old English sheepdog	Old English sheepdog	Working dogs
Papillon	Papillon	Toys
Parson Russell terrier	Parson Russell terrier	Terrier
Pekingese	Pekingese	Toys
Peruvian hairless dog	Peruvian hairless dog	Hounds
Petit basset griffon vendeen	Petit basset griffon vendeen	Hounds
Pharaoh hound	Pharaoh hound	Hounds
Pit bull terrier	Pit bull terrier	Terrier
Pig dog	Cross breed	Terrier
Pointer	Pointer	Gundogs
Polish lowland sheepdog	Polish lowland sheepdog	Working dogs
Pomeranian	Pomeranian	Toys
Poodle	Poodle	Non sporting
Poodle toy	Poodle	Non sporting
Poodle miniature	Poodle	Non sporting
Poodle standard	Poodle	Non sporting
Poodle × Shih Tzu	Poodle	Non sporting
Portugese podengo	Portugese podengo	Hounds
Portuguese water dog	Portuguese water dog	Utility
Pug	Pug	Toys
Puli	Puli	Working dogs
Prague ratter	Cross breed	UB
Pyrenean mastiff	Pyrenean mastiff	Utility
Pyrenean mountain dog	Pyrenean mountain dog	Utility
Rhodesian ridgeback	Rhodesian ridgeback	Hounds
Rottweiler	Rottweiler	Utility
Rottweiler × mastiff	Rottweiler	Utility
Russian black terrier	Russian black terrier	Utility
Saint bernard	Saint bernard	Utility
Saluki	Saluki	Hounds
Samoyed	Samoyed	Utility
Sarplaninac	Sarplaninac	Utility
Schipperke	Schipperke	Non sporting
Schnauzer	Schnauzer	Utility
Schnauzer, miniature	Schnauzer	Utility
Schnauzer, standard	Schnauzer	Utility
Schnauzer, giant	Schnauzer	Utility
Scottish terrier	Scottish terrier	Terrier
Sealyham terrier	Sealyham terrier	Terrier
Shar pei	Shar pei	Non sporting

Table A2. *Cont.*

Breed Reported	Associated Listed Breed	Breed Group
Shar Pei cross	Shar pei	Non sporting
Shetland sheepdog	Shetland sheepdog	Working dogs
Shiba Inu	Shiba Inu	Utility
Shih tzu	Shih tzu	Non sporting
Shih tzu × maltese	Shih tzu	Non sporting
Siberian husky	Siberian husky	Utility
Skye terrier	Skye terrier	Terrier
Sloughi	Sloughi	Hounds
Small terrier cross	Terrier	Terrier
Smithfield cattle dog	Cross breed	UB
Soft coated wheaten terrier	Soft coated wheaten terrier	Terrier
Spaniel	Spaniel	Gundogs
Spanish water dog	Spanish water dog	Gundogs
Spitz	Spitz	Non sporting
Spoodle	Cocker spaniel	Gundogs
Staffordshire bull terrier	American Staffordshire bull terrier	Terrier
Staffordshire bull terrier × labrador	American Staffordshire bull terrier	Terrier
Staghound	Staghound	UB
Swedish vallhund	Swedish vallhund	Working dogs
Tenterfield terrier	Tenterfield terrier	Terrier
Terrier	Terrier	Terrier
Thai ridgeback	Thai ridgeback	Hounds
Tibetan mastiff	Tibetan mastiff	Utility
Tibetan spaniel	Tibetan spaniel	Toys
Tibetan terrier	Tibetan terrier	Non sporting
Timber shepherd	Cross breed	UB
Weimaraner	Weimaraner	Gundogs
Welsh springer spaniel	Springer spaniel	Gundogs
Welsh terrier	Welsh terrier	Terrier
West highland white terrier	West highland white terrier	Terrier
Whippet	Whippet	Hounds
White Swiss shepherd dog	White Swiss shepherd dog	Working dogs
Wirehaired fox terrier	Fox terrier	Terrier
Xoloitzcuintle	Xoloitzcuintle	Non sporting
Yorkshire terrier	Yorkshire terrier	Toys

References

1. Animal Medicines Australia. Pet Ownership in Australia 2016. Available online: http://animalmedicinesaustralia.org.au/wp-content/uploads/2016/11/AMA_Pet-Ownership-in-Australia-2016-Report_sml.pdf (accessed on 26 March 2018).
2. McGreevy, P.D.; Starling, M.; Branson, N.J.; Cobbc, M.L.; Calnon, D. An overview of the dog–human dyad and ethograms within it. *J. Vet. Behav.* **2012**, *7*, 103–117. [CrossRef]
3. O'Haire, M.E.; McKenzie, S.J.; Beck, A.M.; Slaughter, V. Animals may act as social buffers: Skin conductance arousal in children with autism spectrum disorder in a social context. *Dev. Psychobiol.* **2015**, *57*, 584–595. [CrossRef] [PubMed]
4. Levine, G.N.; Allen, K.; Braun, L.T.; Christian, H.E.; Friedmann, E.; Taubert, K.A.; Thomas, S.A.; Wells, D.L.; Lange, R.A. Pet ownership and cardiovascular risk: A scientific statement from the American heart association. *Circulation* **2013**, *127*, 2353–2363. [CrossRef]
5. Gácsi, M.; Maros, K.; Sernkvist, S.; Faragó, T.; Miklósi, Á. Human analogue safe haven effect of the owner: Behavioural and heart rate response to stressful social stimuli in dogs. *PLoS ONE* **2013**, *8*, e58475. [CrossRef] [PubMed]

6. Monsalve, S.; Hammerschmidt, J.; Izar, M.L.; Marconcin, S.; Rizzato, F.; Polo, G.; Garcia, R. Associated factors of companion animal neglect in the family environment in Pinhais, Brazil. *Prev. Vet. Med.* **2018**, *157*, 19–25. [CrossRef] [PubMed]
7. Becker, F.; French, L. Making the links: Child abuse, animal cruelty and domestic violence. *Child Abus. Rev.* **2004**, *13*, 399–414. [CrossRef]
8. Queensland Government—Department of Agriculture and Fisheries. Animal Care and Protection Act 2001. Available online: https://www.legislation.qld.gov.au/view/pdf/inforce/current/act-2001-064 (accessed on 22 April 2019).
9. Mogbo, T.C.; Oduah, F.N.; Okeke, J.J.; Ufele, A.N.; Nwankwo, O.D. Animal cruelty: A review. *J. Nat. Sci. Res.* **2013**, *3*, 94–98.
10. Tallichet, S.E.; Hensley, C. Rural and urban differences in the commission of animal cruelty. *Int. J. Offender Ther.* **2005**, *49*, 711–726. [CrossRef]
11. Newberry, M. Associations between different motivations for animal cruelty, methods of animal cruelty and facets of impulsivity. *Psychol. Crime Law* **2018**, *24*, 500–526. [CrossRef]
12. McEwan, A.; Skandakumar, K. The welfare of greyhounds in Australian racing: Has the industry run its course? *Aust. Anim. Prot. Law J.* **2011**, *6*, 52–74.
13. Burley, S. My dog's the champ: An analysis of young people in urban settings and fighting dog breeds. *Anthropol. Matters J.* **2008**, *10*, 1–18.
14. Fatjó, J.; Bowen, J.; García, E.; Calvo, P.; Rueda, S.; Amblás, S.; Lalanza, J.F. Epidemiology of dog and cat abandonment in Spain. *Animals* **2015**, *5*, 426–441. [CrossRef]
15. Srinivasan, K. The biopolitics of animal being and welfare: Dog control and care in the UK and India. *Trans. Inst. Br. Geogr.* **2012**, *38*, 106–119. [CrossRef]
16. Bennett, P.C.; Perini, E. Tail docking in dogs: A review of the issues. *Aust. Vet. J.* **2003**, *81*, 208–218. [CrossRef]
17. Beatson, R.; Loughnan, S.; Halloran, M. Attitudes toward animals: The effect of priming thoughts of human-animal similarities and mortality salience on the evaluation of companion animals. *Soc. Anim.* **2009**, *17*, 72–89. [CrossRef]
18. Lockwood, R.; Arkow, P. Animal abuse and interpersonal violence: The cruelty connection and its implications for veterinary pathology. *Vet. Pathol.* **2016**, *53*, 910–918. [CrossRef]
19. Ascione, F.R.; Weber, C.V.; Thompson, T.M.; Heath, J.; Maruyama, M.; Hayashi, K. Battered pets and domestic violence: Animal abuse reported by women experiencing intimate violence and by nonabused women. *Violence Against Women* **2007**, *13*, 354–373. [CrossRef]
20. Baldry, A.C. Animal abuse and exposure to interparental violence in Italian youth. *J. Interpers. Violence* **2003**, *18*, 258–281. [CrossRef]
21. Ascione, F.R.; Friedrich, W.N.; Heath, J.; Hayashi, K. Cruelty to animals in normative, sexually abused, and outpatient psychiatric samples of 6- to 12-year-old children: Relations to maltreatment and exposure to domestic violence. *Anthrozoös* **2003**, *16*, 194–212. [CrossRef]
22. Becker, K.D.; Stuewig, J.; Herrera, V.M.; McCloskey, L.A. A study of firesetting and animal cruelty in children: Family influences and adolescent outcomes. *J. Am. Acad. Child Adolesc. Psychiatry* **2004**, *43*, 905–912. [CrossRef]
23. Henry, B.C. The relationship between animal cruelty, delinquency, and attitudes toward the treatment of animals. *Soc. Anim.* **2004**, *12*, 185–207. [CrossRef]
24. Benetato, M.A.; Reisman, R.; McCobb, E. The veterinarian's role in animal cruelty cases. *J. Am. Vet. Med. Assoc.* **2011**, *238*, 31–34. [CrossRef]
25. Tarver, E.C. The dangerous individual('s) dog: Race, criminality and the 'Pit Bull'. *Cult. Theory Crit.* **2014**, *55*, 273–285. [CrossRef]
26. Kim, Y.M.; Kim, S.A.; Lee, S.M.; Choi, Y.J.; Kim, B.J.; Shin, N.S. Canine behavioral problems and their effect on relinquishment of the Jindo dog. *J. Vet. Sci.* **2010**, *11*, 345–350. [CrossRef]
27. Shih, H.Y.; Paterson, M.B.A.; Phillips, C.J.C. Breed group effects on complaints about canine welfare made to the Royal Society for the Prevention of Cruelty to Animals (RSPCA) Queensland, Australia. *Animals* **2019**, *9*, 390. [CrossRef]
28. Shih, H.Y.; Paterson, M.B.A.; Phillips, C.J.C. A retrospective analysis of complaints to RSPCA Queensland, Australia, about dog welfare. *Animals* **2019**, *9*, 282. [CrossRef]

29. Flynn, C. Acknowledging the zoological connection: A sociological analysis of animal cruelty. *Soc. Anim.* **2001**, *9*, 71–87. [CrossRef]
30. Carter, J.; Taylor, C.S. Socio-economic factors in companion animal relinquishment on the Sunshine Coast, Australia. *Soc. Anim.* **2017**, 1–19. [CrossRef]
31. Gullone, E. An evaluative review of theories related to animal cruelty. *J. Anim. Ethics* **2014**, *4*, 37–57. [CrossRef]
32. Smith, R. Investigating financial aspects of dog-fighting in the UK. *J. Financ. Crime* **2011**, *18*, 336–346. [CrossRef]
33. Dotson, M.J.; Hyatt, E.M.; Clark, J.D. Traveling with the family dog: Targeting an emerging segment. *J. Hosp. Mark. Manag.* **2011**, *20*, 1–23. [CrossRef]
34. Pink, B. Socio-economic Indexes for Areas (SEIFA), 2011. Available online: https://www.abs.gov.au/AUSSTATS/abs@.nsf/DetailsPage/2033.0.55.0012011?OpenDocument (accessed on 30 July 2019).
35. Australian National Kennel Council LTD. Breeds. Available online: http://ankc.org.au/Home (accessed on 19 November 2018).
36. New Zealand Kennel Club. Recognised Pedigree Breeds. Available online: https://www.dogsnz.org.nz/home/home (accessed on 1 August 2018).
37. American Kennel Club. Dog Breeds. Available online: https://www.akc.org/dog-breeds/ (accessed on 1 August 2018).
38. United Kennel Club. Breed Standards. Available online: https://www.ukcdogs.com/breed-standards (accessed on 1 August 2018).
39. Bursac, Z.; Gauss, C.H.; Williams, D.K.; Hosmer, D.W. Purposeful selection of variables in logistic regression. *Source Code Biol. Med.* **2008**, *3*, 17. [CrossRef]
40. Neumann, S.L. Animal welfare volunteers: Who are they and why do they do what they do? *Anthrozoös* **2010**, *23*, 351–364. [CrossRef]
41. Degeling, C.; Burton, L.; McCormack, G.R. An investigation of the association between socio-demographic factors, dog-exercise requirements, and the amount of walking dogs receive. *Can. J. Vet. Res.* **2012**, *76*, 235–240.
42. Saunders, J.; Parast, L.; Babey, S.H.; Miles, J.V. Exploring the differences between pet and non-pet owners: Implications for human-animal interaction research and policy. *PLoS ONE* **2017**, *12*, e0179494. [CrossRef]
43. Australian Bureau of Statistics. ABS Maps, Filtered by Postcode and Significant Urban Areas. Available online: https://itt.abs.gov.au/itt/r.jsp?absmaps (accessed on 26 June 2019).
44. Taylor, N.; Signal, T.D. Community demographics and the propensity to report animal cruelty. *J. Appl. Anim. Welf. Sci.* **2006**, *9*, 201–210. [CrossRef]
45. Lockwood, R. Cruelty toward cats: Changing perspectives. In *The State of the Animals III: 2005*, 1st ed.; Salem, D.J., Rowan, A.N., Eds.; Humane Society Press: Washington, DC, USA, 2005; pp. 15–26.
46. Haidt, J.; Koller, S.H.; Dias, M.G. Affect, culture, and morality, or is it wrong to eat your dog? *J. Pers. Soc. Psychol.* **1993**, *65*, 613–628. [CrossRef]
47. Kasperbauer, T.J. Animals and the expanding moral circle. In *Subhuman: The Moral Psychology of Human Attitudes to Animals*; Oxford University Press: New York, NY, USA, 2018; pp. 152–158.
48. Mueller, M.K.; Chubb, S.; Wolfus, G.; McCobb, E. Assessment of canine health and preventative care outcomes of a community medicine program. *Prev. Vet. Med.* **2018**, *157*, 44–49. [CrossRef]
49. Freiwald, A.; Litster, A.; Weng, H.Y. Survey to investigate pet ownership and attitudes to pet care in metropolitan Chicago dog and/or cat owners. *Prev. Vet. Med.* **2014**, *115*, 198–204. [CrossRef]
50. Woodhead, J.K.; Feng, L.C.; Howell, T.J.; Ruby, M.B.; Bennett, P.C. Perceptions of dog breeding practices, breeding dog welfare and companion dog acquisition in a self-selected sample of Australian adults. *Anim. Welf.* **2018**, *27*, 357–368. [CrossRef]
51. Hsu, Y.; Severinghaus, L.L.; Serpell, J.A. Dog keeping in Taiwan: Its contribution to the problem of free-roaming dogs. *J. Appl. Anim. Welf. Sci.* **2003**, *6*, 1–23. [CrossRef]
52. Knobel, D.L.; Laurenson, M.K.; Kazwala, R.R.; Boden, L.A.; Cleaveland, S. A cross-sectional study of factors associated with dog ownership in Tanzania. *BMC Vet. Res.* **2008**, *4*, 5. [CrossRef]
53. Lambert, K.; Coe, J.; Niela, L.; Dewey, C.; Sargeant, J.M. A systematic review and meta-analysis of the proportion of dogs surrendered for dog-related and owner-related reasons. *Prev. Vet. Med.* **2015**, *118*, 148–160. [CrossRef]

54. Weiss, E.; Slater, M.; Garrison, L.; Drain, N.; Dolan, E.; Scarlett, J.M.; Zawistowski, S.L. Large dog relinquishment to two municipal facilities in New York City and Washington, DC.: Identifying targets for intervention. *Animals* **2014**, *4*, 409–433. [CrossRef]
55. Scarlett, J.; Salman, M.D.; New, J.J.G.; Kass, P.H. Reasons for relinquishment of companion animals in U.S. animal shelters: Selected health and personal issues. *J. Appl. Anim. Welf. Sci.* **1999**, *2*, 41–57. [CrossRef]
56. Weng, H.Y.; Kass, P.H.; Hart, L.A.; Chomel, B.B. Risk factors for unsuccessful dog ownership: An epidemiologic study in Taiwan. *Prev. Vet. Med.* **2006**, *77*, 82–95. [CrossRef]

animals

MDPI

Article

Early Recognition of Behaviour Problems in Shelter Dogs by Monitoring them in their Kennels after Admission to a Shelter

Liam Clay [1,*], Mandy Paterson [1,2], Pauleen Bennett [3], Gaille Perry [4] and Clive Phillips [1]

1 Centre for Animal Welfare and Ethics, University of Queensland, Gatton, Queensland 4343, Australia; mpaterson@rspcaqld.org.au (M.P.); c.phillips@uq.edu.au (C.P.)
2 Royal Society for the Prevention of Cruelty to Animals Queensland, Brisbane, Queensland 4076, Australia
3 School of Psychology and Public Health, La Trobe University, Bendigo, Victoria 3552, Australia; pauleen.bennett@latrobe.edu.au
4 Delta Society, Summer Hill, New South Wales 2130, Australia; perrygaille@gmail.com
* Correspondence: liam.clay@uqconnect.edu.au; Tel.: +61-422-706-076

Received: 28 September 2019; Accepted: 23 October 2019; Published: 28 October 2019

Simple Summary: Canine behaviour assessments are commonly used in shelters to identify behaviour problems in dogs prior to adoption. The aim of this study was to evaluate whether kennel monitoring of dogs could identify early signs of behaviour problems. Kennel behaviour was monitored for 38 dogs in their first five days in kennels at a shelter in Brisbane, Australia. This was compared to a formal assessment of exploratory, handling, play, run/freeze, and food guarding behaviour, as well as stranger and fake toddler interactions, and behaviour when the dog was alone, conducted five days after shelter admission. Kennel behaviours associated with fear, anxiety, and arousal in dogs were significantly correlated with the same behaviours in the formal assessment. With respect to outcomes, dogs that displayed more whining, tense body posture, standing leaning forward, panting, ears forward, less barking, lowered body and balanced/relaxed body posture, standing still, and standing by the wall had increased odds of failing the behaviour assessment. The study demonstrates that monitoring kennel behaviour could detect early signs of behaviour problems.

Abstract: Canine behaviour assessments are commonly used in shelters to identify behaviour problems in dogs prior to adoption. The aim of this study was to evaluate whether kennel monitoring of dogs could identify early signs of behaviour problems, thereby facilitating early intervention and better management of dogs displaying behaviour problems. Kennel behaviour was monitored for dogs ($n = 38$) in their first five days in kennels at a shelter in Brisbane, Australia. This was compared to a formal assessment of exploratory, handling, play, run/freeze, and food guarding behaviour, as well as stranger and fake toddler interactions, and behaviour when the dog was alone, conducted five days after shelter admission. Kennel behaviours associated with fear, anxiety, and arousal in dogs were significantly correlated with the same behaviours in the formal assessment. Positional correlations were also evident. With respect to outcomes, dogs that displayed more whining, tense body posture, standing leaning forward, panting, ears forward, less barking, lowered body and balanced/relaxed body posture, standing still, and standing by the wall had increased odds of failing the behaviour assessment. Over the five days in the kennel, the frequency and duration of fear-related behaviours decreased, suggesting a reduction in arousal as the dog became accustomed to the shelter environment. The study demonstrates that monitoring kennel behaviour could detect early signs of behaviour problems.

Keywords: behaviour; problems; assessment; canines; shelters; predict

1. Introduction

The largest Australian animal welfare organisation, the Royal Society for Prevention of Cruelty to Animals (RSPCA), received 40,286 surrendered dogs in the 12 months from July 2017 to June 2018 [1]. Reasons for dog relinquishment commonly include behaviour problems, e.g., inappropriate toileting, barking, digging, separation anxiety, fear, or aggression [2–4]. Entry to a novel shelter environment, plus alienation from its former owner, home, and routine, is likely to result in a potentially stressful form of social isolation in a surrendered dog [5]. Dogs experience fear and anxiety upon relinquishment to a shelter, with overt signs of stress sometimes persisting for several weeks after relinquishment [5,6]. Furthermore, as the length of time in a shelter increases, the detrimental impact on dogs' emotional state worsens [7–9]. Coping capacity differs considerably between individual dogs, with variable habituation to the environment and the same stressor being experienced as neutral or aversive [10–12]. Therefore, in order to reliably and effectively assess and monitor the mental well-being of surrendered dogs, it is important that early interactions with the novel environment are recorded to identify signs of negative affect, e.g., separation anxiety, which occur with high frequency in adopted dogs from shelters [13].

Behaviour assessments are used in shelters globally, assessing adoption suitability, identifying behaviour problems, and matching dogs with the most suitable adoptees [14]. Veterinarians also implement a variety of testing procedures for quality of life assessments in animals with medical and behavioural issues [15]. However, behaviour assessments in shelters have been recently criticised, due to both the nature and consequences of pass or fail assessment procedures and doubt about their ability to accurately predict behaviour problems [16]. It is claimed that they cannot accurately determine the frequency of false positive (identification of a behavioural problem that does not really exist e.g., aggression, which renders the dog unfit for adoption) or false negatives (failure to detect a behavioural problem during the test). Usually, dogs are removed from their kennel to undertake the test in a standard facility, through which many other dogs have passed. This single context assessment is likely to present a stressful situation for the dog, which is unlikely to replicate the best environment to examine their anticipated behaviour in the home in which they are adopted. For example, the presence of excreta from previous dogs, or potentially even odours from dogs previously tested, can affect the outcome of tests [17].

An alternative is to observe behaviour in their kennel (hereafter kennel behaviour), handler interactions, and interspecies behaviour, allowing them to be tested in the environment into which they are becoming settled. Kennel monitoring has been used previously in shelters to identify behaviour problems [18–20]. Furthermore, kennel behaviour monitoring could potentially be automated, using for example motion sensing or by programming computers to recognise specific behaviour patterns, e.g., escape attempts [15].

There is a need for better observational tools for assessment in shelters [21]. These could include assessing behaviour longitudinally in shelters, to account for plasticity, and the greater predictability of behaviour when measured over a period of time [20]. Therefore, the aim of this study was to compare the manifestation of behaviours in a structured assessment with behaviours observed in their kennel over the first five days in a shelter.

2. Materials and Methods

2.1. Ethical Approval

This study was granted ethical approval from the University of Queensland Animal Ethics Committee (AE04214). All dogs were owner-surrendered, and permission was obtained from the owners to enrol their animals into the study.

2.2. Subjects

Criteria for dogs to enter the study were that they were between six months and 10 years of age, had no predisposed medical conditions and had not been previously admitted to the shelter. Thirty-eight dogs (18 male, 20 female) of mean age 3.1 years (SEM 0.37 years) and weight 20.3 kg (SEM 1.43) that had been surrendered to the RSPCA Queensland's Animal Care Facility over a three month period were enrolled into the study. They represented the following 20 different breeds: Bull terrier (n = 9), Kelpie Cross (n = 6), Mastiff (n = 4), Beagle cross (n = 2), Staffordshire Bull Terrier (n = 2), and one each of Jack Russell cross, Alaskan Malamut, American Bulldog, Australian Cattle Dog, Australian Shepherd Cross, Border Collie, Boxer, Bull Arab cross, German Shepard cross, Husky cross, Labrador Retriever, Papillon, Poodle Cross, Portuguese Podengo, and Spoodle. All had been privately surrendered, with owners being required to declare the reasons for surrender.

2.3. Housing and Feeding

Dogs were housed in a single block of kennels, which held 16 dogs in individual kennels. Each kennel had a floor area of 3.5 m^2 (120 cm × 180 cm), concrete floors and two solid walls separating each kennel and a gate opening into the kennel block, a fence opening out toward a garden area, a separate sleeping area with a raised bed, soft bedding, and toys. The dogs were fed twice daily with a combination of dry and wet food and had access to fresh water. Each dog received walks twice a day at 09:00 and 15:00 by shelter staff or volunteers.

2.4. Behaviour Monitoring

2.4.1. Kennel

Dogs were observed on days 1–5, following surrender on day 1, for 60 min (07:30–08:30, before interactions with volunteers). Data were collected using two video surveillance cameras (KOBI CCD video cameras, Model: K-32HCVF, Taipei, Taiwan) placed in each individual kennel at a height of 3 m.

2.4.2. Standard Behaviour Test

The standard RSPCA Qld behaviour assessment (RSPCA, 2018) was conducted on day 6, i.e., the day after the five days of kennel observations, as used by Queensland RSPCA shelters in each state to assess adoption suitability in shelter dogs. The assessment comprised a series of 10 tests of increasing provocation. Dog responses were scored based on frequency and durations of a variety of behaviours as described below. The tests were performed over 15 min with the following aids: a 1.8 m leash, tennis ball, plush squeaky toy, rope, plastic hand on a extend pole, bowl, raw hide or bone, and combination of wet and dry dog food.

The assessments were performed in a room (3 × 5 m) 20–30 m from the kennels, with one window and two half frosted doors, and a concrete floor with hospital-grade non-slip painted covering. All dogs were moved on lead from their kennel block to the assessment room. A single lead was attached to the wall for a 1.8 m leash to restrain the dog. During the assessment, one researcher acted as the handler, and a second person helped in observer interaction and implementing two tests requiring two people (Stranger and Fake toddler tests, described below). Data for all the following tests were recorded using a video recorder (Digital Video Recorder 1.1, Model: XQ-L400H, Manufacture: Kobi, Seoul, Korea).

Exploring the Room, One Minute

The handler entered the room, dropped the lead attached to the dog, and sat in the centre on a chair. Then, the observer started a timer and waited for 1 min without any interaction with the dog by either person.

Sociability to Handler

At the end of test 1, the handler called the dog to them in a friendly voice, remaining in the chair with no other body movement. If there was no response, a second attempt was made, and if still no response the handler clapped their hands on their lap and said 'come here' in the direction of the dog, trying at least three times to call the dog to them. When the dog came (at the first, second, or third call), the handler picked up the leash and then stroked the dog from the base of neck to tail three times. If the dog did not respond to the first, second, or third, call the handler approached the dog, picked up the leash, and gave the dog three strokes from the base of neck to tail. Following each stroke, the observer and handler counted 10 s, with behaviours exhibited noted.

Tolerance to Handling

The handler dropped the leash and held the dog's collar. With the dog standing, the other handler (in the standing position, or crouching if a small breed of dog) picked up the dog's rear inside foot, then the front inside foot, then reached over its back to pick up its rear outside foot, and finally the front outside foot. Each foot was held for 2 s. After picking up all four paws in this manner, the handler stood for 10 s with no dog interaction and finally removed the dog's leash.

Toy Interactions

A tennis ball, squeaky toy, and tugging rope were shown to the dog and gently thrown across the room, and the handler verbally engaged the dog in play. If the dog picked up the ball, the handler waited to see if it returned to the handler without encouragement. If it did not, the handler encouraged the dog to bring the ball back by calling his/her name and saying "come". If the dog still did not return, the handler went to the dog.

In both situations, the handler waited 10 s to see if the dog dropped the ball. If it did not, he/she asked the dog to "drop it". If the dog did not respond, then a second command was given, "give", and if necessary, a third attempt, "out", was tried. If the dog did not respond to these commands, the handler approached the dog carefully and removed the ball from the dog's mouth. These steps were repeated for a second throw, and after completion, the handler waited 10 s with no interaction before moving on to the next test.

Tag (Run and Freeze)

The run and freeze test was used to mimic a tag game. The handler gently moved the dog to the opposite end of the room and left it standing against the wall. Then, he gently moved one hand over its head, down toward the back to gently tap the rump area, and then ran across the room, laughing and waving arms, followed by suddenly stopping, folding his arms, and ignoring the dog. The tap, run, and freeze series was repeated a second time. The handler waited for 10 s after the run and freeze, ignoring the dog, before moving onto the next test. The dog was then placed back on the leash.

Resource Guarding

The handler tethered the dog to the wall for safety reasons, and proceeded to give the dog wet canned food, smeared in a bowl. The bowl was then placed near the dog at the end of the leash perimeter, allowing the dog to begin eating for 2 s. The handler then proceeded with a plastic hand, walking to the side of the dog while it was eating. Using the fake hand, the handler patted the dog on the head, continuing to stroke down its back and body twice. The fake hand was then placed 5 cm in front of the bowl and moved around in a semi-circle. The hand was then placed on the inside edge of the bowl and moved around the edge of the bowl next to the dog's face, without touching it. Finally, the bowl was pulled away from the dog using the fake hand. The bowl was then returned to the dog, which was observed for 10 s.

The handler then gave the dog a pig's ear or bone, depending on dog's food interest, and it was allowed to chew it for 30 s. The steps above with wet food were repeated; then, the handler attempted to retrieve the food, asking the dog to "drop it", "leave it", or "give" before attempting to retrieve it by offering a higher value treat/food, e.g., the pig's ear.

Stranger Interaction

The handler placed the dog on a leash as the observer exited the room and returned dressed in a reflective vest, large brimmed hat and using a walking stick. The observer entered the room, and bent down to extend an open flat hand as if to pat the dog on the head. The observer then talked to the dog normally and stopped for 3 s, allowing the dog to approach. If the dog approached, the observer patted the dog on the top of its head for 3 s. If the dog did not approach, it was observed for 10 s, with an emphasis on any interaction between the handler and/or the observer.

Fake Toddler Interaction

The handler stood and held the dog's leash while the observer exited the area and returned carrying a toddler doll simulating a small child. Once the toddler was within the leash perimeter from the dog, the observer placed the doll on the floor facing the dog, with the doll's arm extended toward the dog. The handler allowed the dog to approach if it desired. If the dog did not approach the observer, it was observed for 20 s.

Time Alone

The handler and observer removed the leash from the dog and left the room for 2 min, with a video camera in the front of the room monitoring behaviour and vocalisations. Then, the handler and observer re-entered through the same door.

Behaviour with Another Dog

This test was conducted in a yard (10–20 m), allowing adequate space between the test dog and another dog, both with handlers. Each dog had a handler, who interacted with their dog by giving treats and ignoring the other handler and dog. The handler had a short, 1 m, leash, so that the dog walked close to the handler. At the start, both handlers walked parallel to each other, 5 m apart, with the dogs on the outside. If one or both dogs were reactive and pulled toward each other, the distance between the handlers was increased. If both dogs were relaxed and focused on their handler, the handlers moved the dogs to an exercise circle. If the dogs did not breach a minimum distance of 5 m between them, they were introduced on opposite sides of a fence. There followed a circling activity, which required one handler to stand still with their dog on no more than 1.5 m of leash while the other handler and their dog completed a circle around the handler. Handlers then swapped places and repeated the circling activity. If no adverse behaviours were displayed, the handler in the middle of the circle remained at that location, ensuring that the only tension on the leash was from the dog. The other handler identified the leash threshold of the dog in the centre and moved close enough to allow the dogs to be nose to nose, also ensuring that the only tension on their leads was caused by the dog pulling, not them pulling against the dog. Once the leads became loose, and the dogs stopped pulling against the handler, the handlers took a step closer to each other, allowing the dogs to interact if they chose. Leashes remained loose. If there were signs of adverse reactions or aggression, dogs were then separated by increasing the threshold.

2.5. Behaviour Scoring

Following preliminary observation of dogs in their kennel and during the formal behaviour assessment, an ethogram with 48 behaviours, classified as either long duration behaviours (for which the duration was recorded) or events (for which the number of occurrences was recorded) was devised. The behaviours focused on eight components: activities of the mouth, body, tail, tail movement, ears, eyes, position, and movement (Table 1). Descriptions of each behaviour are presented in Table 2 and their connection to emotions (Anxiety, Fear, Friendliness, Arousal, Aggression) [22–26] in Table 3. Kennel behaviours were continuously recorded over a 1 h period (07:00–08:00), and the formal behaviour assessments were recorded for all tests. Behaviour recording was assisted by coding software (BORIS) [27]. The following behaviour variables with no or only one occurrence were discarded: Squint, Whale eyes.

Table 1. Canine behaviours recorded for each body part, as well as positions and movement types.

Mouth	Body	Tail	Tail Movement	Ears	Eyes	Position	Movement
Open/Closed	Weight forward	Low	Wagging	Alert	Soft	Front	Pacing
Panting	Weight back	Med	Fast	Back	Hard	Bed/Sleep	Sit/Lay
Mouthing	Balanced	High	Stiff	Forward	Direct	Wire	Stand
Lip Lick	Relaxed	Tucked	Slow	Open	Squinty	Wall	Still
Snap	Tense				Whale Eyes		
Bite	Lowered				Dilated		
Whining	Play bow				Targeting		
Barking	Jumping up				Diverting		
Growl	Lowered head						
Howling	piloerect						

Table 2. Behaviours measured, their descriptions and mean values (± SEM) for duration and frequency during kennel observations.

Behaviour	Description	Duration (s/days)	Frequency (no./days)
Mouth			
Open/Closed	Mouth is open or close, no visual signs of panting	3017 ± 568.0	4.35 ± 0.83
Panting	Increased respiration, deep gasps, and salivation	8314 ± 654.0	22.2 ± 5.35
Mouthing	Nipping or play biting	0	0
Lip Lick	Licking of the upper lip	21.9 ± 21.60	0.17 ± 0.040
Snap	Rapid open and close mouth, possible baring teeth, growl, bark, lunge	0	0
Bite	Closure the teeth on victim causing a wound	0	0
Whining	A sustained, high pitched, plaintive sound	0.464 ± 0.360	0.10 ± 0.060
Barking	Brief vocalization	952 ± 299.0	19.2 ± 5.86
Growl	Low guttural prolonged vocalisation	6.99 ± 6.990	0.12 ± 0.060
Howling	Raise muzzle perpendicular to ground, vocalise over extended period, open jaws	0	0
Body			
Weight forward	Body weight forward while standing still	204 ± 82.7	4.5 ± 2.03
Weight back	Body weight back while standing still	3371 ± 466.0	17.7± 2.823
Balanced	Balanced body posture standing still	6546 ± 633.0	37.6 ±4.960
Relaxed	Body posture relaxed in movement	0	0
Tense	Body Posture is still and tense in association to stimulus	669 ± 147.2	3.98 ±0.922
Lowered	Body lowered to ground	728 ± 183.1	4.85 ±1.1676
Play bow	Stretching front legs out in front, leaning down on its elbows	14.1 ± 11.70	0.625 ±0.2454
Jumping up	Jumping in air	568 ± 115.1	31.9 ± 5.21
Lowered head	Lowered head as body posture is high	0	0
Piloerect	Hackles rise	0	0
Tail			
Low	Tail positioned low	6438 ± 469.6	25.4 ±3.56
Med	Tail positioned in line with spine	2945 ± 404.7	23.6 ± 4.34
High	Tail high or above spine	871 ± 325.2	6.85 ± 1.86
Tucked	Tail positioned underneath body	1721 ± 395.1	3.91 ± 0.6460

Table 2. *Cont.*

Behaviour	Description	Duration (s/days)	Frequency (no./days)
Tail Movement			
Wagging	Relaxed tail movement	0	0
Fast	Movement of tail fast	29.8 ± 16.10	1.88 ± 1.2500
Stiff	Still and no movement in tail	9531 ± 397.1	38.4 ± 3.68
Slow	Slow movement of the tail	2030 + 264.1	35.9 +- 4.17
Ears			
Alert	Ears forward and directed at an object, stimulus, or sound	243 + 169.0	1.88 + 0.8610
Back	Ears positioned back and flat	4470 + 617.0	18.0 + 3.07
Forward	Ears positioned forward	3066 ± 621.0	9.6 ± 1.760
Open	Ears neutral	4221 ± 473.0	17.0 ± 2.22
Eyes			
Soft	Relaxed eyes	275.6 ± 80.5	1.85 ± 0.361
Hard	Hard focused stare	0	0
Direct	Directed at object	5832 ± 516.0	11.4 ±- 0.997
Squinting	Eyes not fully open	0	0
Whale Eyes	Showing whites of eyes	0	0
Dilated	Pupil dilation	219 ± 137.0	1.28 ± 0.699
Targeting	Constricted pupils and targeting object or stimulus	0	0
Diverting	Eyes moving and not maintaining eye contact	5585 ± 484.0	11.1 ± 0.94
Position			
Front	At the front of the kennel/front of room	4705 ± 388.0	136.9 ± 12.70
Bed/Sleep	In bed	1022 ± 224.0	19.8 ± 3.69
Wire	At wire	5065 ± 334.0	134.8 ± 12.50
Wall	At wall of kennel or in behaviour assessment room	1303 ± 237.0	18.7 ± 2.46
Movement			
Pacing	Repeated movement in a regular pattern	3540 ± 308.0	128.8 ± 9.62
Sit/Lie	Sitting position	4290 ± 352.0	62.4 ± 5.19
Stand	Standing on all fours	4242 ± 295.0	119.5 ± 9.45
Still	Motionless	0	0
Walking	Progressive locomotion with at least three legs on floor at one time	0	0

The RSPCA staff classified the dogs for adoption suitability following the formal behaviour assessment: (1) pass and ready for adoption, (2) some behaviour issues which should be addressed in a behaviour modification program, and (3) fail due to extreme behaviour problems. However, in the current study no dogs were classified under category 2.

Table 3. The behaviours contributing to the emotional states Fear, Anxiety, Aggression, Arousal, and Friendliness.

Behaviour Number															
1	2	3	4	5	6	7	8	9	10	11	12	13	14	15	16
Fear															
Diverting	Ears Back	Lip Licking	Lowered Body	Lowered head	Shiver	stiff tail	Tail Low	Tail tucked	Tense Body Posture	Weight back	Yawn				
Anxiety															
Fast tail	High tail	Jumping	Licking	Lip licking	Medium	Pacing	Panting	Stiff Tail	Tense body	Weight back	Weight forward	Whining			
Aggression															
Biting	Ears Forward	Growling	High tail	Lip Licking	Lowered head	Mediuam tail	Snapping	Standing	Stiff tail	Still tail	Targetting	Vertical Lip Raise			
Arousal															
Barking	Diverting Gaze	Fast tail	High Tail	Jumping up	Jump off	Licking	Medium Tail	Mouthing	Pacing	Panting	Weight forward	whining			
Friendliness															
Balanced	Body Curve	Direct eye	Ears forward	Ears open	Fast tail	Handler interaction	Jump	Medium Tail	Play	Relaxed body	slow	Sniff	Soft	Tail loose	Walking

2.6. Statistical Analysis

Results were analysed using Minitab 17, Lead technology Inc., Pennsylvania State University, Pennsylvanina, USA. Behaviours were entered as the percentage of the total observation time or percentage of the frequency of occurrence during their period in the kennel and during the behaviour assessment. These two were compared using multivariate general linear models with the following factors: reason for surrender, age, weight, animals, days since entry, and outcome (adopted/euthanized). Residuals were checked for normal distribution using the Anderson Darling test. Spearman's rank order correlations were computed between kennel and formal behaviour assessment variables. As comparisons with 38 other behaviours were made for each behaviour in each test of the behaviour assessment, results were corrected for false discovery using the Benjamini-Hochberg procedure [28]. The Bonferroni correction was rejected as it assumes independence in the individual tests. The Benjamini-Hochberg procedure ranks the P values for each test and compares P values to critical values [(rank/no. tests) x false discovery rate (selected as 0.20 as recommended by McDonald, 2014)]. All P values up to the critical one were considered to indicate a significant difference [28]. Correlations were further analysed on tests of the sample split according to owner surrender information, sex, adopted vs euthanasia, and daily behaviours. Linear and Binary Logistics Regressions were conducted to compare dog behaviour with RSPCA classification of outcomes and comparing behaviours over days for different tests. Two tests, Time Alone and Exploration of the Room, were subjected to additional logistic regression because of their predictive ability for kennel behaviour.

3. Results

3.1. Reasons for Dog Surrender

The reasons for surrender were moving away or insufficient time to care for the dog (22.2%); dog being aggressive or escaping, or family issues (8.3%); medical concerns (5.5%); and arousal, barking, chasing, destruction, owner's death, resource guarding, or separation anxiety (2.8%).

3.2. Emotional Characteristics of Dogs in Their Kennels That Were or Were Not Subsequently Euthanased

3.2.1. Emotional States of Dogs in Their Kennel

Over the first five days, dogs spent most time and had the highest frequencies of the following behaviours (Table 2): weight back, balanced body, and jumping up. Tail movement and position were spent in tail low and medium with still or slow movement, not wagging (Table 2). Ear position was most commonly ears back, then ears open, and finally ears forward. Eye direction was most commonly direct and diverting. In regards to position, dogs spent the most of the time in a kennel at the wire or front and the least amount of time in bed/sleeping or at the wall. Movement patterns were commonly standing, sit/lay, and pacing (Table 2). Over the five-day period, dogs spent 36% of their time in friendly behaviours, 25% displaying fear, 13% displaying anxiousness, 15% in high arousal, and 7% displaying aggression. Dogs' frequency of emotions differed from duration, with 33% of occurrences being high arousal, 25% friendliness, 24% anxiousness, 16% fear, and 2% aggression. Thus, friendliness and fear were displayed less frequently but for a longer duration compared with arousal and anxiousness, which were of short duration but more frequent.

Over the five-day period, there was a significant reduction in the frequency of fear-related behaviours, including tense body posture ($p < 0.05$), tail tucked ($p < 0.05$), and alert response in ears ($p < 0.05$) (Figure 1). There was increases in stiff and slow tail movement ($p < 0.05$) (Figure 1) and the duration of time spent at the front of the kennel ($p = 0.016$), wire of the kennel ($p = 0.008$), and in bed/sleep ($p = 0.0019$) (Figures 1 and 2).

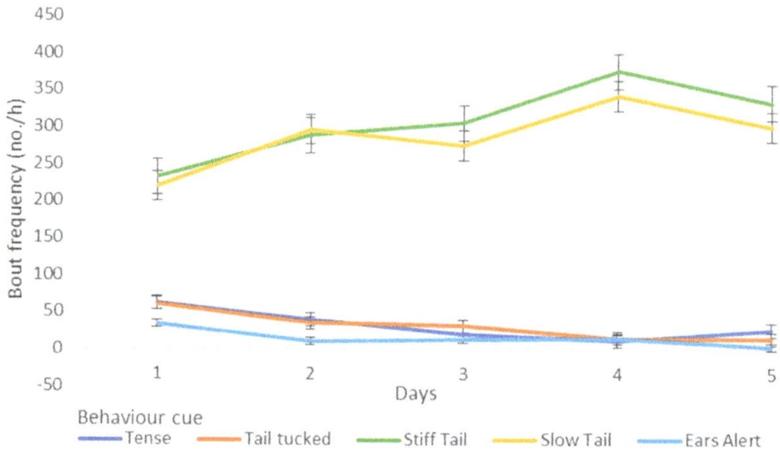

Figure 1. The frequency of fear-related behaviours, alert ears, and tail behaviours over the first five days that dogs (n = 38) spent in a shelter.

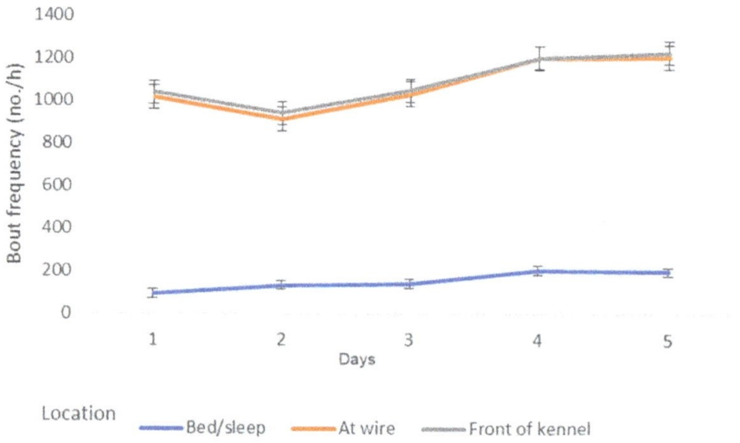

Figure 2. The frequency of position over the first five days that dogs (n = 38) spent in a shelter.

There were reductions in time spent panting ($p < 0.001$) (and corresponding increase in mouth open or closed, $p < 0.001$), a reduction in lowered ($p < 0.008$) and tense body posture ($p < 0.001$), and reductions in tucked tail and stiff tail movement, and a corresponding increase in slow tail movement ($p < 0.05$) (Figure 3).

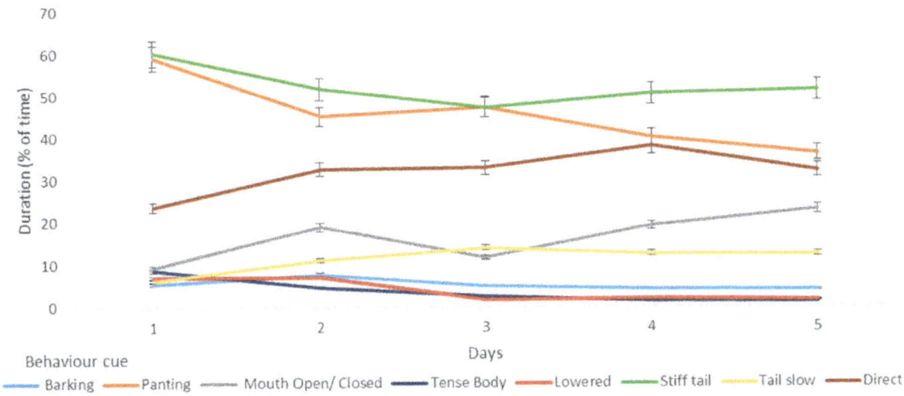

Figure 3. The duration of fear-related behaviours, arousal behaviours, and tail behaviours over the first five days that dogs (n = 38) spent in a shelter.

3.2.2. Relationship between Kennel Behaviour and Outcome for the Dogs

Comparing behavioural characteristics of dogs that were adopted or euthanized, the latter had an increased duration of tense body posture overall, but inspection of changes over time revealed that this was mainly on the first day, with this behaviour declining over time in both sets of dogs (p = 0.001) (Table 4, Figure 4). Conversely, dogs that were adopted, which generally exhibited more mouth open/closed behaviour, had similar levels to euthanased dogs by day 5. Dogs that were adopted had a greater frequency of balanced/relaxed posture, but this declined over time, in contrast to euthanased dogs, which had little evidence of decline over time (p = 0.004). Jumping kennel was more common in euthanased dogs, and this declined over time in both euthanased and adopted dogs (p = 0.03) (Figure 5).

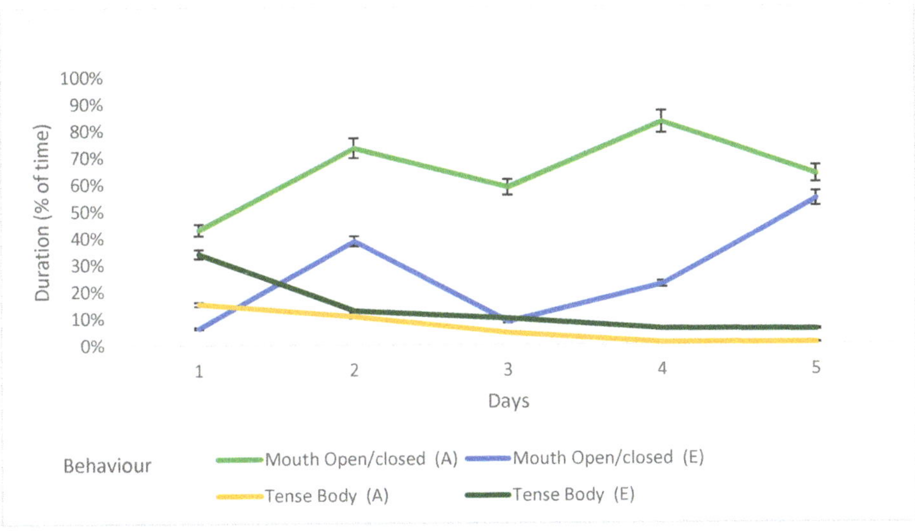

Figure 4. The duration of behaviours over the first five days that adopted or euthanased dogs (n = 38) spent in a shelter.

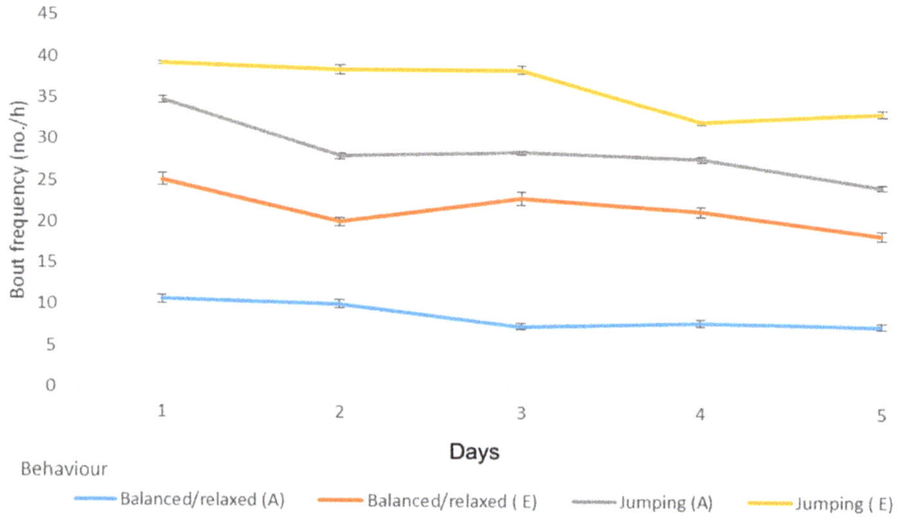

Figure 5. The frequency of behaviours over the first five days that adopted or euthanased dogs (n = 38) spent in a shelter.

Table 4. Differences in kennel behaviour between dogs that were euthanased and adopted, either overall or on certain days.

F/D	Behaviours	Interaction	p-Value
D	*Tense body*	*Day 1*	*0.001*
F	Balance/relaxed	Overall	0.004
F	Jumping in kennel	Overall	0.03

D = Duration, F = Frequency.

3.3. Emotional Characteristics of Dogs in the Behavioral Assessment That Were or Were Not Subsequently Euthanased

Behaviour of Dogs in Formal Behaviour Assessment

In the behaviour assessment, dogs spent 39% of their time in friendly behaviours, 17% displaying fear, 17% displaying anxiousness, 24% in high arousal, and 3% displaying aggression. Considering the frequency of behaviours, 26% were incidences of high arousal, 41% friendliness, 19% anxiousness, 12% fear, and 2% aggression.

Total scores for each behaviour were obtained from the formal behavioural assessment and categorised into emotional domains (Anxiety, Fear, Friendliness, Aggression, and Arousal). See Table 5 for Pearson's correlations of scores, with significance levels corrected using the Benjamini-Hochberg procedure.

Table 5. Spearman Rank correlation coefficients between behaviours recorded in the formal behaviour assessment in shelter dogs (n = 38) (Numbered behaviours relate to those presented in Table 3).

	Fear												
		1	2	3	4	5	6	7	8	9	10	11	12
Fear	1		*0.40*	*0.32*	−0.22	*0.33*	0.19	*0.40*	0.17	0.27	−0.11	0.09	−0.18
	2			**0.61**	**0.48**	**0.60**	−0.26	−0.03	0.21	0.02	**0.49**	**0.43**	*0.38*
	3				0.02	0.02	0.12	*0.35*	*−0.35*	−0.10	0.31	0.19	0.00
	4					0.08	−0.18	**0.55**	*0.32*	**0.56**	−0.06	**0.61**	0.00
	5						−0.18	−0.04	0.03	*0.34*	*0.36*	*0.32*	0.00
	6							0.32	0.15	−0.08	0.21	−0.22	0.00
	7								**0.41**	**0.60**	*0.48*	**0.59**	0.00
	8									**0.65**	0.18	**0.45**	0.00
	9										**0.43**	**0.42**	0.00
	10											−0.15	**0.54**
	11												0.00

	Anxiety														
		1	2	3	4	5	6	7	8	9	10	11	12	13	14
Anxiety	1		0.18	**0.51**	*0.35*	−0.05	**0.77**	0.32	**0.47**	**−0.64**	−0.13	*−0.32*	**0.42**	−0.03	*0.36*
	2			−0.11	−0.14	0.00	*−0.35*	−0.20	−0.20	−0.17	−0.14	−0.17	*0.41*	0.17	0.00
	3				−0.12	−0.08	*0.37*	**0.40**	**0.39**	−0.16	−0.12	−0.08	**0.53**	0.15	0.00
	4					−0.01	−0.20	**0.56**	**0.36**	−0.02	**0.69**	−0.07	**0.64**	**0.39**	0.00
	5						−0.05	0.13	**0.56**	**0.44**	**0.63**	0.19	**0.81**	−0.14	0.00
	6							**0.53**	**0.37**	**−0.37**	0.07	−0.07	**0.38**	−0.23	0.00
	7								0.56	**0.44**	**0.63**	**0.48**	**0.47**	0.09	0.00
	8									0.04	−0.01	0.05	**0.47**	−0.07	0.00
	9										**0.48**	**0.59**	−0.29	−0.16	0.00
	10											−0.15	**0.72**	**0.43**	**0.53**
	11												*0.35*	0.22	0.00
	12													0.03	**0.49**
	13														0.00

Table 5. *Cont.*

	Aggression													
		1	**2**	**3**	**4**	**5**	**6**	**7**	**8**	**9**	**10**	**11**	**12**	**13**
Aggression	**1**		*0.33*	**0.90**	−0.08	*0.31*	0.24	−0.01	**0.72**	0.07	*0.29*	**0.79**	−0.03	0.00
	2			0.25	0.21	−0.19	−0.16	−0.15	0.18	0.16	0.12	0.16	−0.11	*0.33*
	3				−0.03	−0.01	**0.55**	*0.34*	**0.97**	**0.59**	0.05	**0.71**	−0.04	**0.90**
	4					−0.33	**−0.37**	−0.19	−0.01	0.04	0.08	0.22	**0.44**	−0.08
	5						*0.31*	0.16	−0.08	−0.02	0.35	0.27	0.10	*0.31*
	6							0.24	**0.51**	**0.41**	0.13	0.06	−0.17	0.24
	7								*0.36*	**0.44**	*−0.37*	0.02	0.05	−0.01
	8									**0.60**	**0.55**	−0.02	**0.72**	0.01
	9										−0.02	0.06	0.07	0.07
	10											*0.32*	0.24	*0.29*
	11												**0.70**	**0.70**
	12													−0.03
	Arousal													
		1	**2**	**3**	**4**	**5**	**6**	**7**	**8**	**9**	**10**	**11**	**12**	**13**
Arousal	**1**		0.05	−0.23	**0.45**	−0.10	−0.09	−0.20	−0.24	−0.19	−0.19	−0.11	−0.09	−0.16
	2			*0.35*	0.27	*0.33*	−0.03	*0.31*	0.26	*0.34*	0.23	**0.37**	*0.31*	0.03
	3				−0.08	**0.51**	0.09	*0.32*	**0.77**	*0.32*	0.24	**0.47**	**0.41**	−0.16
	4					−0.06	−0.07	*0.33*	−0.19	−0.03	0.26	0.06	0.20	0.02
	5						−0.02	0.02	*0.37*	**0.43**	**0.40**	**0.38**	**0.52**	0.13
	6							0.09	−0.02	−0.15	**0.46**	**0.41**	0.15	−0.16
	7								0.16	**0.44**	**0.56**	**0.36**	**0.64**	**0.39**
	8									**0.51**	0.16	*0.37*	**0.38**	−0.01
	9										0.15	*0.34*	**0.42**	*0.31*
	10											**0.56**	**0.81**	0.10
	11												**0.47**	−0.21
	12													*0.30*

Table 5. *Cont.*

	Friendliness																
		1	2	3	4	5	6	7	8	9	10	11	12	13	14	15	16
Friendliness	1		−0.03	0.30	**0.39**	**0.36**	0.08	−0.16	*−0.31*	0.18	0.18	**−0.47**	**0.54**	0.05	−0.04	0.11	**0.51**
	2			**0.36**	−0.10	−0.17	*0.36*	*0.31*	0.20	*0.35*	0.03	*−0.36*	0.15	**−0.39**	**−0.47**	−0.25	0.13
	3				0.09	0.01	*0.35*	0.14	−0.22	**0.39**	−0.01	−0.15	0.26	−0.18	**−0.40**	−0.10	0.08
	4					**−0.54**	*−0.37*	−0.01	−0.28	−0.15	0.26	0.11	0.20	*0.36*	0.03	0.14	0.01
	5						0.21	*−0.06*	0.08	0.11	*0.31*	0.09	0.17	0.25	0.15	0.06	0.24
	6							0.30	**0.51**	**0.77**	−0.04	−0.25	0.04	−0.25	−0.25	−0.12	0.02
	7								*0.30*	0.32	0.10	*0.33*	0.00	−0.20	0.25	0.12	−0.24
	8									*0.37*	**0.43**	−0.24	−0.18	*0.39*	**0.38**	−0.15	−0.08
	9										0.07	−0.13	0.27	−0.21	−0.14	−0.01	−0.08
	10											**0.53**	0.21	**0.48**	*0.34*	−0.04	−0.03
	11												0.05	*0.36*	**0.39**	−0.06	−0.21
	12													−0.18	−0.19	0.04	**0.44**
	13														**0.47**	0.20	−0.05
	14															**0.65**	−0.24
	15																−0.01

Almost all correlations were statistically significant but ranged from weak to strong for both positive and negative correlations. There were positive correlations between the following behaviours that we associated with Fear: ears back, lip licking, lowered body, lowered head, shiver, tail low, tail tucked, tense body posture, weight back, and yawning; Anxiousness: fast, high tail, jumping, licking, lip licking, medium tail, pacing, panting, stiff tail, tense body posture, weight back, weight forward, and yawning. There were positive correlations between the following behaviours that we associated with Aggression: biting, ears forward, growling, high tail, lip licking, lowered head, medium tail, snapping, standing, stiff tail and still tail, and targeting gaze. There were positive correlations between the following behaviours that we associated with Arousal: barking, diverting gaze, fast and high tail, jumping up and off, licking, medium tail, mouthing, pacing, panting, weight forward, and whining. There were positive correlations between the following behaviours that we associated with Friendliness: balanced body posture, body curve, direct eye contact, ears forward and open, fast tail, handler interaction, jumping, medium tail, play behaviour, relaxed body, slow tail movement, sniffing, soft eye contact, wag loose, and walking.

3.4. Relationship between Kennel Behaviour and Formal Behaviour Assessment

There were positive correlations between anxiety, fear, and arousal behaviours displayed in kennels and in the formal behaviour assessment: whining, diverting eye contact, lip licking, panting, barking, jumping up, ears alert and forward, ears back, lowered body and tense body posture, tail tucked and stiff, and body weight back ($p < 0.02$) (Table 6). In addition, there were positive correlations between position in the kennel (at wall, wire, and at front door) and locations in behaviour assessment (at wall, window, and door) ($p < 0.02$) (Table 7).

Table 6. Significant ($p < 0.05$) Spearman Rank correlation coefficients between behaviours recorded in kennel and the formal behaviour assessment of shelter dogs ($n = 38$), listed for the emotional states of Arousal, Fear and Anxiety.

Arousal		**Fear**		**Anxiety**	
Barking	**0.57**	Diverting	*0.34*	Ears back	**0.57**
Diverting Gaze	*0.30*	Ears Back	**0.46**	Fast tail	**0.40**
Fast tail	**0.40**	Lip Licking	**0.42**	High tail	**0.63**
High Tail	**0.63**	Lowered Body	**0.44**	Jumping	*0.35*
Jumping up	**0.53**	Lowered head	**0.45**	Licking	*0.31*
Jump off	*0.35*	Shiver	**0.41**	Lip licking	*0.29*
Licking	*0.31*	Stiff tail	*0.33*	Medium tail	**0.45**
Lip licking	*0.29*	Tail Low	**0.45**	Pacing	*0.42*
Medium Tail	**0.45**	Tail tucked	*0.25*	Panting	*0.25*
Mouthing	*0.59*	Tense Body Posture	*0.28*	Stiff tail	*0.33*
Pacing	*0.31*	Weight back	**0.41**	Tense body	0.28
Panting	**0.42**	Yawn	*0.33*	Weight back	**0.41**
Weight forward	**0.38**			Weight forward	*0.38*
Whining	**0.36**			Whining	*0.36*

$p < 0.01$, *$p < 0.05$*.

Table 7. Significant ($p < 0.05$) Spearman Rank correlation coefficients between locations recorded in kennel and formal behaviour assessment of shelter dogs ($n = 38$).

Location	**Behaviour Assessment**			
Kennel assessment	Door	Front of room	Wall	Window
Front of kennel	**0.45**	−0.08	*−0.36*	−0.11
Wall	−0.22	0.00	**0.49**	−0.23

$p < 0.01$, *$p < 0.05$*

3.4.1. Exploration of Room

Comparing exploration of the room in the behaviour assessment with kennel behaviours, there were significant correlations between many duration and frequency behaviours in the anxiety, arousal, and fear emotional states (Table 8). Nearly all correlations were positive, demonstrating that for most behaviours recorded in the kennel were related to those exhibited in the behavioural assessment. Only two—whining and lip licking—were negatively related, suggesting that these are not reliable indicators of the room exploration test.

Table 8. Spearman Rank correlation coefficients between behaviours recorded in kennel and behaviours exhibited during the 'exploration of room test' in the behaviour assessment of shelter dogs ($n = 38$) within the emotional domains of arousal, fear, and anxiety.

Arousal	Title	Fear	Title	Anxiety	Title
Barking	*0.40*	Ears Back	**0.59**	Ears back	**0.59**
Diverting Gaze	*0.35*	Lip Licking	*−0.12*	Fast tail	**0.38**
High Tail	**0.69**	Lowered Body	*0.33*	High tail	**0.59**
Jumping up	**0.45**	Lowered head	**0.46**	Jumping	**0.28**
Jump off	*0.33*	Shiver	**0.52**	Licking	*0.34*
Licking	*0.35*	Stiff tail	**0.39**	Medium tail	*0.36*
Lip licking	0.27	*Tail Low*	*0.25*	Pacing	**0.44**
Medium Tail	*0.36*	Tail tucked	*0.25*	Panting	**0.46**
Pacing	*0.25*	Tense Body	*0.30*	Stiff tail	**0.39**
Panting	**0.46**	Weight back	**0.42**	Tense body	**0.36**
Weight forward	*0.33*			Weight back	**0.49**
Whining	*−0.50*				

$p < 0.01$, *$p < 0.05$*.

3.4.2. Time Alone Assessment

Similarly, comparing the time alone assessment with kennel behaviours, there were also significant correlations between many duration and frequency behaviours in the anxiety, arousal and fear emotional states (Table 9). Nearly all correlations were positive, demonstrating that most behaviours recorded in the kennel were related to those exhibited in the time alone assessment. Only three—whining, fast tail, and direct eyes—were negatively related, suggesting that these are not reliable indicators of the time alone test. There were also positive correlations between locations (Table 10).

Table 9. Significant ($p < 0.05$) Spearman Rank correlation coefficients between behaviours recorded in kennel and behaviours exhibited during the time alone test in the behaviour assessment of shelter dogs ($n = 38$) within the emotional domains of arousal, fear, anxiety, and friendliness.

Arousal		Fear		Anxiety		Friendliness	
Barking	**0.54**	Diverting	**0.52**	Ears back	**0.64**	Direct eye	*−0.32*
Fast tail	*−0.24*	Ears Back	**0.45**	Fast tail	*0.36*	Ears forward	**0.61**
High Tail	**0.61**	Lowered Body	*0.28*	High tail	**0.61**	Ears open	*0.29*
Jumping up	**0.38**	Lowered head	*0.30*	Jumping	**0.38**	Fast tail	**0.50**
Licking	**0.47**	Tail Low	**0.42**	Lip licking	*0.28*	Mouth open	**0.48**
Lip licking	**0.46**	Tail tucked	**0.43**	Medium tail	**0.46**	Medium tail	**0.46**
Medium Tail	**0.43**	Tense Body Posture	*0.23*	Pacing	*0.36*	Relaxed body	**0.51**
Pacing	**0.41**	Weight back	**0.47**	Panting	**0.41**	slow	**0.52**
Weight forward	*0.31*			Stiff tail	**0.50**	Sniff	*0.35*
Whining	*−0.23*			Tense body	*0.33*	Stand	**0.51**
				Weight back	**0.47**	Walking	**0.39**
				Weight forward	*0.31*		
				Whining	*−0.23*		

$p < 0.01$, *$p < 0.05$*.

Table 10. Significant ($p < 0.05$) Spearman Rank correlation coefficients between locations recorded in kennel and the time alone test of shelter dogs ($n = 38$).

Kennel Assessment	Time Alone Test		
	Door	Wall	Window
Front	*0.34*		
Wall		**0.54**	
Window		**−0.45**	*0.31*

$p < 0.01$, *$p < 0.05$*.

3.4.3. Relationship between Outcomes for the Dogs and Summarised Behaviour Results

Comparing the time spent in the various behaviours for dogs that were adopted with those that were euthanased, dogs that displayed more barking, balanced or lowered posture, or positioned by the wall in the kennel assessment, or balanced/lowered posture or pacing in the behaviour assessment, or balanced posture or jumping up in the time alone test had an increased likelihood of being adopted (Table 11). Those that displayed more tense body posture in the kennel test or sitting/lying in the behavioural assessment were more likely to be euthanased.

Table 11. Time spent in behaviours in the kennel, the formal assessment and time alone test of dogs ($n = 38$) that were adopted or euthanased, with Odds Ratio and Confidence Interval (CI) tested by binary logistic regression.

K/B/T	Behaviour	Adopted (% time)	Euthanased (% time)	Odds Ratio	95% CI
K	Barking	5.58	1.30	**1.47**	0.98–2.21
K	Balanced	44.06	22.34	**1.23**	1.03–1.49
K	Lowered	3.25	4.30	**4.22**	0.98–18.16
K	Tense	1.50	6.49	**0.09**	0.01–1.07
K	Wall	7.99	6.60	*1.53*	0.96–2.42
K	Sit/Lay	21.51	24.93	**1.52**	0.44–1.00
B	Balanced	66.41	44.60	**1.67**	0.97–2.87
B	Lowered	7.97	16.04	**1.72**	0.84–3.48
B	Pacing	37.59	30.94	**1.58**	0.98–2.51
T	Panting	59.21	68.99	*0.95*	0.89–1.00
T	Balanced	78.67	44.66	**1.50**	1.10–2.04
T	Jump up	18.35	30.21	**1.44**	1.07–1.92

$p < 0.01$, *$p < 0.05$*. K: Kennel B: Behaviour assessment T: Time alone.

Comparing the frequency of the various behaviours for dogs that were adopted with those that were euthanased, dogs that displayed more barking in the kennel assessment or balanced posture in the kennel or behaviour assessment or the time alone test had an increased likelihood of being adopted (Table 12). Those that displayed more panting in the kennel assessment, lowered head, or scanning in the behaviour assessment were more likely to be euthanased.

Table 12. Frequency of behaviours in the kennel, behaviours in the formal assessment and time alone test of dogs ($n = 38$) that were adopted or euthanased, together with the significance of the difference tested by binary logistic regression.

K/B/T	Behaviour	Adopted (% Frequency)	Euthanased (% Frequency)	Odds Ratio	95% CI
K	Barking	19.32	8.50	*1.09*	1.01–1.19
K	Panting	54.64	65.46	*0.95*	0.91–1.00
K	Balanced	45.24	26.13	**1.19**	1.05–1.35
B	Balanced	40.89	27.72	**1.48**	1.11–1.98
B	Lowered Head	12.29	12.77	**1.25**	0.99–1.55
B	Scanning	3.08	4.66	**0.65**	0.44–0.95
T	Balanced	54.68	35.58	**1.34**	1.11–1.62

$p < 0.01$, *$p < 0.05$*; K: Kennel B: Behaviour assessment T: Time alone.

4. Discussion

One solution to increasing adoptability of shelter dogs is the early detection of behaviour problems followed by modification programs aimed at helping dogs develop desired behaviours. Longitudinal monitoring of behaviours using both kennel and formal behaviour assessment information to help create comprehensive insight of the dog's behaviour can help achieve this aim [20]. Recent studies have pointed to the uncertainty of single behaviour assessments [16], but the work of Goold and Newberry and this current research clearly demonstrate the benefit of continual monitoring. Continual monitoring allows correct identification of behavioural cues associated with separation-related behaviours, anxiety, fear, arousal, and friendliness. To identify these behavioural cues using monitoring tools in the first five days allows behaviour modification to be implemented to help these dogs to cope effectively in a socially isolating environment. Using a formal behaviour assessment, as customarily practiced in shelters, as a single context assessment of a dog's behaviour creates an ineffective profile of stable behavioural tendencies.

4.1. Behaviour in the Five Days after Surrender

This study focused on behaviour observations in the first five days after admission to a shelter and compared these to behaviour identified in a formal behaviour assessment. Over the first five days after admission, dogs displayed decreasing tense body and tucked tail, which are probably the best indicators of fear in the dogs. Previous studies that found that over the first five days after relinquishment to a shelter dogs will experience social isolation due to the breaking of social bonds with previous companions/owners [5,6]. Prior studies report numerous contradictory indications of the extent to which shelter dogs adapt over time, displaying behavioural and physiological indicators of positive and negative stress [29]. Some studies report a reduction in stress and fear related behaviours over time in shelters [6,10,30], whereas others indicate that dogs display acute signs of negative stress and fear due to the high novelty of the shelter environment [29,31]. Although environmental factors influence these behaviours, including new olfactory, auditory, and sensory stimulation, dogs can either have a positive or negative coping style, thereby demonstrating effective or ineffective ability to cope in a new environment [21,29,32]. These diverse results are likely to be due to differences in resources offered by shelters.

The ability to monitor kennel behaviours associated with positive and negative stress or coping styles can help identify changes in the quality of life (QoL) of dogs in shelters [15,33]. Identifying dogs that have a deterioration in positive behaviours allows early treatment. Interestingly, dogs that were deemed not suitable for adoption had higher durations of tense body posture in-kennel and increased frequency of jumping behaviour in kennel. Conversely, positive behaviours, including a balanced/relaxed body posture, had lower frequency of occurrence in dogs suitable for adoption.

Another interesting finding in the present study is the association between positive behaviours that include friendliness in dogs in the first five days, which agrees with previous studies [6,10,20,30]. These findings highlight the benefit of longitudinal monitoring of behaviour in shelter kennels to identify stable behaviours that included docility and friendliness [20].

4.2. Behaviours in Assessment

Anxiousness, arousal, and fear tendencies correlated with its corresponding emotional domain in the behaviour assessment (Table 5), indicating a positive relationship with the domains identified in kennel and behaviour in the standardized assessment. Previous research by Mornement [26] in behaviour assessments in Australian shelters indicated fear and friendliness were the only behaviours that were predictive. Other research using similar test protocols with social (stranger and toddler interactions) and non-social stimuli reported fear related behaviours as found in this research [34,35]. As stated previously, the effect of acute stress and social isolation in dogs when relinquished to a novel environment have the ability to dramatically change behaviour. Thus, the result of increased fear, arousal, and anxious behaviour found in the kennel and at assessment (Table 4) suggest time-independent coping mechanisms that a dog may implement to help respond to the changing environment [21,36,37]. The results go beyond the previous study, suggesting that if coping mechanisms are ineffective at helping the dog cope with the environment, then those behavioural tendencies can manifest into behaviour problems that can be identified in an assessment.

4.3. Comparison between Kennel and Behaviour Assessment

The comparison of kennel behaviour and the formal behaviour assessment indicates that kennel behavioural cues associated with fear, anxiety, and arousal were confirmed in the formal behaviour assessment (Table 6). Furthermore, in the analysis of the position in kennel, we confirmed that position in the behaviour assessment was associated with front of kennel, door, and wall in each situation (Table 7).

Once the formal assessment was separated into component parts, specifically exploration of room and time alone, there were associations between behaviours found in these tests and kennel behaviours reflecting separation-related behavioural cues, anxiousness, arousal, and fear (Tables 8 and 9). Separation related behaviours are associated with increased whining, pacing, excessive salivation, barking, jumping in orientation of owner's departure, and escaping behaviour [38]. Studies show that separation-related behaviours can be correctly identified in video analysis of dogs in their time alone once the owner has left [18]. Furthermore, a study by Blackwell et al. [39] into the identification of separation-related behaviours in shelters showed the importance of using a time-alone test to assess dogs with behaviour problems. The results clearly demonstrate the positive predictive value of the time alone test to identify separation related behaviours [39]. Separation-related behaviours have been identified as a common problem post adoption [13]. Therefore, to identify these issues early is the key to early treatment, which could lead to an increase in the likelihood of successful adoption and therefore decreasing euthanasia. The findings with respect to fear are consistent with that of Mornement [26], who identified its predictive validity. Research by Tiira et al. [40] outlined high comorbidity between different anxieties, showing that fearful dogs had significantly higher noise sensitivity and separation anxiety.

Dogs with behaviours associated with separation-related problems, such as arousal and fear, were less likely to be deemed adoptable (Tables 11 and 12). Dogs that displayed friendly, low arousal, and docile behaviours were more likely to be adopted (Tables 11 and 12). Behavioural issues that have been linked to reasons for relinquishment of dogs include separation-related behaviours, arousal, and fear [41–46]. In contrast, behaviours that adoptees look for in dogs are associated with friendliness toward people, docility, and low arousal [47]. Thus, increasing positive behaviours and decreasing separation-related behaviours, fear, and high arousal are critical to increase adoptability, thereby decreasing euthanasia. Early recognition of ineffective behaviours and coping mechanisms allows shelters to implement behaviour management programs before behavioural problems manifest [48,49]. Behaviour assessments are comprised of numerous tests that allow for a snapshot of a dog's behaviour that is multifactorial. Therefore, a paradigm shift should occur in shelters to implement assessments as continuous tools to monitor a dogs' behaviour over time. Once unsuitable or problem behaviours are identified, shelters can create effective modification plans to allow issues to be solved before manifesting into serious behavioural problems. Using assessments in shelters to identify past behaviours in the previous home or to predict future behaviour is difficult. However, using assessments as a tool to understand the behaviour of dogs in conjunction with continual kennel monitoring and everyday interaction may allow identification of behavioural issues and ineffective coping mechanisms. Further research into monitoring of behaviours associated with the manifestation of behavioural problems in shelters is warranted.

Some limitations are associated with this research that future studies should consider. To allow for comprehensive behaviour analysis of dogs, previous home environment could be taken into consideration. Therefore, we should try to more accurately represent behaviour in the home. Our sample size was relatively small, but due to the nature of the study, which identified changes in behaviours over time on single dogs, it is not seen as a major restriction. Finally, the limitation of variability between each shelter should be taken into consideration and warrants further study.

5. Conclusions

Previous research suggests that behaviour assessments are ineffective, focusing on the lack of their accurate predictability of behaviour. However, in this study, we found that behaviour assessment information can be related to behaviour over the previous days since relinquishment to the novel environment. Effectively monitoring kennel behaviour allows early recognition of problems. Numerous authors have recommended continual monitoring procedures to help identify key behavioural problems as early as possible. This research has demonstrated numerous correlations between kennel behaviour and that displayed during formal assessments. We suggest that shelters should use continuous monitoring techniques at the same time as supporting automated behaviour problem recognition. Continuing to use formal assessments and incorporating longitudinal monitoring of behaviour to help identify dogs unable to cope effectively in shelter environments may also provide useful additional information of dog behaviour problems. Such monitoring allows early implementation of training modification, thereby increasing adoptability of dogs that once would be deemed unadoptable.

Author Contributions: L.C., M.P., P.B., G.P., and C.P. conceived the project. L.C. drafted the paper and all authors had input into modifying it into the present format.

Funding: This research received no external funding.

Acknowledgments: The authors acknowledge the assistance of RSPCA Queensland.

Conflicts of Interest: Mandy Paterson declares that she works for the RSCPA Qld. Liam Clay declares that RSPCA Qld funds his studentship. Apart from this, no other author has any conflict.

References

1. RSPCA. *RSPCA Australia National Statistics*. 2017. Available online: https://www.rspca.org.au/sites/default/files/RSPCA%20Australia%20Annual%20Statistics%202017-2018.pdf (accessed on 27 May 2018).
2. Salman, M.D.; Hutchison, J.; Ruch-Gallie, R. Behavioral reasons for relinquishment of dogs and cats to 12 shelters. *J. Appl. Anim. Welf. Sci.* **2000**, *3*, 93–106. [CrossRef]
3. Marston, L.C.; Bennett, P.C.; Coleman, G.J. What Happens to Shelter Dogs? An Analysis of Data for 1 Year From Three Australian Shelters. *J. Appl. Anim. Welf. Sci.* **2004**, *7*, 27–47. [CrossRef] [PubMed]
4. Shore, E.R. Returning a recently adopted companion animal: Adopters' reasons for and reactions to the failed adoption experience. *J. Appl. Anim. Welf. Sci.* **2005**, *8*, 187–198. [CrossRef] [PubMed]
5. Hennessy, M.B.; Davis, H.N.; Williams, M.T.; Mellott, C.; Douglas, C.W. Plasma cortisol levels of dogs at a county animal shelter. *Physiol. Behav.* **1997**, *62*, 485–490. [CrossRef]

6. Stephen, J.M.; Ledger, R.A. A longitudinal evaluation of urinary cortisol in kennelled dogs, canis familiaris. *Physiol. Behav.* **2006**, *87*, 911–916. [CrossRef]
7. Wells, D.L.; Graham, L.; Hepper, P.G. The influence of length of time in a rescue shelter on the behaviour of kennelled dogs. *Anim. Welf.* **2002**, *11*, 317–325.
8. Hewson, C.J.; Hiby, E.F.; Bradshaw, J.W.S. Assessing quality of life in companion and kennelled dogs: A critical review. *Anim. Welf.* **2007**, *16*, 89–95.
9. Dalla Villa, P.; Barnard, S.; Di Nardo, A.; Iannetti, L.; Podaliri Vulpiani, M.; Trentini, R.; Serpell, J.A.; Siracusa, C. Validation of the socially acceptable behaviour (sab) test in a centralitaly pet dog population. *Vet. Ital.* **2017**, *53*, 61–70.
10. Hiby, E.F.; Rooney, N.J.; Bradshaw, J.W. Behavioural and physiological responses of dogs entering re-homing kennels. *Physiol Behav.* **2006**, *89*, 385–391. [CrossRef]
11. Rooney, N.J.; Gaines, S.A.; Bradshaw, J.W.S. Behavioural and glucocorticoid responses of dogs (canis familiaris) to kennelling: Investigating mitigation of stress by prior habituation. *Physiol. Behav.* **2007**, *92*, 847–854. [CrossRef]
12. Titulaer, M.; Blackwell, E.J.; Mendl, M.; Casey, R.A. Cross sectional study comparing behavioural, cognitive and physiological indicators of welfare between short and long term kennelled domestic dogs. *Appl. Anim. Behav. Sci.* **2013**, *147*, 149–158. [CrossRef]
13. Serpell, J.A.; Hsu, Y. Development and validation of a novel method for evaluating behavior and temperament in guide dogs. *Appl. Anim. Behav. Sci.* **2001**, *72*, 347–364. [CrossRef]
14. Mornement, K.; Coleman, G.; Toukhsati, S.R.; Bennett, P.C. Evaluation of the predictive validity of the behavioural assessment for re-homing k9's (bark) protocol and owner satisfaction with adopted dogs. *Appl. Anim. Behav. Sci.* **2015**, *167*, 35–42. [CrossRef]
15. Barnard, S.; Calderara, S.; Pistocchi, S.; Cucchiara, R.; Podaliri-Vulpiani, M.; Messori, S.L.; Ferri, N. Quick, Accurate, Smart: 3D Computer Vision Technology Helps Assessing Confined Animals' Behaviour. *PLoS ONE* **2016**, *11*, e0158748. [CrossRef] [PubMed]
16. Patronek, G.J.; Bradley, J. No better than flipping a coin: Reconsidering canine behavior evaluations in animal shelters. *J. Vet. Behav. Clin. Appl. Res.* **2016**, *15*, 66–77. [CrossRef]
17. Poulsen, A.; Lisle, A.; Phillips, C.J.C. An evaluation of a behaviour assessment to determine the suitability of shelter dogs for rehoming. *Vet. Med. Int.* **2010**, *2010*, 523781. [CrossRef]
18. Palestrini, C.; Minero, M.; Cannas, S.; Rossi, E.; Frank, D. Video analysis of dogs with separation-related behaviors. *Appl. Anim. Behav. Sci.* **2010**, *124*, 61–67. [CrossRef]
19. Konok, V.; Dóka, A.; Miklósi, Á. The behavior of the domestic dog (Canis familiaris) during separation from and reunion with the owner: A questionnaire and an experimental study. *Appl. Anim. Behav. Sci.* **2011**, *135*, 300–308. [CrossRef]
20. Goold, C.; Newberry, R.C. Modelling personality, plasticity and predictability in shelter dogs. *R. Soc. Open Sci.* **2017**, *4*, 170618. [CrossRef]
21. Rayment, D.J.; De Groef, B.; Peters, R.A.; Marston, L.C. Applied personality assessment in domestic dogs: Limitations and caveats. *Appl. Anim. Behav. Sci.* **2015**, *163*, 1–18. [CrossRef]
22. Diesel, G.; Brodbelt, D.; Pfeiffer, D.U. Reliability of assessment of dogs' behavioural responses by staff working at a welfare charity in the uk. *Appl. Anim. Behav. Sci.* **2008**, *115*, 171–181. [CrossRef]
23. Dowling-Guyer, S.; Marder, A.; D'Arpino, S. Behavioral traits detected in shelter dogs by a behavior evaluation. *Appl. Anim. Behav. Sci.* **2011**, *130*, 107–114. [CrossRef]
24. Hennessy, M.B.; Voith, V.L.; Mazzei, S.J.; Buttram, J.; Miller, D.D.; Linden, F. Behavior and cortisol levels of dogs in a public animal shelter, and an exploration of the ability of these measures to predict problem behavior after adoption. *Appl. Anim. Behav. Sci.* **2001**, *73*, 217–233. [CrossRef]
25. Valsecchi, P.; Barnard, S.; Stefanini, C.; Normando, S. Temperament test for re-homed dogs validated through direct behavioral observation in shelter and home environment. *J. Vet. Behav. Clin. Appl. Res.* **2011**, *6*, 161–177. [CrossRef]
26. Mornement, K.M.; Coleman, G.J.; Toukhsati, S.; Bennett, P.C. Development of the behavioural assessment for re-homing k9's (b.A.R.K.) protocol. *Appl. Anim. Behav. Sci.* **2014**, *151*, 75–83. [CrossRef]
27. Olivier, F.; Marco, G. Boris: A free, versatile open-source event-logging software for video/audio coding and live observations. *Methods Ecol. Evol.* **2016**, *7*, 1325–1330.
28. McDonald, J.H. *Handbook of Biological Statistics*, 3rd ed.; Sparky House Publishing: Baltimore, MD, USA, 2014.

29. Protopopova, A. Effects of sheltering on physiology, immune function, behavior, and the welfare of dogs. *Physiol. Behav.* **2016**, *159*, 95–103. [CrossRef]
30. Stephen, J.M.; Ledger, R.A. An audit of behavioral indicators of poor welfare in-kenneled dogs in the United Kingdom. *J. Appl. Anim. Welf. Sci.* **2005**, *8*, 79–96. [CrossRef]
31. Kis, A.; Klausz, B.; Persa, E.; Miklósi, Á.; Gácsi, M. Timing and presence of an attachment person affect sensitivity of aggression tests in shelter dogs. *Vet. Rec.* **2014**, *174*, 196. [CrossRef]
32. Taylor, K.D.; Mills, D.S. The effect of the kennel environment on canine welfare: A critical review of experimental studies. *Anim. Welf.* **2007**, *16*, 435–447.
33. Kiddie, J.; Collins, L. Identifying environmental and management factors that may be associated with the quality of life of kennelled dogs (canis familiaris). *Appl. Anim. Behav. Sci.* **2015**, *167*, 43–55. [CrossRef]
34. Stellato, A.C.; Flint, H.E.; Widowski, T.M.; Serpell, J.A.; Niel, L. Assessment of fear-related behaviours displayed by companion dogs (Canis familiaris) in response to social and non-social stimuli. *Appl. Anim. Behav. Sci.* **2017**, *188*, 84–90. [CrossRef]
35. De Meester, R.H.; Pluijmakers, J.; Vermeire, S.; Laevens, H. The use of the socially acceptable behavior test in the study of temperament of dogs. *J. Vet. Behav. Clin. Appl. Res.* **2011**, *6*, 211–224. [CrossRef]
36. Bateson, M.; Brilot, B.; Nettle, D. Anxiety: An Evolutionary Approach. *Can. J. Psychiatry* **2011**, *56*, 707–715. [CrossRef]
37. Hohoff, C. Anxiety in mice and men: A comparison. *J. Neural Transm.* **2009**, *116*, 679–687. [CrossRef]
38. Ogata, N. Separation anxiety in dogs: What progress has been made in our understanding of the most common behavioral problems in dogs? *J. Vet. Behav. Clin. Appl. Res.* **2016**, *16*, 28–35. [CrossRef]
39. Blackwell, E.J.; Bradshaw, J.W.S.; Casey, R.A. Fear responses to noises in domestic dogs: Prevalence, risk factors and co-occurrence with other fear related behaviour. *Appl. Anim. Behav. Sci.* **2013**, *145*, 15–25. [CrossRef]
40. Tiira, K.; Sulkama, S.; Lohi, H. Prevalence, comorbidity, and behavioral variation in canine anxiety. *J. Vet. Behav.* **2016**, *16*, 36–44. [CrossRef]
41. Herron, M.E.; Lord, L.K.; Husseini, S.E. Effects of preadoption counseling on the prevention of separation anxiety in newly adopted shelter dogs. *J. Vet. Behav. Clin. Appl. Res.* **2014**, *9*, 13–21. [CrossRef]
42. Hemy, M.; Rand, J.; Morton, J.; Paterson, M. Characteristics and outcomes of dogs admitted into queensland rspca shelters. *Animals* **2017**, *7*, 67. [CrossRef]
43. Neidhart, L.; Boyd, R. Companion Animal Adoption Study. *J. Appl. Anim. Welf. Sci.* **2002**, *5*, 175–192. [CrossRef] [PubMed]
44. New, J.C.; Salman, M.D.; King, M.; Scarlett, J.M.; Kass, P.H.; Hutchison, J.M. Characteristics of Shelter-Relinquished Animals and Their Owners Compared With Animals and Their Owners in U.S. Pet-Owning Households. *J. Appl. Anim. Welf. Sci.* **2000**, *3*, 179–201. [CrossRef]
45. Patronek, G.J.; Glickman, L.T.; Beck, A.M.; McCabe, G.P.; Ecker, C. Risk factors for relinquishment of dogs to an animal shelter. *J. Am. Vet. Med. Assoc.* **1996**, *209*, 572–581. [PubMed]
46. Salman, M.D.; New, J.J.G.; Scarlett, J.M.; Kass, P.H.; Ruch-Gallie, R.; Hetts, S. Human and Animal Factors Related to Relinquishment of Dogs and Cats in 12 Selected Animal Shelters in the United States. *J Appl. Anim.Welf. Sci.* **1998**, *1*, 207–226. [CrossRef] [PubMed]
47. King, T.; Marston, L.C.; Bennett, P.C. Describing the ideal australian companion dog. *Appl. Anim. Behav. Sci.* **2009**, *120*, 84–93. [CrossRef]
48. Sherman, B.L.; Mills, D.S. Canine Anxieties and Phobias: An Update on Separation Anxiety and Noise Aversions. *Vet. Clin. N. Am. Small Anim. Pract.* **2008**, *38*, 1081–1106. [CrossRef] [PubMed]
49. Takeuchi, Y.; Houpt, K.A.; Scarlett, J.M. Evaluation of treatments for separation anxiety in dogs. *J. Am. Vet. Med. Assoc.* **2000**, *217*, 342–345. [CrossRef]

animals

MDPI

Article

Forgotten, But Not Lost—Alloparental Behavior and Pup–Adult Interactions in Companion Dogs

Péter Pongrácz * and Sára S. Sztruhala

Department of Ethology, Eötvös Loránd University, Pázmány Péter sétány 1/c, 1117 Budapest, Hungary; sztruhalasara@hotmail.com

* Correspondence: peter.pongracz@ttk.elte.hu

Received: 30 October 2019; Accepted: 14 November 2019; Published: 21 November 2019

Simple Summary: Companion dogs are vastly popular animals; however, we know surprisingly little about their natural parental behaviors. Meanwhile, although wolves, dingoes, and, to an extent, even free-ranging dogs show several forms of alloparental behaviors, the parental care among companion dogs is thought to be solely provided by the mother. We circulated an international survey for dog breeders, asking them about the forms of alloparental behaviors they observed among their dogs, as well as further interactions between the puppies and other adult dogs at home. Our results show that allonursing and feeding of the pups by regurgitation is a widespread phenomenon among companion dogs. The behavior of young puppies regarding, for example, their reaction to other dogs' barking was also influenced by the timing of their access to the other dogs at the breeder's home. Based on the breeders' observations, sexual status and age of the other dogs affected the way they interacted with the puppies, and also the way the puppies' mother interacted with them. These results highlight the importance of dog–puppy interactions during the early weeks of life, an often neglected area compared to the well-known elements of puppy socialization with human beings.

Abstract: Socialization with humans is known to be a pivotal factor in the development of appropriate adult dog behavior, but the role and extent of dog–dog interactions in the first two months of life is rarely studied. Although various forms of alloparental behaviors are described in the case of wild-living canids, the social network of companion dogs around home-raised puppies is almost unknown. An international online survey of companion dog breeders was conducted, asking about the interactions of other dogs in the household with the puppies and the pups' mother. Based on the observations of these breeders, our study showed an intricate network of interactions among adult dogs and puppies below the age of weaning. Alloparental behaviors (including suckling and feeding by regurgitation) were reportedly common. Independent of their sex, other household dogs mostly behaved in an amicable way with the puppies, and in the case of unseparated housing, the puppies reacted with lower fear to the barks of the others. Parousness, sexual status, and age of the adult dogs had an association with how interested the dogs were in interacting with the puppies, and also with how the mother reacted to the other dogs. Our study highlights the possible importance of dog–dog interactions during the early life of puppies in forming stable and low-stress interactions with other dogs later in life.

Keywords: behavior; dog; alloparental care; puppies; breeders

1. Introduction

The species-specific traits of dogs (*Canis familiaris*) are mostly inseparable from the process of domestication. The evolutionary changes in the socio-cognitive capacities [1,2], behavior [3,4], anatomy and physical appearance [5,6], as well as the physiology [7,8], are most often evaluated with regard to the differences between dogs and their closer or more distant wild-living relatives.

More recently, researchers concentrated on the behavioral and cognitive features [9,10], both as proximate and ultimate factors behind the adaptation of dogs to the anthropogenic niche, with only a few exceptions—for example, genetic changes that could affect the carbohydrate metabolism in dogs were also highlighted as assumed key factors behind domestication [11,12]. Although differences in the reproductive biology of dogs (e.g., switching from being monestrous to a mainly diestrous cycle), when compared to their closest wild relative, the grey wolf (*Canis lupus*), are also apparent among the crucial changes, and have also been modeled by the well-known silver fox project conducted in Novosibirsk, Russia [13], the reproductive behavior of companion (or "family") dogs is rarely discussed in scientific literature, apart from various issues covered by veterinary science (e.g., [14,15]). As the reproduction of companion (and working) dogs is mainly planned, supervised, and restricted by human caretakers [16,17], this segment of dog behavior remains almost untouched by ethologists. Furthermore, alloparental behavior and paternal caretaking of the young, two factors which are considered uniquely typical for a wide selection of canid species [18], are literally unknown among companion dogs, and have only recently been discovered in the free-ranging dog populations [19,20].

Free-ranging dogs are often considered to be the "ecologically most successful" variants of domestic dogs due to their vast number (according to some estimations, around 800 million worldwide—[21]) and ubiquitous presence in and around human settlements. Subsequently, it is assumed that free-ranging dogs provide the best opportunity to understand the biology of dogs [21], as free-ranging dogs have been adapting to their environment for many generations without excessive artificial selection by humans. However, when it comes to parental behavior, seemingly, there is a considerable difference, even among free-ranging dogs. This has led to such widely differing observations, which have either stated that lack of alloparental care is one of the reasons why mortality rate of young pups is very high among free-ranging dogs (in Italy—[22]; in Mexico—[23]), or described more or less sporadic, but existing paternal/alloparental caretaking (in India—[20,24]). In their exhaustive review on canid reproduction, Lord and colleagues [25] assumed that since domestic dogs became dependent on human resources, they mostly lost the need for alloparental caretaking (i.e., because they have a stable food supply); meanwhile, the same ecologically predictable food resources made it possible that the sexual behavior of dogs also mostly lost the strict seasonality that is typical to wild canids. This feeding ecology-based theory gains further (indirect) support from the observations made with dingoes—feral dogs in Australia that became isolated from other Southeastern Asian dog populations around 3.5–5 thousand years ago [26]. These dogs sustain themselves mainly by hunting large prey—consequently, they have retained many typical features of the reproductive behavior of wolves (e.g., alloparental care—[27], monestrus—[18]), because these seem to support the lifestyle of apex canid predators.

Very little is known about the natural interactions of juvenile (pre- and around weaning period) dogs and their older canine companions (kin and non-kin) in the case of companion (pet) and working dogs. Contrary to dog–human interactions during puppyhood, which were recently investigated from multiple aspects and considered to be a crucial part of the "process of proper socialization" [3,28–30], the behavior and effect of adult dogs on puppies in the home environment have received much less interest from investigators. Among the most likely reasons for this is the difficulty of conducting observations at the owners' home, or the highly variable social environment (i.e., there is no standardized or "natural" social structure at breeders' homes that would include roughly the same kinds of adult dogs around the puppies). Consequently, although there are data about the interactions of dog puppies with other dogs at public areas [31], as well as pup–pup interactions within the litter (e.g., the ontogeny of playful behavior [32]), our knowledge of pups' interactions with familiar, but not necessarily related adult dogs from the household is very limited. The exception is the interaction with the mother—as the extent and style of maternal care was found to have fundamental effects on later behavior in working dogs (e.g., drug seeking dogs—[33]; police dogs—[34]). However, except for the work of [35] regarding the feeding of the pups by regurgitation, we do not know of any studies on alloparental behavior within pet dog groups, and indications of paternal care are missing as well.

In this study, we conducted a detailed, international internet questionnaire about alloparental behavior and pup–adult dog interactions within companion dog groups that live at the homes of dog breeders. We surveyed not only the existence of alloparental nursing and feeding of pups by regurgitation, but we also covered such behaviors as the mother dog's reaction to other adult dogs around her offspring, and the pups' reaction to other household dogs' barking. We analyzed whether the aforementioned behavioral variables were dependent on the circumstances of how the pups were kept—especially with regard to their isolation from the other dogs in the home. Based on the literature about the reproductive biology of free-ranging dogs [36] and dingoes (e.g., [18]) (both of which are not under direct human control), it can be assumed that alloparental behavior emerges in dogs as a functional response of the temporal predictability and ease of access to food [25]. As companion and working dogs in private or professional kennels are steadily provided with easily accessible food, one could assume that the need for alloparental and paternal care is minimal to non-existent. However, as an alternative hypothesis, one could expect that even companion dog populations have retained the capacity of providing alloparental care, as we do not know about active selection that would go against this capacity in dogs under human management.

2. Materials and Methods

2.1. Ethical Approval

This study was carried out in accordance with national and international ethical guidelines (e.g., American Psychological Association, Hungarian Psychological Association). Participation was voluntary; we handled all data obtained confidentially, and anonymized the questionnaires after data collection. The Ethical Committee of Eötvös Loránd University reviewed and approved the study. Ethical permission number: PEI/2016/003.

2.2. Development of the Survey

Two questionnaires were created in Hungarian and English languages, and both versions were disseminated via social media and email. Questionnaire 1 (https://goo.gl/forms/u7Q4ti2jglUHfKy63) was the main endeavor with a complex set of items, while Questionnaire 2 (https://forms.gle/ceRXWfrw4tEmtAUZ6) served to collect additional information regarding adult dogs' food regurgitation to puppies. It took approximately 10–15 min to complete Questionnaire 1, and 5 min to complete Questionnaire 2. Data from Questionnaire 1 were recorded between October 2017 and April 2018. Questionnaire 2 (focusing on solely the regurgitating behavior) was circulated between 8–19 of October 2018.

Nonprobability convenience sampling was used, as the questionnaires were distributed via social media, predominantly by sharing them in various Facebook groups, dedicated to the breeding of specific dog breeds, or breeding of purebred dogs in general. We repeated the call for participation weekly on the social media platforms along the course of the survey. Additionally, the surveys were also sent via email to dog breeders on the basis of personal acquaintances of the authors—however, this resulted in only about 10% of the total sample.

2.3. Subjects

Participation in the survey was voluntary and anonymous; however, participants could provide their name and/or contact address on a non-mandatory basis. No form of incentive was offered for the participation.

Our first sample consisted of 77 dog breeders from 11 countries, who reported their observations of 45 dog breeds. Our second sample from Questionnaire 2 consisted of the observations of 36 dog breeders from 3 countries and 28 dog breeds (see Table 1).

Table 1. Distribution of breeders according to countries and participation in the questionnaire surveys. In the column marked with "Total", the numbers represent the sum of individual respondents (of each breed) who completed either Questionnaire 1 or 2. FCI = Fédération Cynologique Internationale.

Breed	FCI Breed Group	Total	Questionnaire 2 only	Questionnaires 1 and 2	Country
Australian Shepherd	FCI-1	4	2		Hungary, Germany
Belgian Shepherd Dog/Malinois	FCI-1	2	1	1	Hungary, UK
Border Collie	FCI-1	1	1		Hungary
Bouvier des Flandres	FCI-1	1		1	Hungary
Collie Rough	FCI-1	1	1		Hungary
Collie Smooth	FCI-1	1			Germany
German Shepherd Dog	FCI-1	4			Hungary
Mudi	FCI-1	5	1	1	Hungary
Old English Sheepdog	FCI-1	1			Poland
Pumi	FCI-1	1			Sweden
Shetland Sheepdog	FCI-1	2			Hungary
Cane Corso	FCI-2	1			Hungary
Doberman	FCI-2	4	2	2	Hungary, Germany
Great Dane	FCI-2	2			Hungary
Kangal	FCI-2	1		1	Hungary
Rottweiler	FCI-2	3	1	1	Hungary, US
Schnauzer—Giant	FCI-2	3			Hungary
Schnauzer—Middle	FCI-2	1	1		Hungary
Shar Pei	FCI-2	1			Hungary
Am. Pitbull Terrier	FCI-3	3	1		Hungary
Am. Staffordshire Terrier	FCI-3	2			Hungary
Biewer Yorkshire Terrier	FCI-3	1			Hungary
Jack Russell Terrier	FCI-3	1			Hungary
Yorkshire Terrier	FCI-3	1			Hungary
Dachshund Mini	FCI-4	1			Finland
Dachshund Short Haired	FCI-4	1			Belgium
Akita Inu	FCI-5	1	1		Hungary
German Spitz Klein	FCI-5	2			Hungary, UK
Hokkaido	FCI-5	1			Hungary
Keeshond (Wolfspitz)	FCI-5	1			Germany
Shiba Inu	FCI-5	1			Lithuania
Siberian Husky	FCI-5	3	1		Hungary
Rhodesian Ridgeback	FCI-6	1		1	Germany
Braque d'Auvergne	FCI-7	1			Hungary
Hungarian Vizsla	FCI-7	2		1	Hungary
English Cocker Spaniel	FCI-8	1	1		Hungary

Table 1. *Cont.*

Breed	FCI Breed Group	Total	Questionnaire 2 only	Questionnaires 1 and 2	Country
Flatcoated Retriever	FCI-8	1		1	Finland
Labrador Retriever	FCI-8	5	1		Hungary, UK, Italy
Chinese Crested Dog	FCI-9	3		1	Hungary
Bichon Frise	FCI-9	1		1	Hungary
Bolognese	FCI-9	2		1	Hungary, Slovenia
Coton de Tulear	FCI-9	1		1	Hungary
French Bulldog	FCI-9	2		1	Hungary
Havanese	FCI-9	4		1	Hungary
Lhasa Apso	FCI-9	1			Hungary
Poodle—all sizes	FCI-9	3		1	Hungary
Shih Tzu	FCI-9	1			Hungary
Tibetan Terrier	FCI-9	3	1	1	Hungary
Borzoi	FCI-10	1	1		Hungary
Whippet	FCI-10	3		1	Hungary, Austria

Breeders who completed the questionnaire had to meet the following criteria: They raised at least one litter of puppies a priori the completion of the questionnaire; they keep at home a minimum of one more dog of any age, in addition to the mother of the puppies. We requested that the breeders answer the questions regarding their observations on their "entire experience" of their past litters bred.

2.4. Variables

The questionnaires covered the first 8 to 12 weeks of the puppies lives which are spent in their breeders' home. During this period, the puppies normally experience various new stimuli (both social and asocial). Towards the end of this period, puppies start to show an almost complete set of social behaviors, including playful, agonistic, and communicative interactions with not only their kin [37], but also towards humans [38].

Beyond the demographic details, the questionnaires contained items that were aimed at the following interactions between the mother, other dogs, and the puppies:

- Alloparental behaviors (apart from feeding by regurgitation); e.g., when another female dog nursed, or attempted to nurse the puppies. (Multiple choice, where the possible answers were: No alloparental behaviors were observed; puppies were nursed by—the daughter, the mother, the sister of the mother dog; an unrelated adult female nursed the puppies.)
- Feeding with regurgitation (we asked the participants to indicate separately whether the mother of the puppies and/or another dog regurgitated food for the puppies). Multiple choice item.
- Puppies' reaction to adult dogs' barking: Breeders could select one of three options: Puppies show no reaction, puppies become frightened (run and hide, become quiet), or puppies start to bark when they hear the adults' barking.
- Adult dogs' reaction to the puppies' whining: We again used a three grade scale (single choice): Adults show no reaction to puppies' whining, they sniff the puppies, or the adults show alloparental behavior (mother-like caretaking behavior towards the puppies: Licks them, lays down next to them, plays with them).
- Adult dogs' reaction to puppies in general situations: During play, eating, drinking, and chewing on toys (single choice). We got the majority of answers from the following two categories: Adults show no reaction to the puppies, or adults are friendly with the puppies. Other categories, such as

adults start barking, show aggression towards the puppies, puppies show aggression towards adults, were excluded due to the very sporadic or missing answers to these options.

- Mother dogs' reaction to adult dogs around her puppies (single choice). Breeders could again indicate their answers on a three-grade scale: She shows no reaction, she is aggressive with other dogs, or she is friendly with other dogs.

We analyzed the possible associations among the aforementioned parameters and the following factors:

- The way the puppies are separated from the other dogs. Three alternatives were given: Puppies and adult dogs are kept together, puppies are separated with a fence or barrier, or puppies are separated completely from adult dogs. (i.e., they live in a separate room of the house).
- Age of the puppies when they are separated from adult dogs. Participants could choose from three main options: The puppies and adult dogs are kept together, the puppies are kept separated only until a defined age, or the puppies are separated completely from other adult dogs the entire time they are at the breeder.
- Aggression towards puppies. The breeders were asked whether they experienced any aggression towards the puppies in general situations (yes/no).

In the case of the items that included the puppies' or their mother's interaction with "another dog", there were separate options for the following categories of dogs:

- The age of the other dog: Younger than one year, between one and eight years, and over eight years old.
- Paternal status of the other dogs (whether it was the father or not of the current puppies).
- Previous parental experience of adult dogs (i.e., has the dog already reproduced, in the case of both male and female dogs); yes or no answers could be selected.
- Sexual status of the adult dogs: Breeders could select whether the particular dog was neutered/spayed or intact.

The following categories were used as fixed factors in the statistical analysis.

Dog breeds categorized according to the FCI (Fédération Cynologique Internationale) system. For the statistical analysis, we had a large enough sample size from only three FCI breed groups: Group 1 "Sheepdogs and Cattledogs", Group 2 "Pinscher and Schnauzer—Molossoid, Swiss Mountain, and Cattledogs", and Group 9 "Companion and Toy Dogs".

Dog breeds were categorized according to their genetic distance from the wolf (*Canis lupus*) (based on [39]). Three out of the ten groups had a large enough sample size for statistical analysis: "Working Dogs", "Herding Dogs", and "Mastiffs".

2.5. Data Analysis

For the statistical analysis, we used the IBM SPSS statistical program (version 22.0, Armonk, NY, USA). Raw data are available online as a Supplementary Materials.

The alloparental behavior was handled as a binary variable (presence/absence) and its association with the fixed factors was analyzed by generalized linear models (GzLM) with the binary logistic method. The same method was used when we analyzed the adult dogs' behavior to puppies in general situations. The binomial test was used to analyze the occurrence of regurgitating behavior (presence/absence). The puppies' reaction to adult dogs' barking was analyzed by ordinal regression. We used GzLM with ordinal logistic to find out whether there were differences in the adult dogs' behavior towards the puppies' whining and in the mother's reaction towards the adult dogs. We used sex, previous parental experience, sexual status, and age of the adult dogs as independent variables.

In the case of models where there were two or more independent variables, the two-way interactions were also included in the analysis. Alpha was set at 0.05 in each test.

3. Results

3.1. Alloparental Behaviors (Including Nursing, Licking, Cleaning, But Not Regurgitation)

From the 77 responses to Questionnaire 1, we found in 61% of the cases that breeders observed the presence of alloparental nursing behaviors. We found no significant association between the adult dogs' caretaking behavior and any of the fixed factors (GzLM with binary logistic—housing method of puppies $\kappa^2(2) = 3.022$, $p = 0.221$; timing of separation of the puppies $\kappa^2(2) = 2.349$, $p = 0.309$; aggressive behavior with the puppies $\kappa^2(1) = 0.245$, $p = 0.621$; FCI breed groups $\kappa^2(2) = 0.294$, $p = 0.863$; and genetically clustered breed groups $\kappa^2(2) = 0.539$, $p = 0.764$).

3.2. Feeding of the Puppies with Regurgitation

From the 41 breeders who responded to Questionnaire 2, 34 reported that he/she observed mother dogs regurgitating to their puppies. By setting the chance level to 0.5, according to the binomial test, this ratio is significantly above chance level ($p < 0.001$). However, one could also expect that mothers will almost always feed their litters by regurgitation (parallel with the pariah dogs, [36]), therefore we ran the binomial test again with the chance level set at 0.99. The abovementioned observed ratio is significantly below this ($p < 0.001$). We also investigated the occurrence of alloparental regurgitative feeding. From the 41 responses, 18 breeders reported observations of other (i.e., not the mother) dogs providing food by regurgitation to the puppies. Based on the very sporadic incidence of this behavior, based on the literature about free-ranging dogs [20,36], we set the reference ratio near zero (0.01). The observed ratio was significantly higher than this (binomial test, $p < 0.001$).

3.3. The Puppies' Reaction to the Other Dogs' Barking

We found a significant association with the housing method of puppies (ordinal regression—$\kappa^2(2) = 9.363$, $p = 0.009$). Based on the post-hoc comparisons (Table 2), puppies show the weakest reaction to the barking of adult dogs when they are not separated from the other dogs; meanwhile, they show fear or they join in barking with the others when they are kept partly or fully separated from the other dogs in the household (Figure 1). A similar, but non-significant trend was found in the case of the timing of separation of puppies (ordinal regression—$\kappa^2(2) = 4.631$, $p = 0.099$), where again, puppies that are never separated from the other dogs showed the weakest reaction to other dogs' barking according to the breeders' observations. We found no further significant associations between reaction to other dogs' barking and the fixed factors (ordinal regression—aggressive behavior with the puppies $\kappa^2(1) = 0.628$, $p = 0.428$; FCI breed groups $\kappa^2(2) = 1.112$, $p = 0.573$; and genetically clustered breed groups $\kappa^2(2) = 0.598$, $p = 0.742$).

Table 2. Parameter estimates of the puppies' reaction to other dogs' barking as a function of the method of housing of the puppies (ordinal regression). "Barkreact" 0–2 levels show an increasing intensity of fearful/joining reaction when other dogs bark. "Separation" 1–3 levels show a decreasing extent of separating the puppies from the other dogs during the time spent at the breeder's house. * this parameter is set to 0 because it is redundant.

Parameter		Estimate	Std. Error	Wald	df	Sig.	95% Confidence Interval	
							Lower Bound	Upper Bound
Threshold	[barkreact = 0]	−0.317	0.472	0.450	1	0.502	−1.243	0.609
	[barkreact = 1]	0.893	0.486	3.367	1	0.067	−0.061	1.846
	[barkreact = 2]	1.496	0.504	8.809	1	0.003	0.508	2.484
Location	[separation = 1]	1.461	0.589	6.154	1	0.013	0.307	2.615
	[separation = 2]	1.651	0.592	7.771	1	0.005	0.490	2.812
	[separation = 3]	0 *	.	.	0	.	.	.

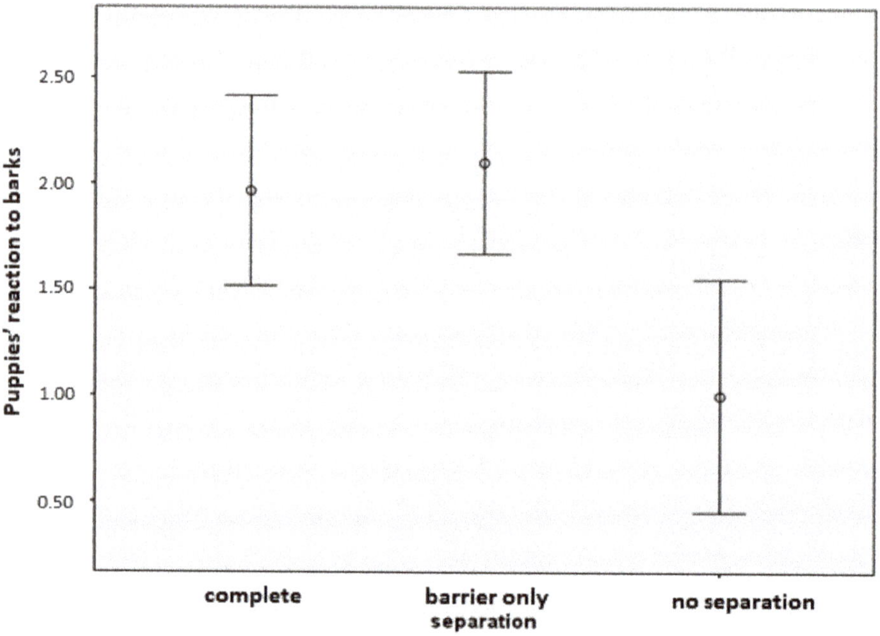

Figure 1. Association between the housing method of the puppies ("separation" 1–3 shows a decreasing level of isolation of the puppies from other dogs) and their reaction to other dogs' barking (mean ± 95% confidence interval (CI)).

3.4. *Adult Dogs' Reaction to the Puppies' Whining*

We found no difference between the reaction of the puppies' father and other adult male dogs (GzLM with ordinal logistic—$\kappa^2(1) = 0.606$, $p = 0.436$). In the case of the adult female dogs' age, reproductive status (intact vs. spayed), and parousness (had offspring vs. did not have offspring previously), we found a significant association between the age of the female dogs and their reaction to the puppies' whining (GzLM with ordinal logistic—age $\kappa^2(2) = 8.564$, $p = 0.014$). According to the post-hoc comparisons (Table 3), female dogs under 1 year of age showed the strongest reaction (including either sniffing, or even more intense, mother-like behaviors, such as licking, lying beside, or attempting to nurse) to the puppies' whining (Figure 2).

Table 3. Parameter estimates of the reactions of adult female dogs to the whining of the puppies as a function of age, reproductive status, and parousness (had offspring or not). GzLM with ordinal logistic. * this parameter is set to 0 because it is redundant.

Parameter		B	Std. Error	95% Wald Confidence Interval		Hypothesis Test		
				Lower	Upper	Wald Chi-Square	df	Sig.
Threshold	[react_whine = 1]	−1.485	0.4545	−2.375	−0.594	10.671	1	0.001
	[react_whine = 2]	0.912	0.4410	0.047	1.776	4.275	1	0.039
[age = below 1 year]		2.113	7824	0.579	3.646	7.292	1	0.007
[age = between 1–8 years]		0.579	0.5375	−0.474	1.633	1.162	1	0.281
[age = older than 8 years]		0 *	.	.	.	.	.	.
[spayed]		−1.152	0.5439	−2.218	−0.086	4.488	1	0.034
[intact]		0	.	.	.	.	.	.
[parous = 0]		−1.140	0.4780	−2.077	−0.203	5.690	1	0.017
[parous = 1]		0 *	.	.	.	.	.	.

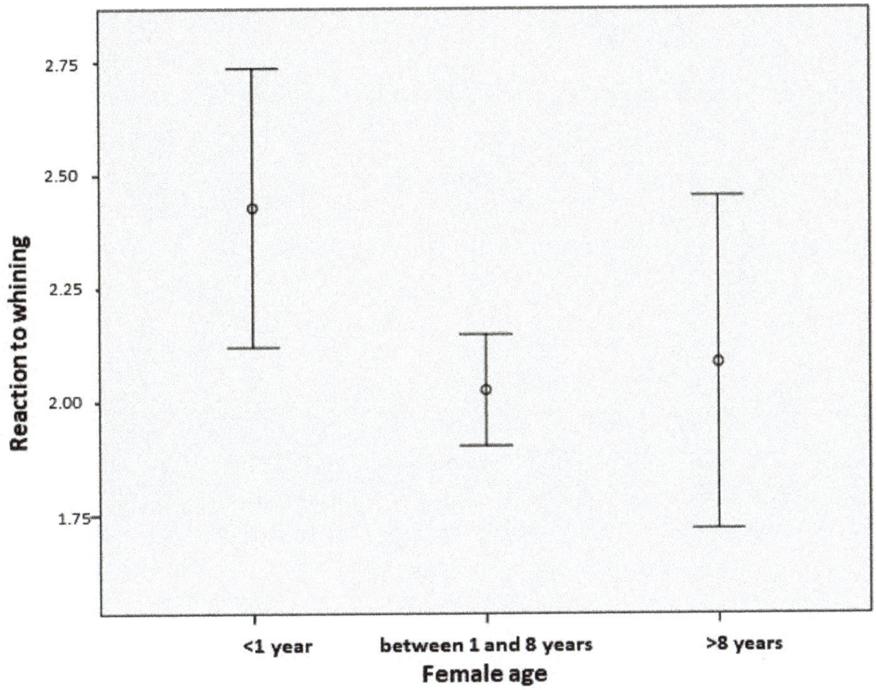

Figure 2. Association between the female dogs' age (1 = younger than one year; 2 = adult; 3 = older than 8 years) and the intensity of their reaction to the puppies' whining (mean ± 95% CI).

Neither the reproductive status ($\kappa^2(1) = 2.442$, $p = 0.118$) nor the parousness ($\kappa^2(1) = 2.341$, $p = 0.126$) of the adult females had significant association with reaction to the puppies' whining. We did not find any significant interactions between the factors. According to the breeders' observations, the parousness of the adult dogs (including both males and females) had a significant association with their reaction to the puppies' whining (Table 4; GzLM with ordinal logistic—$\kappa^2(1) = 5.795$, $p = 0.016$), where parous adult dogs react more intensely than the nulliparous ones (Figure 3). Sex itself did not have a significant effect ($\kappa^2(1) = 2.240$, $p = 0.134$), and we did not find significant interaction between the factors.

Table 4. Parameter estimates of adult dogs' reaction to the whining of puppies, as a function of their sex and parousness (0 = did not have puppies; 1 = had puppies). GzLM with ordinal logistic. * this parameter is set to 0 because it is redundant.

Parameter		B	Std. Error	95% Wald Confidence Interval		Hypothesis Test		
				Lower	Upper	Wald Chi-Square	df	Sig.
Threshold	[react_whine = 1]	−2.322	0.3886	−3.083	−1.560	35.694	1	0.000
	[react_whine = 2]	0.396	0.3295	−0.250	1.041	1.441	1	0.230
[male]		−0.963	0.4643	−1.873	−0.053	4.302	1	0.038
[female]		0 *	.	.	.	.	.	.
[parous = 0]		−1.273	0.4984	−2.250	−0.296	6.527	1	0.011
[parous = 1]		0 *	.	.	.	.	.	.

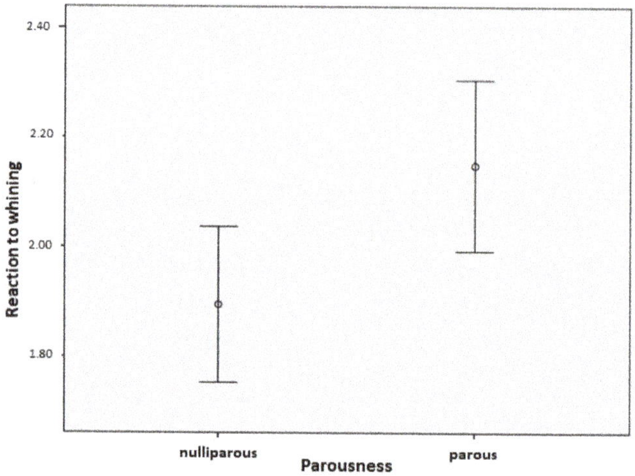

Figure 3. Adult (male + female) dogs' reaction (mean ± 95% CI) to the whining of puppies in association with the parousness of the adults (0 = did not have puppies; 1 = had puppies before).

3.5. Adult Dogs' Reaction to Puppies in General Situations

Adult males' reactions did not show a difference between the puppies' father and other males (GzLM with binary logistic—$\kappa^2(1) = 0.000$, $p = 1.000$). The age (GzLM with binary logistic—$\kappa^2(2) = 8.300$, $p = 0.016$) and reproductive status ($\kappa^2(1) = 6.293$, $p = 0.012$) of female dogs showed a significant association with their reaction to the puppies—according to the post-hoc comparisons (Table 5), less than 1 year old females and the intact females were more likely to react with playful interest to the puppies' presence than the older or spayed females (Figure 4). The parousness of the females did not have a significant effect on this parameter ($\kappa^2(1) = 0.271$, $p = 0.603$). The sex ($\kappa^2(1) = 0.451$, $p = 0.502$) and parousness ($\kappa^2(1) = 0.164$, $p = 0.686$) of the other adult dogs in the household, as well as the interaction of these factors, did not show a significant association with the reaction to the puppies in general encounters.

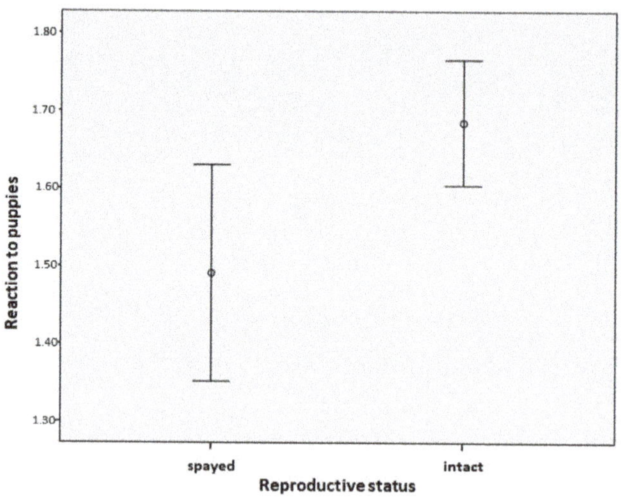

Figure 4. The association between the amicable reaction (mean ± 95% CI) of female dogs to puppies as a function of their reproductive status (0 = spayed, 1 = intact).

Table 5. Parameter estimates of the female dogs' behavior with the puppies in association with their age, reproductive status, and parousness (did not have puppies or had puppies before). GzLM with binary logistics. * this parameter is set to 0 because it is redundant.

Parameter	B	Std. Error	95% Wald Confidence Interval		Hypothesis Test		
			Lower	Upper	Wald Chi-Square	df	Sig.
(Intercept)	0.241	0.4029	−0.549	1.031	0.358	1	0.549
[age = below 1 year]	−2.044	0.7752	−3.564	−0.525	6.955	1	0.008
[age = 1–8 yeas]	−1.259	0.5064	−2.252	−0.266	6.181	1	0.013
[age = older than 8 years]	0 *	.	.	.	.	.	.
[spayed]	0.935	0.3727	0.204	1.666	6.293	1	0.012
[intact]	0	.	.	.	.	.	.
[parous = 0]	0.194	0.3723	−0.536	0.923	0.271	1	0.603
[parous = 1]	0 *	.	.	.	.	.	.

3.6. The Reaction of the Mother of the Puppies to Other Dogs in the Household

In the case of other adult male dogs, we did not find a significant difference between the reaction of the mother to the father of the puppies or other adult male dogs (GzLM with ordinal logistic—$\kappa^2(1) = 1.091$, $p = 0.296$). When we analyzed the reaction of the mother to other females in the household, according to the breeders' observations, there was a significant association with the other females' age (GzLM with ordinal logistic—$\kappa^2(2) = 13.090$, $p = 0.001$), and we also found a significant interaction between the reproductive status and parousness of the other females ($\kappa^2(1) = 12.183$, $p < 0.001$). According to the post-hoc comparisons, mother dogs are less friendly with other adult females than with juveniles or older females. Furthermore, mother dogs are the least friendly with nulliparous, intact females (Table 6). This result was strengthened by another analysis, where we tested the association with the sex and parousness of other dogs in the household. Besides the two significant main effects (GzLM with ordinal logistic—sex $\kappa^2(1) = 12.784$, $p < 0.001$; parousness $\kappa^2(1) = 15.090$, $p < 0.001$), we also found a significant interaction ($\kappa^2(1) = 5.732$, $p = 0.017$). According to this, mother dogs are the least friendly with nulliparous females by the observations of the breeders.

Table 6. Parameter estimates of the mother dogs' reaction (agonistic, neutral, friendly) to other females in the household as a function of the other females' age, reproductive status, and parousness. Significant interaction (*) between reproductive status and parousness is included. GzLM with ordinal logistic. * this parameter is set to 0 because it is redundant.

Parameter		B	Std. Error	95% Wald Confidence Interval		Hypothesis Test		
				Lower	Upper	Wald Chi-Square	df	Sig.
Threshold	[react_female = −1]	−2.019	0.3824	−2.768	−1.269	27.876	1	0.000
	[react_female = 0]	−0.751	0.3556	−1.448	−0.054	4.463	1	0.035
[age = less than 1 year]		1.744	0.6716	0.428	3.061	6.743	1	0.009
[age = 1–8 years]		−0.195	0.3999	−0.979	0.588	0.239	1	0.625
[age = older than 8 years]		0 *	.	.	.	.	.	.
[spayed]		−0.646	0.4022	−1.434	0.142	2.579	1	0.108
[intact]		0 *	.	.	.	.	.	.
[parous = 0]		−2.003	0.4294	−2.844	−1.161	21.748	1	0.000
[parous = 1]		0 *	.	.	.	.	.	.
[reproductive = 0] * [parous = 1]		2.117	0.6065	0.928	3.306	12.183	1	0.000

4. Discussion

In this study, we surveyed an international sample of active dog breeders, with questions aimed at the various forms of alloparental caretaking of the puppies, the interactions between puppies and other dogs apart from the mother, plus the interactions between the mother dog and other dogs at the same home. Based on the breeders' reports, alloparental nursing can be considered widespread in companion dogs, although its presence is significantly less than 100%. Other forms of nurturing the puppies (i.e., regurgitation) is typical for the mothers, and additionally provided by other dogs in slightly less than 50% of the cases as well. The presence of alloparental caretaking was not associated with the type of breed, or with the way the puppies were separated (or not) from the other dogs. The puppies showed a stronger reaction to other dogs' barking if they were kept at least partially separated from other dogs in the household. Paternity (including both the father of the puppies or an unrelated male) did not have significant association with the dogs' reaction to the puppies' whining and with the dogs' playful behavior with the puppies. Mother dogs also reacted similarly with the father of the puppies and other males. Young female, parous dogs showed stronger interest towards the whining of the puppies; and the young/sexually intact females behaved more playfully with them. In turn, mother dogs were reported as being the least friendly with other females that were either adult, or intact, nulliparous ones.

Before the detailed discussion of the results, we should consider some limitations of our study. As it was a questionnaire survey, we entirely relied on the expertise and experiences of the breeders who reported on the behavior of their dogs. Therefore, in the future, targeted behavioral tests would be useful for validating the key findings about the various dog–dog interactions—however, we should also keep in mind that conducting experiments in the homes of dog owners/breeders has its difficulties/limitations, too. The sample size was relatively low in our study, partly caused by the rather strict rules of participation (only breeders were invited with actual experience of raising at least one litter of puppies in the past with other adult dogs around). In any case, a more comprehensive survey would be beneficial (preferably involving more participants from non-European countries), especially which includes a more complete set of dog breeds from preferably each breed groups based both on the artificial (FCI) and genetic clustering. Finally, the scope of our survey was limited by the way in which the questionnaire was distributed—the utilization of internet and social media has its benefits (one can reach participants quickly and technically with no geographic limitations); however, potentially knowledgeable participants can also be left out because they do not use these means of communication.

Compared to the detailed comparative work done on the parental behavior of wild canids [25], and even on free-ranging dogs [36], empirical studies are noticeably lacking for companion dogs. Companion and working/service dogs are usually bred with close human supervision (e.g., [40,41]); therefore, the circumstances and, many times, even the process of parental care can be considered as more or less artificial compared to wild dogs. Apart from veterinary and animal breeding texts (e.g., [17,42]), scientifically accumulated information is surprisingly sparse regarding the maternal (see review [43]) and alloparental behaviors, as well as the interaction between the puppies and other adult dogs, in the first two months of life.

Our results show that alloparental caretaking behaviors (allonursing, regurgitating of food) are widespread among dogs that are kept by hobby breeders. Fostering of the young has a solid ecological basis in such species as the super-social members of the *Canidae* family (gray wolf; African wild dog *Lycaon pictus*; dhole *Cuon alpinus*). Among others, Riedman [44] lists the following factors that, along the course of evolution, could facilitate alloparental caretaking to develop: "(1) Prolonged or energetically intensive parental investment; (2) small groups with tight kinship bonds; (3) highly social or cooperative group structure; and (4) young that are raised in high density breeding colonies". From these, the first three conditions are typically true of the abovementioned wild canids—however, they are harder to interpret in the case of companion, or even-free ranging dogs. If we consider the latter (pariah, or village dogs) as the most valid ecotype of domestic dogs [21], we should notice that the change in feeding ecology (being mainly a scavenger instead of a hunter of large prey, [45]) could be the main driving force behind the alteration of reproductive and parental behavior in the majority of dogs. Scavenging does not require adults to act cooperatively—neither during the hunt, nor when provisioning the lactating mother and the young. Furthermore, in free-ranging dogs, the freshly weaned juveniles become mostly a competitor for the adults [21]; therefore, their role as "helpers" for the next generation is limited or non-existent. Interestingly, our results, as well as in the earlier study of [35], show a considerably common occurrence of alloparental behaviors among companion dogs. The apparent difference compared to the infrequently observed foster-behaviors among free-ranging dogs [19,20] could be the result of the complex effect of different levels of food competition, human intervention, and, in part, the density of animals around the breeding mother/puppies (see condition #iv by Riedman [44]). Based on this, we can hypothesize that the capacity for alloparental caretaking is steadily present in the domestic dog, even in the population of pets and working dogs where the conditions for both the breeding and caretaking of the young are highly artificial. The competition for resources is rather strong among free-ranging dogs [46], thus alloparental behaviors may be less adaptive (given that scavenging is the main type of food procurement). In companion dogs, alloparental behavioral tendencies are rarely discouraged—according to our results, its occurrence was not affected by the method of puppy-raising that the breeders chose. Somewhat surprisingly, we did not find a breed effect; however, this could be explained by the low sample size of individual breed groups. One could hypothesize that those breeds that are more closely related to wolves would show stronger alloparental tendencies. Although basal (or "ancient") breeds were not represented well enough in our sample, the fact that we did not find a difference between the occurrence of alloparental behaviors among breed groups, with such widely varying functions as herding vs. toy/companion or belonging to the mastiffs vs. herding dogs, it shows that alloparental behaviors were probably less affected by recent functional selection of dogs.

Keeping conditions reportedly affected the puppies' reactivity to other dogs' barking—the breeders observed the weakest reaction (i.e., lower levels of fearfulness, or tendency to join the others barking) in those puppies that were kept together with the other dogs in the household without any restriction. Although the first couple of months are considered crucial in the proper socialization of young dogs [47], and successful socialization is unequivocally considered a key factor in avoiding problem behaviors in dogs (e.g., [48]), still, the majority of scientific studies concentrate on the events of socialization that typically follow the puppies' departure from the breeder's home (e.g., [29,30]). Relatively few papers target the early interactions between the living environment and puppies still with their mother (e.g., [49]). In a few earlier studies (such as [50]), it was shown that early isolation of the puppies resulted in a decreased level of interactivity later with conspecifics. However, most studies have focused on dog–human interactions (e.g., [3,51]), because, in the case of companion and working dogs, the main measure of their success as adults depends on their fit to the human environment. Therefore, the observations reported in our study have a relevance from the aspect that the chance to interact uninterruptedly with other dogs from a very early age can improve young dogs' behavior regarding dog–dog interactions. It is important to see, however, that the familiarity between the young puppies and the other adult dogs they interact with also can play a crucial role in the behavioral development of the juvenile dogs. Earlier, it was found [52] that the younger the puppy was when its new owners introduced it to other dogs, the higher the chance became that later the dog showed undesired aggression towards conspecifics. Adding these findings to our results, one could conclude that the predominantly amicable interactions (including alloparental care) with familiar dogs at the breeder's home could provide the best experience for young puppies with their conspecifics; meanwhile, owners should be careful with the early exposure of the puppies to possibly negative experiences with unknown dogs at public areas [52]. So far, this particular aspect of environmental effects on behavioral ontogeny has only been marginally covered by other studies (e.g., [33,53]), where the focus has mainly been on other behaviors, such as interactions with the physical environment or with humans.

We did not find a difference between the behavior of adult male dogs with the puppies, whether the particular dog was the father of the puppies or not, and the mother dog's reaction to the males was also independent from the paternal status of the males. This result is in parallel with the reports on free-ranging dogs, where sometimes, more than one male dog was observed to be loosely associated with particular litters [36].

Friendly interactions were reportedly facilitated with the puppies if the other dog was a young, sexually intact female. Sexual status was also important in the case of the other dogs' reaction to the whining of the puppies—besides the young females, intact and parous dogs showed stronger interest towards whining puppies. Interestingly, in a recent laboratory study, Lehoczki et al. [54] found no association between the sexual status, parousness, or sex of adult dogs and their responsiveness to playbacks of separation calls of puppies. Besides the option that breeders may misinterpret their dogs' behavior, it is also possible that dogs behave differently in an artificial laboratory setting compared to a realistic situation at home when puppies are truly present. In the case of highly social canid species, fostering (alloparental helping) behaviors are facilitated by several factors [18]: Monestrum (which prevents deliberate further pregnancies in most females along the season) and an unusually long diestrous (pseudopregnancy) phase (which puts the non-pregnant, usually young females into a hormonal state that facilitates maternal behaviors). As monestrum and alloparental helper behavior are definitely present in some dogs (dingoes, [27]), as well as pseudopregnancy, which is a pronounced feature even in companion dog females [55], our results (i.e., young females are among the most attentive to whining puppies and respond to puppies in the friendliest manner) are in line with the literature. Breeders also reported that older, intact, and parous dogs showed heightened interest towards whining puppies and were more likely to play with the puppies—this behavior can be expected, according to Schradin et al. [56], who, besides the neoteny-helper hypothesis, highlighted another way of alloparental care: The parent–helper scenario. According to the latter, even adult members of the group can show helping behavior with the puppies, if their endocrine system mimics the changes otherwise typical to the lactating mother.

Finally, breeders reported some level of discrimination by the mother dog regarding her interactions with other dogs in the household. Mothers showed the most agonistic behaviors with other females, if those were either adult, or intact, nulliparous ones. One should take into consideration, of course, that the distribution of sexes and age cohorts were uncontrolled in our sample, which could have an effect on the statistical reliability. On the other hand, there are reports of protective mother dogs (in free-ranging dogs, [57]), with an emphasis on the most frequent agonistic interactions among the adult females [58]. Paul and colleagues [19], however, showed that in free-ranging dogs, grandmothers may provide help with their daughters' puppies. As in our study, the mother dogs showed higher aggression against intact, nulliparous females, the observations made on companion dogs and on free-ranging dogs (that were accepting help from their mothers) are not excluding each other. However, in general, the reportedly higher occurrence of agonistic behaviors in dog mothers against other (intact) females can be explained as a result of avoiding possible resource competition, and even a probability of infanticide (which is observed among female wolves, [59]).

5. Conclusions

Based on the observations of companion dog breeders, our study shows an intricate network of interactions among adult dogs of the household and puppies below the age of weaning. Alloparental behaviors and amicable interactions from the adult dogs dominated the scene, with an eventual stress-reducing effect on the behavior of puppies in the case of alarm barks of the adults. The role of dog–dog interactions during the first two months of life might be an important factor for proper socialization and later problem-free behavior with future canine partners.

Supplementary Materials: The following are available online at http://www.mdpi.com/2076-2615/9/12/1011/s1, raw data file (puppy_survey)—this file includes the data which were used for the statistical analyses.

Author Contributions: Conceptualization, P.P. and S.S.S.; Methodology, P.P. and S.S.S.; Formal analysis, P.P.; Investigation, S.S.S.; Data curation, P.P. and S.S.S.; Writing—original draft preparation, P.P. and S.S.S.; Writing—review and editing, P.P.; Visualization, P.P.; Supervision, P.P.; Project administration, S.S.S.

Funding: This research received no external funding.

Acknowledgments: The authors are thankful to Celeste Pongrácz for checking the English of this manuscript.

Conflicts of Interest: The authors declare no conflicts of interest.

References

1. Topál, J.; Miklósi, Á.; Gácsi, M.; Dóka, A.; Pongrácz, P.; Kubinyi, E.; Csányi, V. The dog as a model for understanding human social behavior. *Adv. Study Behav.* **2009**, *39*, 71–116.
2. Udell, M.A.; Dorey, N.R.; Wynne, C.D. What did domestication do to dogs? A new account of dogs' sensitivity to human actions. *Biol. Rev.* **2010**, *85*, 327–345. [CrossRef] [PubMed]
3. Topál, J.; Gácsi, M.; Miklósi, Á.; Virányi, Z.; Kubinyi, E.; Csányi, V. Attachment to humans: A comparative study on hand-reared wolves and differently socialized dog puppies. *Anim. Behav.* **2005**, *70*, 1367–1375. [CrossRef]
4. Marshall-Pescini, S.; Virányi, Z.; Range, F. The effect of domestication on inhibitory control: Wolves and dogs compared. *PLoS ONE* **2015**, *10*, e0118469. [CrossRef]
5. Clutton-Brock, J. Origins of the dog: Domestication and early history. In *The Domestic Dog: Its Evolution, Behaviour and Interactions with People*; Serpell, J., Barrett, P., Eds.; Cambridge University Press: Cambridge, UK, 1995; pp. 7–20.
6. Gácsi, M.; McGreevy, P.; Kara, E.; Miklósi, Á. Effects of selection for cooperation and attention in dogs. *Behav. Brain Funct.* **2009**, *5*, 31. [CrossRef]
7. Belyaev, D.K.; Plyusnina, I.Z.; Trut, L.N. Domestication in the silver fox (*Vulpes fulvus* Desm): Changes in physiological boundaries of the sensitive period of primary socialization. *Appl. Anim. Behav. Sci.* **1985**, *13*, 359–370. [CrossRef]
8. Axelsson, E.; Ratnakumar, A.; Arendt, M.-L.; Maqbool, K.; Webster, M.T.; Perloski, M.; Liberg, O.; Arnemo, J.M.; Hedhammar, Å.; Lindblad-Toh, K. The genomic signature of dog domestication reveals adaptation to a starch-rich diet. *Nature* **2013**, *495*, 360. [CrossRef]
9. Miklósi, A.; Topál, J.; Csányi, V. Comparative social cognition: What can dogs teach us? *Anim. Behav.* **2004**, *67*, 995–1004. [CrossRef]
10. Range, F.; Virányi, Z. Tracking the evolutionary origins of dog-human cooperation: The Canine Cooperation Hypothesis. *Front. Psychol.* **2015**, *5*, 1582. [CrossRef]
11. Arendt, M.; Fall, T.; Lindblad-Toh, K.; Axelsson, E. Amylase activity is associated with AMY 2B copy numbers in dog: Implications for dog domestication, diet and diabetes. *Anim. Genet.* **2014**, *45*, 716–722. [CrossRef]
12. Arendt, M.; Cairns, K.M.; Ballard, J.W.O.; Savolainen, P.; Axelsson, E. Diet adaptation in dog reflects spread of prehistoric agriculture. *Heredity* **2016**, *117*, 301. [CrossRef] [PubMed]
13. Belyaev, D.K. Destabilizing selection as a factor in domestication. *J. Hered.* **1979**, *70*, 301–308. [CrossRef] [PubMed]
14. Harrop, A.E. *Reproduction in the Dog*; Williams and Wilkins: Philadelphia, PA, USA, 1960.
15. Hoffman, J.M.; Creevy, K.E.; Promislow, D.E. Reproductive capability is associated with lifespan and cause of death in companion dogs. *PLoS ONE* **2013**, *8*, e61082. [CrossRef] [PubMed]
16. McGreevy, P.D.; Nicholas, F.W. Some practical solutions to welfare problems in dog breeding. *Anim. Welf.* **1999**, *8*, 329–342.
17. Kustritz, M.V.R. Determining the optimal age for gonadectomy of dogs and cats. *J. Am. Vet. Med. Assoc.* **2007**, *231*, 1665–1675. [CrossRef]
18. Asa, C.S.; Valdespino, C. Canid reproductive biology: An integration of proximate mechanisms and ultimate causes. *Am. Zool.* **1998**, *38*, 251–259. [CrossRef]
19. Paul, M.; Majumder, S.S.; Bhadra, A. Grandmotherly care: A case study in Indian free-ranging dogs. *J. Ethol.* **2014**, *32*, 75–82. [CrossRef]
20. Paul, M.; Bhadra, A. The great Indian joint families of free-ranging dogs. *PLoS ONE* **2018**, *13*, e0197328. [CrossRef]
21. Coppinger, R.; Coppinger, L. *What is a Dog?* University of Chicago Press: Chicago, IL, USA, 2016.
22. Matter, H.C.; Daniels, T.J. Dogs, Zoonoses and Public Health. In *Dog Ecology and Population Biology*; McPherson, C.N.L., Meslin, F.X., Wandeler, A.I., Eds.; CABI: Wallingford, UK, 2000; pp. 17–62.
23. Coppinger, R.; Coppinger, L. *Dogs: A New Understanding of Canine Origin, Behavior and Evolution*; University of Chicago Press: Chicago, IL, USA, 2002.

24. Pal, S.K. Reproductive behaviour of free-ranging rural dogs in West Bengal, India. *Acta Theriol.* **2003**, *48*, 271–281. [CrossRef]
25. Lord, K.; Feinstein, M.; Smith, B.; Coppinger, R. Variation in reproductive traits of members of the genus Canis with special attention to the domestic dog (*Canis familiaris*). *Behav. Proc.* **2013**, *92*, 131–142. [CrossRef]
26. Sacks, B.N.; Brown, S.K.; Stephens, D.; Pedersen, N.C.; Wu, J.T.; Berry, O. Y chromosome analysis of dingoes and Southeast Asian village dogs suggests a Neolithic continental expansion from Southeast Asia followed by multiple Austronesian dispersals. *Mol. Biol. Evol.* **2013**, *30*, 1103–1118. [CrossRef] [PubMed]
27. Thomson, P.C. The behavioural ecology of dingoes in north-western Australia. II. Activity patterns, breeding season and pup rearing. *Wildl. Res.* **1992**, *19*, 519–529. [CrossRef]
28. Seksel, K. Puppy socialization classes. *Vet. Clin. N. Am. Small* **1997**, *27*, 465–477. [CrossRef]
29. Gazzano, A.; Mariti, C.; Alvares, S.; Cozzi, A.; Tognetti, R.; Sighieri, C. The prevention of undesirable behaviors in dogs: Effectiveness of veterinary behaviorists' advice given to puppy owners. *J. Vet. Behav.* **2008**, *3*, 125–133. [CrossRef]
30. Howell, T.J.; King, T.; Bennett, P.C. Puppy parties and beyond: The role of early age socialization practices on adult dog behavior. *Vet. Med. Res. Rep.* **2015**, *6*, 143. [CrossRef]
31. Řezáč, P.; Viziová, P.; Dobešová, M.; Havlíček, Z.; Pospíšilová, D. Factors affecting dog–dog interactions on walks with their owners. *Appl. Anim. Behav. Sci.* **2011**, *134*, 170–176. [CrossRef]
32. Ward, C.; Bauer, E.B.; Smuts, B.B. Partner preferences and asymmetries in social play among domestic dog, Canis lupus familiaris, littermates. *Anim. Behav.* **2008**, *76*, 1187–1199. [CrossRef]
33. Slabbert, J.M.; Rasa, O.A.E. Observational learning of an acquired maternal behaviour pattern by working dog pups: An alternative training method? *Appl. Anim. Behav. Sci.* **1997**, *53*, 309–316. [CrossRef]
34. Foyer, P.; Wilsson, E.; Jensen, P. Levels of maternal care in dogs affect adult offspring temperament. *Sci. Rep.* **2016**, *6*, 19253. [CrossRef]
35. Malm, K. Regurgitation in relation to weaning in the domestic dog: A questionnaire study. *Appl. Anim. Behav. Sci.* **1995**, *43*, 111–122. [CrossRef]
36. Pal, S.K. Parental care in free-ranging dogs, *Canis familiaris*. *Appl. Anim. Behav. Sci.* **2005**, *90*, 31–47. [CrossRef]
37. Scott, J.P.; Fredericson, E.; Fuller, J.L. Experimental exploration of the critical period hypothesis. *Personality* **1951**, *1*, 162–183.
38. Gácsi, M.; Győri, B.; Miklósi, Á.; Virányi, Z.; Kubinyi, E.; Topál, J.; Csányi, V. Species-specific differences and similarities in the behavior of hand-raised dog and wolf pups in social situations with humans. *Dev. Psychobiol.* **2005**, *47*, 111–122. [CrossRef]
39. Parker, H.G.; Dreger, D.L.; Rimbault, M.; Davis, B.W.; Mullen, A.B.; Carpintero-Ramirez, G.; Ostrander, E.A. Genomic analyses reveal the influence of geographic origin, migration, and hybridization on modern dog breed development. *Cell Rep.* **2017**, *19*, 697–708. [CrossRef] [PubMed]
40. Wilsson, E.; Sundgren, P.E. The use of a behaviour test for the selection of dogs for service and breeding, I: Method of testing and evaluating test results in the adult dog, demands on different kinds of service dogs, sex and breed differences. *Appl. Anim. Behav. Sci.* **1997**, *53*, 279–295. [CrossRef]
41. King, T.; Marston, L.C.; Bennett, P.C. Describing the ideal Australian companion dog. *Appl. Anim. Behav. Sci.* **2009**, *120*, 84–93. [CrossRef]
42. Kustritz, M.V.R. Reproductive behavior of small animals. *Theriogenology* **2005**, *64*, 734–746. [CrossRef]
43. Lezama-García, K.; Mariti, C.; Mota-Rojas, D.; Martínez-Burnes, J.; Barrios-García, H.; Gazzano, A. Maternal behaviour in domestic dogs. *Int. J. Vet. Sci. Med.* **2019**, *7*, 20–30. [CrossRef]
44. Riedman, M.L. The evolution of alloparental care and adoption in mammals and birds. *Q. Rev. Biol.* **1982**, *57*, 405–435. [CrossRef]
45. Boitani, L.; Ciucci, P.; Ortolani, A. Behaviour and social ecology of free-ranging dogs. In *The Behavioural Biology of Dogs*; Jensen, P., Ed.; CAB International: Wallingford, UK, 2007; pp. 147–165.
46. Pal, S.K.; Ghosh, B.; Roy, S. Dispersal behaviour of free-ranging dogs (*Canis familiaris*) in relation to age, sex, season and dispersal distance. *Appl. Anim. Behav. Sci.* **1998**, *61*, 123–132. [CrossRef]
47. Scott, J.P. The process of primary socialization in canine and human infants. *Monogr. Soc. Res. Child Dev.* **1963**, *28*, 1–47. [CrossRef] [PubMed]

48. Duxbury, M.M.; Jackson, J.A.; Line, S.W.; Anderson, R.K. Evaluation of association between retention in the home and attendance at puppy socialization classes. *J. Am. Vet. Med. Assoc.* **2003**, *223*, 61–66. [CrossRef] [PubMed]
49. Dendoncker, P.A.; De Keuster, T.; Diederich, C.; Dewulf, J.; Moons, C.P. On the origin of puppies: Breeding and selling procedures relevant for canine behavioural development. *Vet. Record.* **2019**, *184*, 710. [CrossRef] [PubMed]
50. Fisher, A.E. The Effects of Differential Early Treatment on the Social and Exploratory Behavior of Puppies. Ph.D. Thesis, Pennsylvania State University, University Park, PA, USA, 1955.
51. Howell, T.J.; Bennett, P.C. Puppy power! Using social cognition research tasks to improve socialization practices for domestic dogs (*Canis familiaris*). *J. Vet. Behav.* **2011**, *6*, 195–204. [CrossRef]
52. Wormald, D.; Lawrence, A.J.; Carter, G.; Fisher, A.D. Analysis of correlations between early social exposure and reported aggression in the dog. *J. Vet. Behav.* **2016**, *15*, 31–36. [CrossRef]
53. Lenkei, R.; Pogány, Á.; Fugazza, C. Social behavior in dog puppies: Breed differences and the effect of rearing conditions. *Biol. Fut.* **2019**, *70*, 134–142. [CrossRef]
54. Lehoczki, F.; Szamosvölgyi, Z.; Miklósi, Á.; Faragó, T. Dogs' sensitivity to strange pup separation calls: Pitch instability increases attention regardless of sex and experience. *Anim. Behav.* **2019**, *153*, 115–129. [CrossRef]
55. Smith, M.S.; Mc Donald, L.E. Serum levels of luteinizing hormone and progesterone during the estrous cycle, pseudopregnancy and pregnancy in the dog. *Endocrinology* **1974**, *94*, 404–412. [CrossRef]
56. Schradin, C.; Vuarin, P.; Rimbach, R. The neoteny-helper hypothesis: When to expect and when not to expect endocrine mechanisms to regulate allo-parental care? *Physiol. Behav.* **2018**, *193*, 127–134. [CrossRef]
57. Pal, S.K. Urine marking by free-ranging dogs (*Canis familiaris*) in relation to sex, season, place and posture. *Appl. Anim. Behav. Sci.* **2003**, *80*, 45–59. [CrossRef]
58. Pal, S.K.; Ghosh, B.; Roy, S. Agonistic behaviour of free-ranging dogs (*Canis familiaris*) in relation to season, sex and age. *Appl. Anim. Behav. Sci.* **1998**, *59*, 331–348. [CrossRef]
59. McLeod, P.J. Infanticide by female wolves. *Can. J. Zool.* **1990**, *68*, 402–404. [CrossRef]

animals

MDPI

Article

Salivary Vasopressin as a Potential Non–Invasive Biomarker of Anxiety in Dogs Diagnosed with Separation–Related Problems

Federica Pirrone [1,*], Ludovica Pierantoni [2], Andrea Bossetti [1], Stefania Uccheddu [3] and Mariangela Albertini [1]

1 Department of Veterinary Medicine, University of Milan, via Celoria 10, 20133 Milan, Italy; andrea.bossetti@gmail.com (A.B.); mariangela.albertini@unimi.it (M.A.)
2 Veterinary Behaviour & Consulting Services at CAN Training Centre, 80128 Naples, Italy; ludovica.pierantoni@gmail.com
3 Vet Ethology, 3090 Overijse, Belgium; uccheddus@gmail.com
* Correspondence: federica.pirrone@unimi.it

Received: 24 October 2019; Accepted: 19 November 2019; Published: 26 November 2019

Simple Summary: Oxytocin and vasopressin have been shown to have opposite effects on the expression of anxiety and fear responses in rodents. In the present study, we analyzed the salivary fluctuations of these neuropeptides in both behaviorally normal dogs and dogs with separation distress in response to a three-minute separation from the owner, in a new environment. Dogs with a previous diagnosis of separation distress showed more anxiety-related behaviors and higher concentrations of vasopressin than control dogs when separated from the owner. Further research is needed on the potential use of salivary vasopressin as an early, non-invasive biomarker of anxiety-related disorders in pet dogs.

Abstract: Physiological biomarkers of canine anxiety have not been extensively investigated to date. To identify new biomarkers in dogs, we compared behaviorally normal dogs (Control group, N = 13) to dogs diagnosed with separation problems (Case group, N = 13) as they were introduced into a novel environment in the presence of two strangers and subjected to a short episode of separation and reunion with the owner. During the separation phase, dogs in the Case group explored significantly less than controls and were significantly more persistent in expressing passive stress-coping strategies aimed at seeking proximity to their owners. When the owners returned, dogs with separation distress spent significantly more time jumping up on the strangers than control dogs did. Salivary oxytocin and vasopressin concentrations did not differ between samples taken before and after the separation. However, vasopressin concentrations immediately after separation were significantly higher in the Case than in the Control group and remained higher, although not significantly so, 10 min later. These results indicated that dogs with separation distress became more anxious than typical dogs when separated from their owner in an unfamiliar environment and provided preliminary support for the use of salivary vasopressin as a possible biomarker for anxiety-related responses in dogs.

Keywords: dogs; separation distress disorder; vasopressin; oxytocin; anxiety

1. Introduction

Both humans and dogs are highly social mammals who develop selective forms of sociality, in which are embedded lasting relationships defined as social bonds or attachments [1]. An attachment bond may also link dogs and owners [2], the latter becoming the animals' reference point in the environment, influencing their welfare [3]. The attachment relationship between dogs and owners is one of the primary reasons why people keep dogs as companion animals [4].

Studies throughout the last decade have explored dogs' attachment-related behavioral reactions to separation from and reunion with the owner [2,3,5], while little is known about concomitant physiological reactions. Physiological changes, which include increased secretion of glucocorticoids and/or adrenocorticotropic hormone (ACTH), as well as altered cardiovascular or immune parameters, have been described in dogs during laboratory testing [4,6]. Increased circulating levels of nerve growth factor (NGF) have been directly associated with psychosocial stress in human and animal models [7,8]. However, in both rodents and humans, the closely related neuropeptides oxytocin (OT) and vasopressin (AVP) have been associated with distinct, partly opposite roles in social behavior, as well as in stress, fear, and anxiety responsiveness following social separation [9–11]. In particular, OT has been found to attenuate anxiety, central fear responses, and neuroendocrine reactivity [12], while stimulation of the AVP system has been shown to lead to augmented anxiety and fear expression [10] and increased neuroendocrine stress response [12].

Recently, some studies have begun to focus on OT in affiliative interactions in dog-human dyads [13,14]. We are also aware of two studies investigating the relationships between AVP and behavior in dogs, both of which revealed positive associations with fear and aggression [15,16].

To the best of our knowledge, there are no published data on the roles of these neuropeptides in anxiety-like behavior in dogs.

This study sought to examine the behavior and OT and AVP fluctuations in dogs in response to a commonly occurring human-dog interaction that might potentially induce anxiety in dogs. Specifically, we compared two groups of pet dogs that were separated from their owners for 3 min and then reunited with them, all while they were in a novel environment and the presence of two unfamiliar people. The Control group consisted of behaviorally normal dogs, while dogs diagnosed with separation-related problems (SRP) formed the Case group. Separation-related problems are described as physical, physiological, and/or behavioral signs of the distress exhibited by the dog only in the absence of, or lack of access to, the owner [17].

Approximately 14%–20% of dog patients [18,19] from general veterinary practices show signs of SRP in their owners' absence [20], and the anxiety emotional system is recognized as one possible cause. However, as with other anxiety disorders in veterinary patients, underdiagnosis of anxiety-dependent separation problems has been previously reported [21] due to misinterpretation of normal and pathological anxiety [22] and the lack of recognition of mild clinical signs by owners [23]. Therefore, unsurprisingly, according to Karagiannis et al. [24], it is suspected that up to 50%–56% of the overall dog population may actually display clinical symptoms of SRP at some point in their life [25,26], which, out of the total population of approximately 160 million dogs in the US and Europe [27,28] represents approximately 85 million dogs with SRP on the two continents. Underdiagnosis of SRP suggests a need to identify putative measurable markers that are specific to these behaviors, which would allow early diagnosis and intervention. Since separation-related problems are one of the primary cited reasons for the relinquishment of dogs to animal shelters [29,30], the relevant practical implications make research in this direction unquestionably worthwhile.

To assess the levels of relevant hormones, we examined their concentrations in saliva. A recent study by MacLean [31] validated salivary measures of OT and AVP in dogs. Because saliva collection is simple and painless, it is better suited for assessments in dogs than other methods, such as using plasma or urine, which present a host of challenges related to design, analysis, and welfare [31]. However, it is currently unknown whether these measures respond dynamically to anxiety-related aspects of human-dog interaction.

From previous findings [4], we hypothesized that separation from the owner in our experimental setting would not lead to different behavioral responses from dogs in the two groups. In fact, using an experimental protocol similar to ours, Parthasarathy and Crowell–Davis [4] found that, regardless of their separation anxiety status, all of the dogs displayed behavioral signs of increased anxiety when their owners left the room. Conversely, from observations in rodent and human models, we expected differences in endocrine responses during the separation phase, with a lower level of OT and a higher concentration of AVP in the Case group than in the Control group.

2. Materials and Methods

The Animal Welfare Committee (OPBA) of the University of Milan approved this research (OPBA_106_2018). All methods were performed in accordance with the current European and Italian legislation.

2.1. Participants and Study Setting

This study was part of a research project designed to discover novel potential biomarkers of stress and anxiety in pet dogs brought by their owner to the Veterinary Behavior and Consulting Services at CAN (Comportamento Animale Napoli) Training Center in Naples, Italy, for signs of fear or anxiety-related conditions. Initially, the owners of all dogs were asked, through a brief telephone interview, the reason for requesting the behavioral visit. Dogs were then tested, and, for this study, 13 dogs were selected by simple random sampling from those who had SRP according to the owners and received a diagnosis that confirmed SRP based on the outcome of the behavioral visit (Case group).

We also selected 13 controls from dogs whose owners attended the training center for the first time to get information on the service of dog's night board and daycare. The target sample size for the study was determined on the basis of a power analysis with 80% power, a large effect size (0.50), and an α value (error rate) of 0.05. The selected controls, who were rated as behaviorally typical based on the outcome of the visit purposely performed to include them in the present study, were similar to the cases with respect to the owners' gender and the dogs' sex, age, origin, and breed type (Supplementary Table S1). The dogs were 12 females (of which 4 were spayed), and 14 males (of which 6 were neutered), either pure or mixed breed. The mean age of the dogs, in years, was 3.7 (SE = 0.4, range: 1–8). Moreover, cases and controls had belonged to their current owners for at least one year to ensure that they had a reasonable amount of time to form a relationship with the owners and were an adult at the time of the study (1–10 years). All dogs underwent a physical examination. The Case group also had blood tests done. Those who were healthy for the veterinarian and not in treatment for behavioral or physical problems at the time of data collection were included in the study. Exclusion criteria included estrus, pregnancy, and nursing, as well as a history or present signs of stranger-directed aggression or fear of strangers or novel environments.

The experimenter explained the general objective of the research to all owners, who signed written informed consent and provided their assent to participate in the study. By agreeing to participate, owners confirmed that they were informed of a cameraperson's presence for the videotaping procedure and acknowledged that, in compliance with the General Data Protection Regulation (GDPR) (EU) 2016/679, their data and video recordings would be stored on the principal investigator's computer at the Department of Veterinary Medicine at the University of Milan and otherwise processed by the investigators of this study, with all ensuring an adequate level of data protection. Informed consent was obtained from the owners of dogs whose information, images, and/or videos would be published in an online journal, which could lead to their identification. Testing took place during the behavioral visit in a 300 m^2 outdoor arena (Figure 1), containing three chairs (for the owner, the veterinary behaviorist, and a cameraperson), a bowl for fresh water, and some toys. The arena had four sides and was enclosed with chain-link fencing approximately 2 m tall; one side of the arena included an entrance gate. The duties of the veterinary behaviorist and the cameraperson were always performed by the same two women, who had never met the dogs before. The arena was located in a larger area. For standardization, and to minimize distractions and background noises, which could have acted as confounders, all dogs were tested in mild weather conditions and the late afternoon (between 1630 and 1830), after the center's planned closure, so no other social stimuli were present.

Figure 1. The spatial arrangement of the test. A moment during the session with one of the dogs. The owner and the two strangers (the veterinary behaviorist and the camerawoman) are visible. Drawing by Valentina Sammartano.

2.2. Study Protocol

Owners were instructed not to provide their dogs with food or exercise 1.5 hours prior to the start of the test. A simplified version of the Ainsworth's strange situation test (ASST) was employed. During the test, the owner and veterinary behaviorist could talk to each other and interact with the dog only if he/she was seeking their attention. The cameraperson was instructed not to interact with the dog or with the veterinary behaviorist and owners to avoid reinforcing attention-seeking behaviors and to control for the possibility that the strangers would act differently around different dogs [4]. The owners were told that if at any point they were concerned about their dogs' welfare or safety, they could stop the test. The entire procedure comprised three steps: An acclimatization phase, a separation phase, and a reunion phase.

Acclimatization phase (10 min). Prior to the short separation task, the owner, the dog, the veterinary behaviorist, and the camerawoman entered the arena. The dog was left unleashed and free to explore the environment, while the owner and the strangers remained seated in the chairs. Saliva was collected from the dog at the end of the tenth minute (T0).

Separation phase (3 min). Immediately afterward, the owner left the arena, where the dog remained in the company of the two strangers. During the ASST, the veterinary behaviorist attempted to engage the dog in friendly interaction, including gently petting him/her and speaking to him/her in a calm tone (Supplementary Video S1). However, the veterinary behaviorist allowed the dog to lead these interactions, and dogs were always free to disengage and move away from her. If a dog exhibited signs of severe distress or anxiety, the owner was asked to come back, and the testing stopped (Supplementary Video S2). At the end of the third minute, the owner returned to the arena, sat in the same chair as before, and made conversation with the veterinary behaviorist, and the second sample of saliva was collected (T1).

Reunion phase (10 min). Immediately upon returning, the owner was allowed to respond to his/her dog's greeting by interacting both verbally and physically in a calm way. After ten minutes, saliva was collected (T2), and the test ended.

2.3. Parameters Recorded: Behavioral Responses

Observer-blind analysis of behavior was carried out with focal animal sampling and continuous recording using the Observer XT software package (Noldus Information Technology, 6702 EA Wageningen, The Netherlands). Another coder, expert in animal behavior but unfamiliar with the aims and conditions of the study, verified the reliability of the coding in 20% of the videos. A Cronbach's α of 0.80 or higher was considered acceptable for this study. According to a study by Mariti et al. [2], the behaviors were divided into social and non–social, and each social behavior towards the owner and the strangers was analyzed (Table 1). Behavior definitions were formulated on the basis of a literature review [2–4,32–37]. As for social behaviors, attention-seeking and proximity were grouped to create the category spontaneous interactions. During the separation phase, we also recorded the interactions solicited by the veterinary behaviorist from the dogs, which included talking to and petting the dogs to comfort them if they showed signs of distress. Behavioral variables were measured in terms of relative frequency (the number of occurrences per minute) and/or duration (time spent on a behavior, expressed in seconds) of occurrence during each observation period.

Table 1. List of behaviors and definitions used in the study. F = frequency (number of occurrences); D = duration (s).

Behaviors	Description	Measured Values (F/D)
Social behaviors		
Jumping up	Both of the dog's forelegs were out of contact with the ground, regardless of the position of the hind legs; the dog was in proximity to a person. The dog might also be entirely on a person's lap	F, D
Spontaneous interactions	Staying close to and seeking attention and physical contact (nuzzling or pawing for attention, soliciting petting) from the owner or the stranger	F, D
Mounting	Sexual mounting of people or inanimate objects	F, D
Non–social behaviors		
Explore	Activity directed towards physical aspects of the environment, including sniffing, visual inspection, and gentle licking	F, D
Individual play	Any behavior performed vigorously or at a galloping gait and directed towards an object when clearly not interacting with any human; these play behaviors included chewing, biting, shaking from side to side, scratching or batting with the paw, chasing rolling balls, and tossing objects using the mouth	F, D
Standing by the fence	Standing close to the fence (<1 m), regardless of whether the face was oriented towards the exit	F, D
Attention oriented towards the fence	Staring fixedly at the fence, either when close to it or from a distance	F, D
Behaviors oriented towards the fence	All activities, resulting in physical contact with the fence, including scratching the gate with the paws, jumping on the fence, and pulling on the fence with the forelegs or mouth.	F, D
Restlessness	A feeling of agitation expressed by continual motion; changing the state of locomotion; digging (scratching the floor with the forepaws in a way that is similar to when dogs are digging holes)	F
Drinking	Taking in fluids by lapping up water from the bowl with the tongue	F
Whining	High-pitched vocalization	F
Pacing	Increased motor activity, walking or running around without exploring the environment	F

2.4. Parameters Recorded: Endocrine Responses

We collected saliva samples from dogs using commercially available swabs (SalivaBio Children's Swab, Salimetrics, Carlsbad, CA, USA). All samples were taken by the veterinary behaviorist. The swab was gently placed into the cheek or under the tongue of the dog for approximately 60 s, without the restraint of the animal. The dog's salivation was stimulated by presenting the odor of food treats. The dog received a food treat only after the last saliva sample was taken because the consumption of food immediately before sample collection has been found to affect OT and AVP measurements [31]. Each sample was replaced in the device tube and closed with a plastic stopper to prevent evaporation. The collected material was refrigerated at 4 °C and then stored at −20 °C immediately after it arrived at the laboratory. At the time of analysis, the samples were thawed at room temperature and centrifuged according to the protocol for salivary samples. The laboratory technician who performed these analyses was blinded to the hypotheses and conditions. All samples were analyzed by enzyme-linked immunosorbent assay (ELISA) following previously validated protocols [31,34]. For the measurement of OT and AVP, we used commercially available enzyme-linked immunosorbent assay kits from Arbor Assays (Ann Arbor, MI, USA) and MyBiosource Inc. (San Diego, CA, USA). Each sample was prepared in duplicate, and concentrations were calculated using a Labisystem Multiskan Ex (Nepean, ON, Canada) microplate reader according to the relevant standard curves.

The mean recovery was 102.8% ± 10.8 for OT and 94.3 ± 2.2% for AVP. The average intra- and inter-assay coefficients of variation, respectively, were 4.7% and 8.8% for OT and 5.7% and 6.5% for AVP. The assay sensitivity was 17 pg/mL and 1 pg/mL for OT and AVP, respectively.

The laboratory technician was blinded to the hypotheses and conditions.

2.5. Statistical Analysis

Due to the number of animals and the distribution of the data, non-parametric statistics were used to analyze the behavioral and hormonal data [38,39]. Differences between groups were analyzed using the Mann–Whitney *U* test, while the Kruskal–Wallis test for multiple comparisons was conducted to compare behaviors within each group. A post hoc Mann–Whitney *U* test with the Bonferroni correction followed the Kruskal–Wallis test in case a significant effect was detected. The Friedman test for paired samples was used to test the difference in endocrine parameters among time points. In addition, OT and AVP concentrations in the two groups of dogs were compared using Mann–Whitney *U* tests. The OT and AVP concentrations and the duration and relative frequency of behaviors were presented as median. *p*-values ≤ 0.05 were deemed statistically significant. Statistical analyses were performed with IBM SPSS Statistics 25.0.

3. Results

3.1. Behavioral Responses

Intra-observer and inter-observer reliability were confirmed, with a Cronbach's α of 0.995 and 0.997, respectively. No significant differences were found in behavioral responses during the acclimatization phase between groups (Mann–Whitney *U* test, $p > 0.05$). As shown in Figures 2 and 3, in this phase, attention directed towards the fence and exploring were the most frequent behaviors and had the greatest total duration in both Case (duration: Kruskal–Wallis test, $\chi^2 = 52.803$, $p = 0.001$; relative frequency: Kruskal–Wallis test, $\chi^2 = 93.552$, $p = 0.001$) and Control dogs (duration: Kruskal–Wallis test, $\chi^2 = 37.895$, $p = 0.001$; relative frequency: Kruskal–Wallis test, $\chi^2 = 97.349$, $p = 0.001$). Behaviors oriented towards the fence and standing by the fence were not sufficiently expressed to be analyzed in this phase.

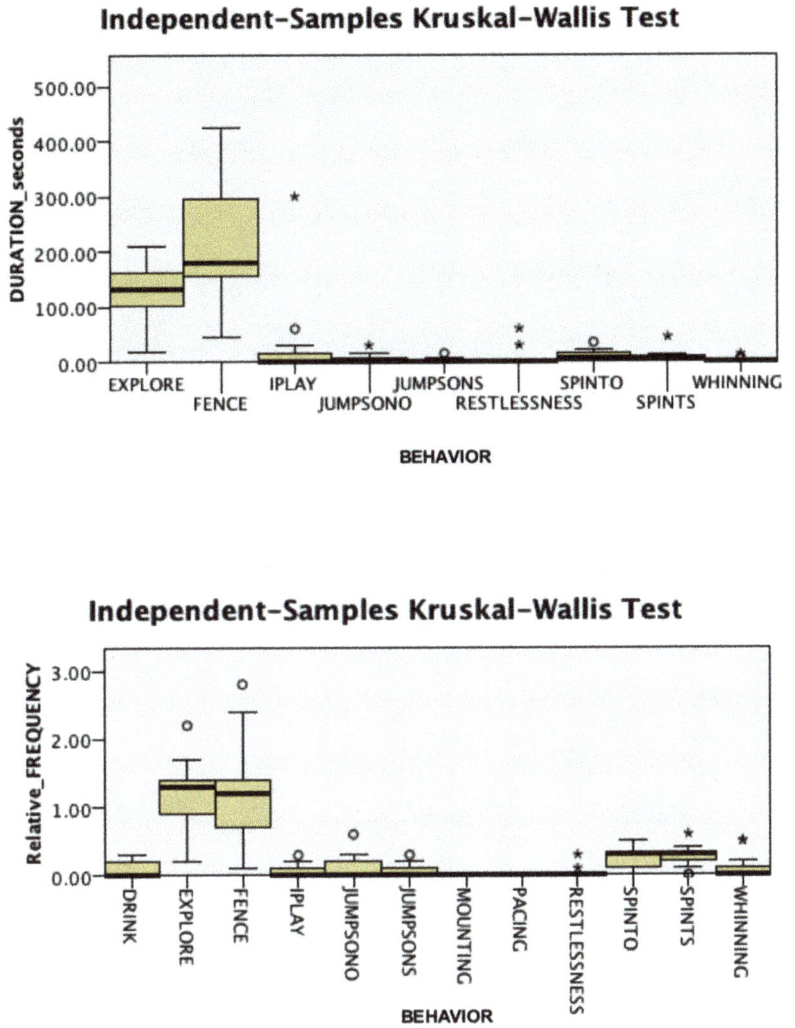

Figure 2. Duration and relative frequency, expressed as the number of occurrences per minute, of the behaviors observed during the acclimatization period in the Case group (N = 13). FENCE: Attention oriented to the fence; IPLAY: Individual play; JUMPSONO: Jumps on owner; JUMPSONS: Jumps on the stranger; SPINTO: Spontaneous interactions with the owner; SPINTS: Spontaneous interactions with a stranger.

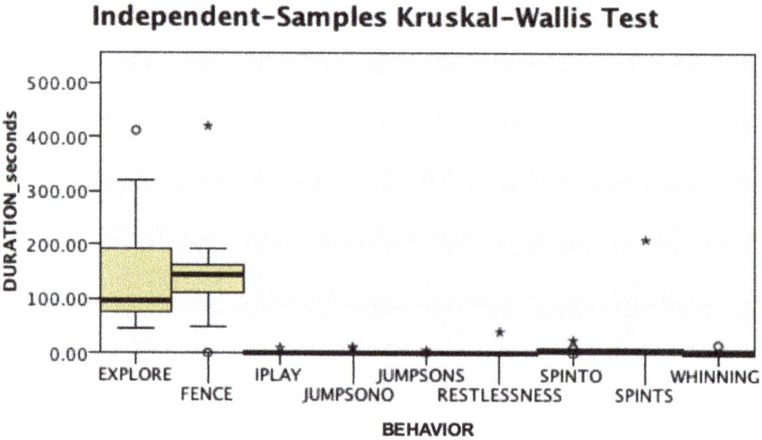

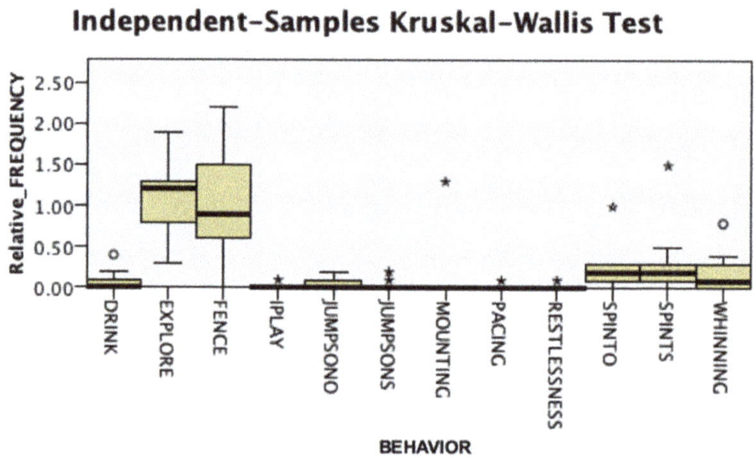

Figure 3. Duration and relative frequency, expressed as the number of occurrences per minute, of the behaviors observed during the acclimatization period in the Control group (N = 13). FENCE: Attention oriented to the fence; IPLAY: Individual play; JUMPSONO: Jumps on owner; JUMPSONS: Jumps on the stranger; SPINTO: Spontaneous interactions with the owner; SPINTS: Spontaneous interactions with a stranger.

The groups differed significantly during the separation phase (Figures 4 and 5). Dogs in the Case group showed significantly less exploration (duration: 0 vs. 6, Case vs. Control; Mann–Whitney U = 122.500, $p = 0.05$; relative frequency: 0 vs. 0.33, Mann–Whitney U = 125.000, $p = 0.039$) and spent significantly more time standing by the fence (duration: 260 vs. 151, Case vs. Control; Mann–Whitney U = 36.000, $p = 0.012$, Figure S1) than Control dogs. In addition, at T1, we recorded significantly more frequent attempts by the veterinary behaviorist to interact with Case dogs (relative frequency: 2 vs. 0.33, Case vs. Control; Mann–Whitney U = 40.000, $p = 0.022$) than with controls. During dog-owner reunion, Case dogs were significantly more persistent than dogs from the Control group in jumping up on the stranger (duration: 10 vs. 0, Case vs. Control; Mann–Whitney U = 185.000, $p = 0.004$; Supplementary Video S3), with the veterinary behaviorist as the only target of this behavior.

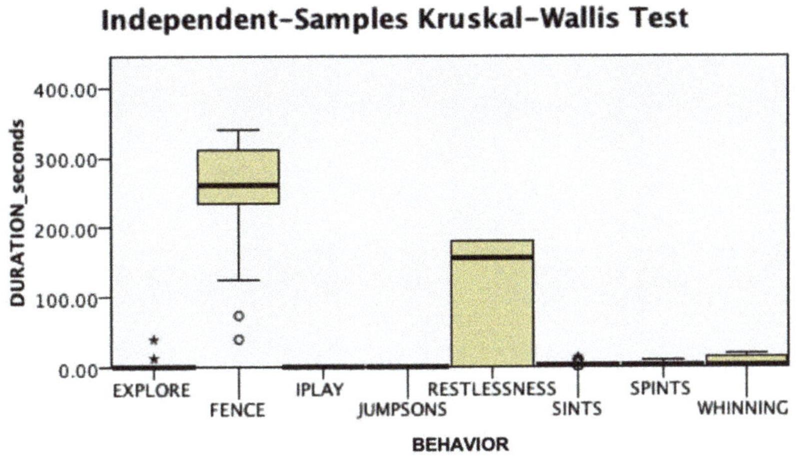

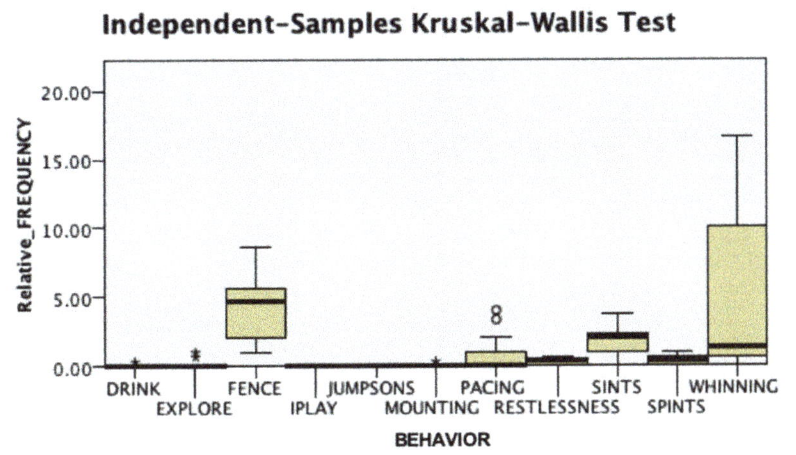

Figure 4. Duration and relative frequency, expressed as the number of occurrences per minute, of the behaviors observed during the separation period in the Case group (N = 13). Fence: Attention oriented to the fence; IPLAY: Individual play; JUMPSONO: Jumps on owner; JUMPSONS: Jumps on the stranger; SPINTO: Spontaneous interactions with the owner; SPINTS: Spontaneous interactions with a stranger.

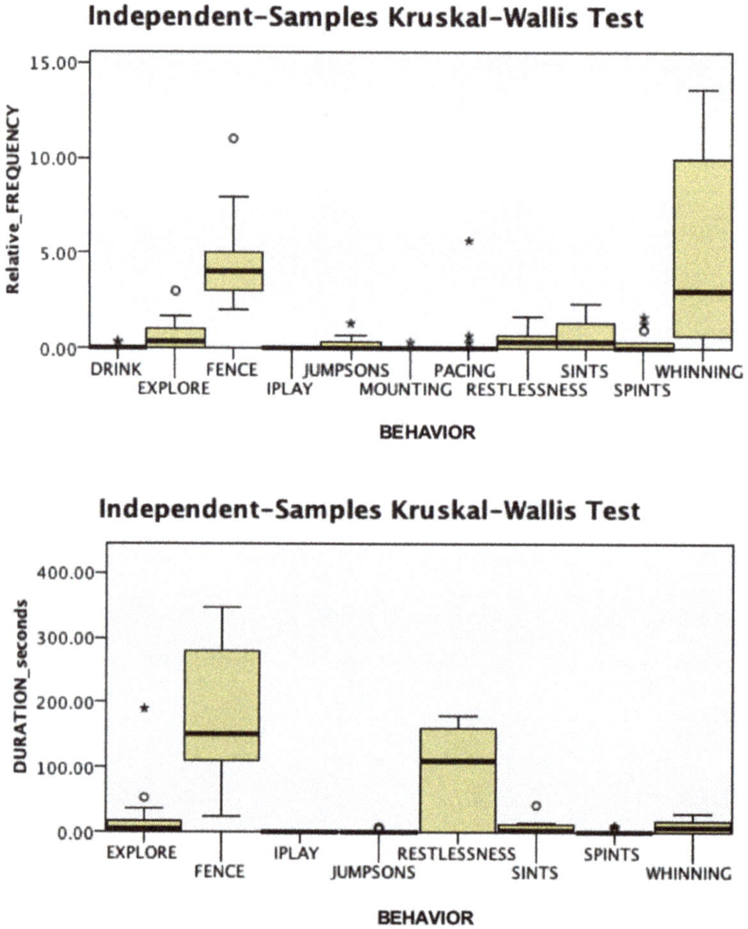

Figure 5. Duration and relative frequency, expressed as the number of occurrences per minute, of the behaviors observed during the separation period in the Control group (N = 13). FENCE: Attention oriented to the fence; IPLAY: Individual play; JUMPSONO: Jumps on owner; JUMPSONS: Jumps on the stranger; SPINTO: Spontaneous interactions with the owner; SPINTS: Spontaneous interactions with a stranger.

3.2. Endocrine Responses

As shown in Table 2 and Figure 6, the analysis of salivary OT and AVP revealed non-significant differences among time points within each group. However, at T1, AVP concentrations were significantly higher in the Case group than in the Control group. In addition, there was a trend, although not statistically significant, towards lower OT concentrations at T1 and T2 and higher AVP concentrations at T2 in Case dogs than in controls.

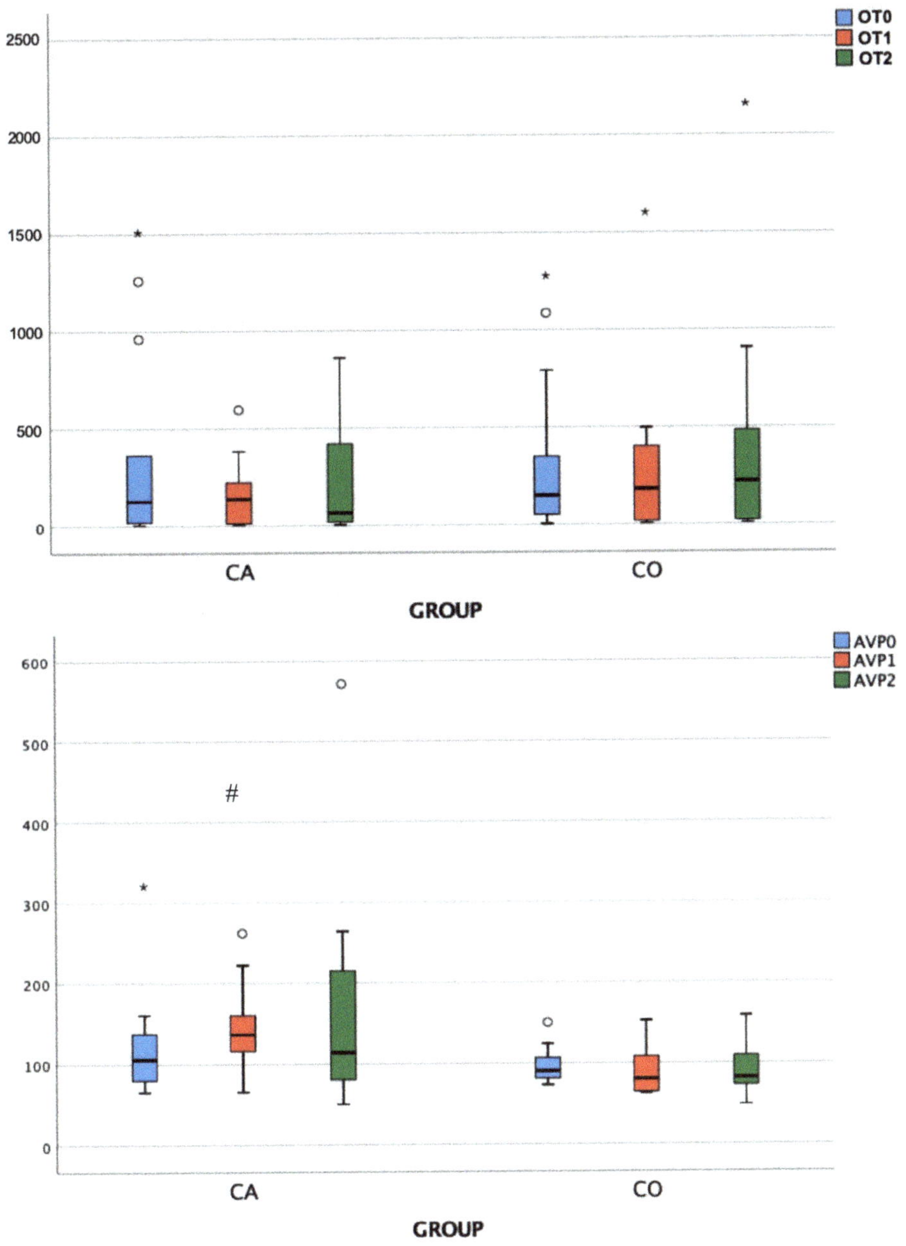

Figure 6. Concentrations of oxytocin (OT) and vasopressin (AVP) measured in saliva before (T0), immediately (T1), and 10 min after (T2) separation from the owner. CA: Case group, N = 13. CO: Control group, N = 13. Salivary AVP concentrations showed a statistically significant difference between groups at T1 ($^{\#}$, $p < 0.05$).

Table 2. Salivary concentrations of oxytocin (OT) and vasopressin (AVP).

Group	OT0	OT1	OT2	Friedman test	
				χ2	P
Case	127.87	138.79	67.04	3.231	0.199
Control	149.99	183	221.60	0.462	0.794
Mann–Whitney U test	86	98	100		
Mann–Whitney U test p	0.960	0.511	0.448		
Group	**AVP0**	**AVP1**	**AVP2**	**Friedman test**	
				χ2	P
Case	105.97	136.61	114	2.923	0.232
Control	90.40	80.12	81.53	3.231	0.199
Mann–Whitney U test	71	29	55		
Mann–Whitney U test p	0.511	0.003	0.139		

4. Discussion

This study aimed to determine whether there were differences in specific behaviors and endocrine responses between dogs affected by separation distress at home and behaviorally normal dogs while experiencing a potentially anxiogenic situation (separation from the owner in a novel environment and the presence of two strangers). We found that during the initial acclimatization phase, when the owner was still present, dogs with SRP behaved similarly to normal dogs. They were mainly explorative, as reflected by the higher frequency and longer duration of time spent in investigating the physical environment investigation time than in the other behaviors. Thus, dogs in both groups exhibited a similar attachment style, showing that they viewed the attachment figure (the owner) as a secure base for exploration of the novel environment [2]. This finding did not conform to the traditional vision, which considers dogs with separation distress as having excessive attachment towards their owners [40,41]. However, it agreed with what was more recently reported by Parthasarathy and Crowell–Davis [4], namely, that separation distress is not correlated with hyper attachment, confirming that these dogs were confident with the novel environment and strangers. Regarding the behavioral responses observed during the separation episodes, contrary to what was expected based on the results by Parthasarathy and Crowell–Davis [4], the patterns of behavior were different for the two groups in our study. Dogs in the Case group were significantly less explorative and more persistent in passive behaviors aimed at maintaining proximity to the owners, such as staying near the fence by the exit, than controls. According to Topál et al. [37], who analyzed the attachment relationship of behaviorally typical dogs with their owners, the fact that, during the separation phase, this behavior was not reduced by the presence of a stranger, despite her attempts to relate positively, might suggest that dogs with SRP have an especially strong preference for their owners in stress situations. This searching response has been observed in children [42] and non-human primates [43,44] and is regarded as aimed at maintaining the comforting bond of attachment. Nevertheless, dogs in the study by Topál et al. [45] showed a tendency to seek and maintain contact with the returning owner but not the stranger. Conversely, in our study, dogs with separation distress tended to show greater persistence in investigating the veterinary behaviorist (as suggested by more time spent jumping up) during post-separation reunion than did Control (CO) dogs, and this could also be seen as indirect evidence that these dogs experienced a higher level of anxiety during the separation. In fact, although we could not exclude the possibility that this was a side effect of dogs with separation distress needing to spend more time in proximity to the owner, who was sitting near the stranger, another possible explanation could be put forward. The veterinary behaviorist was the stranger who had made attempts to relate to dogs during the separation phase when proximity to their attachment figure was no longer possible. Those efforts might have made her a target of exploration for the SRP dogs in the reunion phase when contact with the attachment figure was re-established. This would also explain why the dogs' jumping up behavior was directed only towards the veterinary behaviorist and not the camerawoman. Indeed, even if our experimental design was supposed to lead to calmer dogs compared to that used by Topal et al. [45] (e.g. a 10 vs. 4 minutes acclimatization period; dog vs. stranger initiated contact in the separation phase), dogs in the study by Topal et al. [45] were all behaviorally normal, and therefore less prone to anxiety than our Case dogs. More research is needed to understand whether behavioral reactions of a dog while alone (such as reduced exploration and increased seeking of proximity to the owner), or even after the owner's return, might be used as effective indicators of the presence of separation distress disorders.

Notably, salivary AVP concentrations were significantly higher in Case dogs than Control dogs immediately after the end of the separation period (T1). The timing with which OT and AVP reach saliva is not well understood, but they appear faster than other salivary hormones (e.g., cortisol), which reach peak concentrations in saliva ~10 min after those in blood [31]. Previous studies have shown effects at a minimum time delay of 10 min [13,31]. However, other studies revealed early increases in salivary OT concentrations in nursing mothers and dams [31,46]. In the study by MacLean et al. [31], in particular, a large and statistically significant increase in salivary OT was detected in dams from baseline, at the end of a pre-test separation from their litters, to nursing 3 min later. This effect has been interpreted as reflecting an anticipatory rise in salivary OT. Similarly, it is possible that, in our study, the significant between-group difference observed in salivary AVP at 3 min partially reflected a rapid anticipatory response in SRP dogs. Given that non-social fears, such as fear of novel situations/environments, are common comorbidities of separation anxiety in dogs [17,47], we chose only dogs that did not present signs of stranger-directed aggression or fear of strangers or novel environments. Therefore, jointly with the fact that all dogs acted normal during the acclimatization period, the difference at 3 min was unlikely to be the result of stress at the start of the experiment in a novel environment. It is worth noting that, at T2, Case dogs still had higher concentrations of AVP than Control dogs, although the difference was not statistically significant due to the relatively high standard error of the mean. Although we could not be certain, it is possible that separation from the owner accounted also for this difference in salivary concentrations.

This was an intriguing outcome because central AVP, particularly that released within the amygdala, has been shown to be involved in the generation of passive coping strategies for acute stress in rodent models [37,48] through processes that would be mediated by the V1a and V1b receptors [48].

Both OT and AVP are synthesized in the hypothalamus, primarily in large magnocellular neurons situated in the supraoptic and paraventricular nuclei, and secreted from their axons, which are projected to the neurohypophysis, into the general circulation (for example, during labor or imbalance of water homeostasis) [48]. Vasopressin molecules that have been released in this way are, for the most part, prevented from re-entering the central nervous system via the blood-brain barrier [49]. In parallel, AVP and OT are also secreted within the brain, from the dendrites of the same neurons, in a manner regulated semi-independently of axonal release [50]. Following secretion, these peptides diffuse throughout the extracellular space, serving as neuromodulators for surrounding brain tissue [51]. As above, dendritic secretion of AVP has been shown to be of central importance in animal models of anxiety disorders [48], while axonal secretion has been shown to affect fear responses in mice [52], probably by regulating stress responses through the hypothalamic–pituitary–adrenal axis [53].

In our study, we could detect only an increased peripheral AVP level immediately after separation-induced social stress in SRP dogs compared to normal dogs, but central concentrations of AVP were likely increased as well. In fact, plasma and salivary AVP, measured using ELISA kits, were found to be moderately correlated in humans in one study [54], and plasma AVP concentrations significantly and positively predicted cerebrospinal fluid AVP concentrations in human neonates in another study [55]. Thus, these two studies provided preliminary support for the use of salivary AVP ELISA measurement as a proxy for brain AVP activity, at least, in humans. Future studies are now required to determine the relationship between behavioral measures and AVP concentrations in both the central and peripheral compartments in dogs.

It is worth mentioning that interest in using V1b antagonism to treat anxiety disorders has been investigated. Clinical trials in humans failed, with the V1b receptor antagonist SSR149415 not being useful for the treatment of generalized anxiety disorder [56]. However, two newly synthesized V1b receptor antagonists—TASP0233278 and TASP0390325—have shown potential benefits in rodent models [57]. Future studies could examine whether this therapeutic approach might also benefit dogs diagnosed with SRP. The antidepressants—clomipramine and fluoxetine—which act primarily as serotonin reuptake inhibitors, are currently approved for the treatment of canine separation distress [58]. As more evidence is collected on the role of AVP in the pathophysiology of SRP, future drugs targeting the vasopressinergic system would offer treatment options for canine separation distress therapy.

Finally, although AVP is often anxiogenic, the closely related nonapeptide OT often has anxiolytic effects [59] and may reduce the stress of negative social interactions [60]. The concentrations of OT that we reported here with a sample size of 26 dogs (13 behaviorally normal, 13 with SRP) did not reach statistical significance over time, although a trend towards a reduction (and an increase in AVP concentrations) was observed during and even after the separation from the owner in dogs from the Case group. This could also be a type II error-related false-negative result due to the small sample size. Future research, exploring in a larger sample size whether both OT and AVP respond dynamically to this potentially anxiogenic situation, might help detect a significant effect.

5. Conclusions

In conclusion, this study showed that, when placed in a novel environment, dogs presented different behavioral and endocrine responses to a short separation from the owner, followed by the reunion, depending on whether they suffered from separation distress at home. Dogs with SRP became more distressed than CO dogs when the owner was gone for a short time and left them in the company of two unfamiliar persons. They were less able to mediate their reactions in such a stressful situation, showing more passive coping strategies aimed at seeking proximity to the owner while he/she was absent, and had significantly higher salivary concentrations of AVP at the very end of this phase, which are two responses that have been previously associated in other animal models of social separation. Although these results are preliminary and should, therefore, be interpreted with caution, the differences observed between groups in both behavioral and endocrine responses during and after the separation lead the way to further exploration of the use of salivary AVP as an early, non-invasive biomarker of canine anxiety-related disorders and support AVP antagonism as a potential new mechanism-based therapeutic approach. According to Thielke and Udell [61], improved methods of treating SRP in dogs would not only benefit human-dog relationships but also potentially contribute to the decreased surrender of dogs to shelters by reducing the effort needed to modify this behavior problem successfully.

Supplementary Materials: The following are available online at http://www.mdpi.com/2076-2615/9/12/1033/s1, **Figure S1:** A photo of a dog from the Case group who is standing by the fence during the 3-minute separation phase (T1). Photo by Valentina Sammartano. **Table S1:** Main characteristics of the dogs involved in the study. **Video S1:** A dog is showing signs of stress while the owner is out of the arena and the veterinary behaviorist is engaging the dog in friendly interaction, including gently petting her and speaking to her in a calm tone. **Video S2:** A dog is exhibiting signs of severe distress and anxiety; consequently, the owner is asked to come back, and the test is stopped. **Video S3.** A dog in the Case group persists in jumping up on the veterinary behaviorist during the post-separation reunion phase.

Author Contributions: First Author (F.P.) was responsible for conceptualization, data curation, funding acquisition, investigation, and writing (original draft + review and editing). The second author (L.P.) was responsible for conceptualization and data collection. The third author (A.B.) was responsible for data analysis. Forth author (S.U.) was responsible for conceptualization, reviewing, and editing. The last author (M.A.) was responsible for conceptualization, methodology, interpretation of data, and supervision.

Funding: This research was funded by the UNIVERSITY OF MILAN (LINEA2_FPIRR_AA_2017 grant number 24066).

Acknowledgments: We thank Carla Colombani for her support in conducting laboratory analyses in this study.

Conflicts of Interest: The authors declare no conflict of interest.

References

1. Carter, C.S. The Role of Oxytocin and Vasopressin in Attachment. *Psychodyn. Psych.* **2017**, *45*, 499–517. [CrossRef] [PubMed]
2. Mariti, C.; Ricci, E.; Carlone, B.; Moore, J.L.; Sighieri, C.; Gazzano, A. Dog attachment to man: A comparison between pet and working dogs. *J. Vet. Behav.* **2013**, *8*, 135–145. [CrossRef]
3. Rehn, T.; Handlin, L.; Uvnäs–Moberg, K.; Keeling, L.J. Dogs' endocrine and behavioural responses at reunion are affected by how the human initiates contact. *Physiol. Behav.* **2014**, *124*, 45–53. [CrossRef]
4. Parthasarathy, V.; Crowell–Davis, S.L. Relationship between attachment to owners and separation anxiety in pet dogs (Canis lupus familiaris). *J. Vet. Behav.* **2006**, *1*, 109–120. [CrossRef]
5. Konok, V.; Dóka, A.; Miklósi, Á. The behavior of the domestic dog (Canis familiaris) during separation from and reunion with the owner: A questionnaire and an experimental study. *Appl. Anim. Behav. Sci.* **2011**, *135*, 300–308. [CrossRef]
6. Palestrini, C.; Previde, E.P.; Spiezio, C.; Verga, M. Heart rate and behavioural responses of dogs in the Ainsworth's Strange Situation: A pilot study. *Appl. Anim. Behav. Sci.* **2005**, *94*, 75–88. [CrossRef]
7. Aloe, L.; Bracci–Laudiero, L.; Allevat, E.; Lambiase, A.; Micera, A.; Tirassa, P. Emotional stress induced by parachute jumping enhances blood nerve growth factor levels and the distribution of nerve growth factor receptors in lymphocytes (str/nxet/hormone). *Proc. Natl. Acad. Sci. USA* **1994**, *91*, 10440–10444. [CrossRef]
8. Aloe, L.; Allevat, E.; Bohm, A.; Levi–Montalcini, R. Aggressive behavior induces release of nerve growth factor from mouse salivary gland into the bloodstream (submaxillary salivary gland/adrenal gland). *Proc. Natl. Acad. Sci. USA* **1986**, *83*, 6184–6187. [CrossRef]
9. Campbell, T.; Lin, S.; DeVries, C.; Lambert, K. Coping strategies in male and female rats exposed to multiple stressors. *Physiol. Behav.* **2003**, *78*, 495–504. [CrossRef]
10. Neumann, I.D.; Landgraf, R. Balance of brain oxytocin and vasopressin: Implications for anxiety, depression, and social behaviors. *Trends Neurosci.* **2012**, *35*, 649–659. [CrossRef]
11. Uvnäs–Moberg, K. Oxytocin may mediate the benefits of positive social interaction and emotions. *Psychoneuroendocrinology* **1998**, *23*, 819–835. [CrossRef]
12. Wotjak, C.; Kubota, M.; Liebsch, G.; Montkowski, A.; Holsboer, F.; Neumann, I.; Landgraf, R. Release of vasopressin within the rat paraventricular nucleus in response to emotional stress: A novel mechanism of regulating adrenocorticotropic hormone secretion? *Soc. Neurosci.* **1996**, *16*, 7725–7732. [CrossRef]
13. MacLean, E.L.; Gesquiere, L.R.; Gee, N.R.; Levy, K.; Martin, W.L.; Carter, C.S. Effects of Affiliative Human–Animal Interaction on Dog Salivary and Plasma Oxytocin and Vasopressin. *Front. Psychol.* **2017**, *8*. [CrossRef] [PubMed]
14. Nagasawa, M.; Mitsui, S.; En, S.; Ohtani, N.; Ohta, M.; Sakuma, Y.; Onaka, T.; Mogi, K.; Kikusui, T. Oxytocin–gaze positive loop and the coevolution of human–dog bonds. *Science* **2015**, *348*, 333–336. [CrossRef]
15. Hydbring–Sandberg, E.; von Walter, L.; Höglund, K.; Svartberg, K.; Forkman, L.B.S. Physiological reactions to fear provocation in dogs. *J. Endocrinol.* **2004**, *180*, 439–448. [CrossRef]
16. MacLean, E.L.; Gesquiere, L.R.; Gruen, M.E.; Sherman, B.L.; Martin, W.L.; Carter, C.S. Endogenous Oxytocin, Vasopressin, and Aggression in Domestic Dogs. *Front. Psychol.* **2017**, *8*, 1613. [CrossRef]
17. Overall, K.L. *Manual of Clinical Behavioral Medicine for Dogs and Cats*; Elsevier Saunders: Philadelphia, PA, USA, 2013; ISBN 9780323240659.
18. Guthrie, A. Dogs behaving badly–Canine separation disorder research. *Vet. Pr.* **1999**, *31*, 12–13.
19. Landsberg, G.; Hunthausen, W.; Ackerman, L. Fear and Phobias, Separation Anxiety. In *Handbook of Behavior Problems of the Dog and Cat*, 2nd ed.; Landsberg, G.M., Hunthausen, W.L., Ackerman, L., Eds.; Saunders Company: Philadelphia, PA, USA, 2003; pp. 258–267.
20. Blackwell, E.; Casey, R.; Bradshaw, J. Controlled trial of behavioural therapy for separation–related disorders in dogs. *Vet. Rec.* **2006**, *158*, 551–554. [CrossRef]
21. Overall, K.L.; Dunham, A.E.; Juarbe–Diaz, S.V. Phenotypic determination of noise reactivity in 3 breeds of working dogs: A cautionary tale of age, breed, behavioral assessment, and genetics. *J. Vet. Behav.* **2016**, *16*, 113–125. [CrossRef]
22. Tami, G.A.G. Description of the behaviour of domestic dog (Canis familiaris) by experienced and inexperienced people. *Appl. Anim. Behav.* **2009**, *120*, 159–169. [CrossRef]

23. Mariti, C.; Gazzano, A.; Moore, J.; Baragli, P. Perception of dogs' stress by their owners. *J. Vet. Behav. Clin. Appl. Res.* **2012**, *7*, 213–219. [CrossRef]
24. Karagiannis, C.I.; Burman, O.H.; Mills, D.S. Dogs with separation–related problems show a "less pessimistic" cognitive bias during treatment with fluoxetine (Reconcile^{TM}) and a behaviour modification plan. *BMC Vet. Res.* **2015**, *11*, 80. [CrossRef] [PubMed]
25. Bradshaw, J.; McPherson, J.; Casey, R.; Larter, I. Aetiology of separation–related behaviour in domestic dogs. *Vet. Rec.* **2002**, *151*, 43–46. [CrossRef]
26. Soares, G.; Pereira, J.; Paixão, R. Exploratory study of separation anxiety syndrome in apartment dogs. *Cienc. Rural.* **2010**, *40*, 548–553. [CrossRef]
27. Statista Number of dogs in the United States from 2000 to 2017 (in millions). Available online: https://www.statista.com/statistics/198100/dogs--in--the--united--states--since--2000/ (accessed on 15 July 2019).
28. Statista Number of dogs in the European Union in 2017, by country (in 1000s). Available online: https://www.statista.com/statistics/414956/dog--population--european--union--eu--by--country/ (accessed on 15 July 2019).
29. Diesel, G.; Brodbelt, D.; Pfeiffer, D.U. Characteristics of Relinquished Dogs and Their Owners at 14 Rehoming Centers in the United Kingdom. *J. Appl. Anim. Welf. Sci.* **2010**, *13*, 15–30. [CrossRef]
30. New, J.C., Jr.; Salman, M.D.; Scarlett, J.M.; Kass, P.H.; Vaughn, J.A.; Scherr, S.; Kelch, W.J. Moving: Characteristics of Dogs and Cats and Those Relinquishing Them to 12 U.S. Animal Shelters. *J. Appl. Anim. Welf. Sci.* **1999**, *2*, 83–96. [CrossRef]
31. MacLean, E.; Gesquiere, L.; Gee, N.; Levy, K. Validation of salivary oxytocin and vasopressin as biomarkers in domestic dogs. *J. Neurosci.* **2017**, *293*, 67–76. [CrossRef]
32. O'Farrell, V. Owner attitudes and dog behaviour problems. *Appl. Anim. Behav. Sci.* **1997**, *52*, 205–213. [CrossRef]
33. Pastore, C.; Pirrone, F.; Balzarotti, F.; Faustini, M.; Pierantoni, L.; Albertini, M. Evaluation of physiological and behavioral stress–dependent parameters in agility dogs. *J. Vet. Behav. Clin. Appl. Res.* **2011**, *6*, 188–194. [CrossRef]
34. Pirrone, F.; Ripamonti, A.; Garoni, E.C.; Stradiotti, S.; Albertini, M. Measuring social synchrony and stress in the handler–dog dyad during animal–assisted activities: A pilot study. *J. Vet. Behav. Clin. Appl. Res.* **2017**, *21*, 45–52. [CrossRef]
35. Prato–Previde, E.; Spiezio, C.; Sabatini, F.; Custance, D.M. Is the dog–human relationship an attachment bond? An observational study using Ainsworth's strange situation. *Behaviour* **2003**, *140*, 225–254. [CrossRef]
36. Scaglia, E. Video analysis of adult dogs when left home alone. *J. Vet. Behav. Clin. Appl. Res.* **2013**, *8*, 412–417. [CrossRef]
37. Topál, J.; Gácsi, M.; Miklósi, Á.; Virányi, Z.; Kubinyi, E.; Csányi, V. Attachment to humans: a comparative study on hand–reared wolves and differently socialized dog puppies. *Anim. Behav.* **2005**, *70*, 1367–1375. [CrossRef]
38. Choi, S.W. Life is lognormal! What to do when your data does not follow a normal distribution. *Anaesthesia* **2016**, *71*, 1363–1366. [CrossRef] [PubMed]
39. O'Hara, R.B.; Kotze, D.J. Do not log-transform count data. *Methods Ecol. Evol.* **2010**, *1*, 118–120. [CrossRef]
40. Flannigan, G.; Dodman, N.H. Risk factors and behaviors associated with separation anxiety in dogs. *J. Am. Vet. Med. Assoc.* **2001**, *219*, 460–466. [CrossRef]
41. Takeuchi, Y.; Houpt, K.A.; Scarlett, J.M. Evaluation of treatments for separation anxiety in dogs. *J. Am. Vet. Med. Assoc.* **2000**, *217*, 342–345. [CrossRef]
42. Ainsworth, M.D.S. Individual Differences in Strange–Situational Behaviour of One–Year–Olds. In *Determinants of Infant Behavior (Vol. 4)*; Foss, B.M., Ed.; Methuen: London, UK, 1969; pp. 111–136.
43. Bard, K.A.; Nadler, R.D. The effect of peer separation in young chimpanzees (Pan troglodytes). *Am. J. Primatol.* **1983**, *5*, 25–37. [CrossRef]
44. Kaufman, I.C.; Rosenblum, L.A. Effects of separation from mother on the emotional behavior of infant monkeys. *Ann. N.Y. Acad. Sci.* **1969**, *159*, 681–695. [CrossRef]
45. Topál, J.; Miklósi, Á.; Csányi, V.; Dóka, A. Attachment Behavior in Dogs (Canis familiaris): A New Application of Ainsworth's (1969) Strange Situation Test. *J. Comp. Psychol.* **1998**, *112*, 219–229. [CrossRef]
46. White–Traut, R.; Watanabe, K.; Pournajafi–Nazarloo, H.; Schwertz, D.; Bell, A.; Carter, C.S. Detection of salivary oxytocin levels in lactating women. *Dev. Psychobiol.* **2009**, *51*, 367–373. [CrossRef] [PubMed]
47. Tiira, K.; Lohi, H. Early Life Experiences and Exercise Associate with Canine Anxieties. *PLoS ONE* **2015**, *10*, e0141907. [CrossRef]

48. Ludwig, M.; Leng, G. Dendritic peptide release and peptide–dependent behaviours. *Nat. Rev. Neurosci.* **2006**, *7*, 126–136. [CrossRef] [PubMed]
49. Stevenson, E.L.; Caldwell, H.K. The vasopressin 1b receptor and the neural regulation of social behavior. *Horm. Behav.* **2012**, *61*, 277–282. [CrossRef] [PubMed]
50. Baribeau, D.A.; Anagnostou, E. Oxytocin and vasopressin: Linking pituitary neuropeptides and their receptors to social neurocircuits. *Front. Neurosci.* **2015**, *9*. [CrossRef]
51. Pow, D.; Morris, J. Dendrites of hypothalamic magnocellular neurons release neurohypophysial peptides by exocytosis. *Neuroscience* **1989**, *32*, 435–439. [CrossRef]
52. Knobloch, H.S.; Charlet, A.; Hoffmann, L.C.; Eliava, M.; Khrulev, S.; Cetin, A.H.; Osten, P.; Schwarz, M.K.; Seeburg, P.H.; Stoop, R.; et al. Evoked Axonal Oxytocin Release in the Central Amygdala Attenuates Fear Response. *Neuron* **2012**, *73*, 553–566. [CrossRef]
53. Neumann, R. Involvement of the brain oxytocin system in stress coping: Interactions with the hypothalamo–pituitary–adrenal axis. *Trends Neurosci.* **2002**, *35*, 649–659. [CrossRef]
54. Weisman, O.; Schneiderman, I.; Zagoory–Sharon, O.; Feldman, R. Salivary vasopressin increases following intranasal oxytocin administration. *Peptides* **2013**, *40*, 99–103. [CrossRef]
55. Carson, D. Plasma vasopressin concentrations positively predict cerebrospinal fluid vasopressin concentrations in human neonates. *Peptides* **2014**, *61*, 12–16. [CrossRef]
56. Griebel, G.; Beeské, S.; Stahl, S.M. The Vasopressin V 1b Receptor Antagonist SSR149415 in the Treatment of Major Depressive and Generalized Anxiety Disorders: Results From 4 Randomized, Double–Blind, Placebo–Controlled Studies. *J. Clin. Psychiatry* **2012**, *73*, 1403–1411. [CrossRef] [PubMed]
57. Iijima, M.; Yoshimizu, T.; Shimazaki, T.; Tokugawa, K.; Fukumoto, K.; Kurosu, S.; Kuwada, T.; Sekiguchi, Y.; Chaki, S. Antidepressant and anxiolytic profiles of newly synthesized arginine vasopressin V_{1B} receptor antagonists: TASP0233278 and TASP0390325. *Br. J. Pharmacol.* **2014**, *171*, 3511–3525. [CrossRef] [PubMed]
58. Cannas, S.; Frank, D.; Minero, M.; Aspesi, A.; Benedetti, R.; Palestrini, C. Video analysis of dogs suffering from anxiety when left home alone and treated with clomipramine. *J. Vet. Behav. Clin. Appl. Res.* **2014**, *9*, 50–57. [CrossRef]
59. Heinrichs, M.; von Dawans, B.; Domes, G. Oxytocin, vasopressin, and human social behavior. *Front. Neuroendocr.* **2009**, *30*, 548–557. [CrossRef] [PubMed]
60. Chen, X.; Hackett, P.D.; DeMarco, A.C.; Feng, C.; Stair, S.; Haroon, E.; Ditzen, B.; Pagnoni, G.; Rilling, J.K. Effects of oxytocin and vasopressin on the neural response to unreciprocated cooperation within brain regions involved in stress and anxiety in men and women. *Brain Imaging Behav.* **2016**, *10*, 581–593. [CrossRef]
61. Thielke, L.E.; Udell, M.A.R. The role of oxytocin in relationships between dogs and humans and potential applications for the treatment of separation anxiety in dogs. *Biol. Rev.* **2017**, *92*, 378–388. [CrossRef]

animals

MDPI

Article

Disaster Preparedness among Service Dog Puppy-Raisers (Human Subject Sample)

Sarah E. DeYoung [1,*], Ashley K. Farmer [2], Zoe Callaro [3] and Shelby Naar [4]

1 Disaster Research Center, Department of Sociology and Criminal Justice, University of Delaware, Newark, DE 19716, USA
2 Department of Criminal Justice Sciences, Illinois State University, Normal, IL 61761, USA; akfarme@ilstu.edu
3 Public Administration, University of Colorado Denver, Denver, CO 80204, USA; zoe.callaro@ucdenver.edu
4 Radiological Emergency Preparedness, Georgia Emergency Management and Homeland Security, Atlanta, GA 30316, USA; shelby.naar@gema.ga.gov
* Correspondence: sedeyoun@udel.edu

Received: 20 December 2019; Accepted: 1 February 2020; Published: 4 February 2020

Simple Summary: Before service animals are matched with clients, they undergo training programs for increasing the dogs' ability to navigate public spaces. Increasingly, service dog training programs recruit young adults from universities and college campuses. Little is known, however, how these students prepare for disasters and the ways in which they create plans to keep service dogs in training safe during hazard events. We collected data from service dog puppy raisers in a hurricane-prone region of the United States to understand their concepts and actions of disaster preparedness. People who were raising a service puppy for the first time were more likely to consider evacuating from Hurricane Irma in 2017 than people who had participated in the program before. Additionally, over half of the respondents did not have a disaster preparedness kit. Finally, many respondents in this study indicated that their service dog in training provides a sense of safety and security.

Abstract: Little is known about the ways in which puppy raisers engage in disaster preparedness for their puppies (or "guide dogs in training"). The aim of this research is to understand disaster preparedness among service dog puppy raisers. A web-based survey was distributed to people raising puppies in a service dog training program (n = 53 complete survey responses). Questions in the survey included items about disaster preparedness and plans for canine safety in hazards events. Out of those who said they had an evacuation plan for their puppy in training, 59% stated they would put the dog in their vehicles for evacuating to safety in the event of a hurricane or other disaster. The odds of first-time puppy raisers who considered evacuation for Hurricane Irma in 2017 was 15.3 times the odds of repeat raisers. Over half the raisers reported that they did not have a disaster kit. Additionally, 82% of respondents indicated that having a service puppy in training makes them feel safer. These results can be used as a foundation for service dog organizations in disaster preparedness among their puppy raiser volunteers and in designing recruitment messages for new volunteers.

Keywords: disaster; preparedness; puppy raiser; service dogs; working dogs

---

1. Introduction

The Pets Evacuation and Transportation Standards (PETS) Act was passed on 6 October 2006 after category five Hurricane Katrina devastated the coast of Louisiana and Texas. This new act amended the Robert T. Stafford Disaster Relief and Emergency Assistance Act to address the needs of household companion animal and service animals [1]. However, preparedness information typically disseminated to communities through outreach programs or county emergency managers encourages companion animal guardians to prepare not only for themselves but also for their animals [2]. Individual preparedness—such as being able to care for household animals in the event of an emergency—helps reduce losses in disasters and puts less strain on first responders and resources [3]. Although the PETS Act was an important policy for creating plans for managing pets in disasters, individual preparedness remains extremely important for all companion animal guardians, service dog users, and those with service dogs in training [4].

Globally, pets and companion animals are a critical part of the human landscape. According to 2016 companion animal guardianship statistics, out of Australia's 9.2 million households, 5.7 million housed a companion animal [5]. Companion animal guardianship has grown tremendously among the United States population, serving many different forms of companionship. According to the 2017–2018 American Companion Pet Products Association (APPA) National Companion Pet Guardians Survey, 68% of households within the U.S. own a pet. For context, 68% is equivalent to about 84.6 million homes [6]. The National Pet Survey was first conducted in the year 1988, where it found that about 56% of households owned a companion animal at that time. Besides the typical dog and cat, the survey now includes numbers for birds, horses, freshwater fish, saltwater fish, and even reptiles. In 1994, APPA calculated total U.S. companion animal industry expenditures to be $17 billion [6]. In 2016, that number soared to almost $67 billion, showing the bond between guardian and companion animal has strengthened over the past few decades [6].

While expenditures alone may not capture the overall social landscape of humans and their pets, news media coverage of humans and their pets in disasters is present in almost every disaster [7]. The ongoing concern about pets in disasters reflects the notion that pets are important in emergencies. Less attention has been given to the topic of service animals in disasters and the ways in which people who rear service animals engage in disaster preparedness. In the United States, college students are increasingly volunteering to train guide dog puppies—and many of the campuses in which they live are prone to hurricanes and other hazards. This research is an attempt at understanding the ways in which guide dog trainers prepare their puppies for disasters and emergency scenarios. Specifically, this research focuses on the emerging population of college student puppy raisers and the ways in which they perceive preparedness, safety, and risk for themselves and their puppies in training. We begin the paper by describing the context of the human-animal bond because this may influence decision-making such as disaster preparedness. We then describe research related to college students and the ways in which they interpret risk, followed by literature on the social landscape of animals in disasters. We then provide the method for our study followed by the results, discussion, and implications for future research.

1.1. Human-Animal Bond

Companion animals provide a multitude of benefits for their guardians. Although emotional benefits may be most obvious, companion animals can also have social, physiological, and physical health benefits [8–10]. Companion animal guardianship directly affects cardiovascular health, increases physical activity, enhances the welfare of the elderly, and even serves as a source of support for people experiencing crisis and hardship. These hardships can include homelessness, family stressors, or some type of disaster [9].

Therapeutic organizations have used animals in several areas, including animal-assisted therapy, animal-assisted crisis response, and child psychotherapy [8,11]. The role of pets is crucial, as reflected in some reports, where companion animals were the main reason individuals survived personal crises such as loss of a loved one, a family health crisis, or periods following substance abuse [9]. Benefits created by companion animals are not unique to one specific age, gender, or ethnicity; evidence suggests companion animals to help hospitalized children recover, ease suffering for those in hospice care, and alleviate depression among Acquired Immune Deficiency Syndrome (AIDS) patients [11].

Dogs, although popular companion animals, are also popular in occupational roles. The roles include guiding for the blind, search and rescue, assistance for people with disabilities, and police detection work [12,13]. Despite working dogs serving a different role than companion animals, they may have similar attachment levels to their handlers as pets do to their guardians [14]. Assistance dogs serve important roles for their handlers, especially for handlers who cannot see or are visually impaired [12].

1.2. History and Context of the Guide Dog Occupation

The origins of the guide dog can be traced back to Germany after the end of World War I and have now become more and more common throughout the U.S. and countries around the world [15]. Guide dogs play an important role for their handlers, serving as companions, security detail, and safety officers. After World War I, many German troops returned home with no physical injuries besides the loss of their eyesight [14]. In the U.K., many guide dog users expressed social and psychological benefits from having a guide dog, as well as increased confidence and socialization with other people [16].

The way other people treat those who are visually impaired has also been stated as a benefit among guide dog users; out in public people tend to be friendlier towards those individuals when the dog is present [16]. This particular benefit has been very important because individuals with visual impairment are already prone to social stigma and even discrimination [17]. In South Africa, guide dog users explained the dogs were safer and faster than using a cane, they provided more confidence regarding mobility, and enhanced independence [17]. Japanese guide dog teams expressed that their dogs reduced tension and strain when traveling and had positive effects on mobility and quality of life [18]. Many of these benefits originate from the strong bond and attachment that forms between handler and dog. Additionally, research on guide dog training programs suggests that existence of service animals can be linked with the perceived validity of a persons' functional and access needs status [19].

1.3. Guide Dog-Handler Bond

Guide dogs go through many bonding stages during their training and placement process with a handler. They live and stay with a puppy raiser, which is typically a volunteer through the guide dog school, get assigned to a trainer once they are around sixteen months old, and then are eventually matched with a person who is visually impaired. Each of these stages has been isolated and studied to understand the difference in bond strength between the dogs and their raisers, the dogs and their trainers, and the dogs and their handlers [20,21]. For example, after one year of partnership, guide dogs showed all signs of attachment when separated from their handler, even in the presence of another friendly individual, and remained oriented to the door for their handler to return [20,21].

Some evidence suggests that because handlers who are blind have different situational awareness, some dogs developed signals to more effectively communicate with their handler, such as maintaining physical contact [21]. Because of the process used to train guide dogs, bonds are formed and technically lost after the dogs leave their puppy raisers and then again when they leave their trainers. However, the use of puppy raisers is still important in ensuring the dogs reach their end goal of helping individuals dependent upon their services [22].

1.4. Puppy Raisers

Guide dog schools utilize volunteer puppy raisers as an essential part of their process [22]. Puppy raisers typically have the dogs from around eight weeks old until a little over a year old or until the dog has matured. Because the puppy raisers have the dogs for quite a while, it is understandable that bonds form between the trainer and puppy. Puppy raisers can help contribute to a dog's success or more accurately predict the ability of the dog they are raising to become a guide dog. For example, assessments or questionnaires done by puppy raisers concerning their dogs' chances of success have served as important predictors for the respective guide dog schools [22–24]. This suggests that puppy raisers bond with and know their dog very well over just a year of time. Other benefits can also come with being a puppy raiser, including improved mental, social, and physical health; increased awareness of responsibility; and new friendships with other raisers [25].

Depending on the reach and location of each guide dog school, a diverse array of people volunteer as puppy raisers. In many instances, college students tend to be the most common group of volunteers. To become successful guide dogs, raisers take the puppies with them everywhere they go so that the dogs grow accustomed early on to what they will be doing as working canines. For working adults with full-time jobs, it is harder for them to be accompanied by young puppies every day, and the puppies are also big commitments. College students can more easily take a puppy in training to class, grocery shopping, or to social events [26]. Interestingly, the popularity of having a guide dog puppy is expressed as a trend among college students in some areas of the United States, like fashion and technology branding [27] that signifies identity, social status, and possible wealth. In fact, it is common to see students walking on campuses with the training dogs that wear the signature yellow vests. For the broader puppy-raiser population, training of the dogs is likely to take place by able-bodied persons [28]. In the context of this research, it is important to consider how the able-bodiedness of the handlers influences their ability to engage in activities such as disaster preparedness, as well as perceptions about ease of preparedness by able-bodied handlers [29].

1.5. College Students: Risk Perception

Risk perception is defined as a "highly personal process of decision making, based on an individual's frame of reference developed over a lifetime" [30] (p. 1) and can be directly linked to protective actions taken [30,31] In other words, perception of risk is a very individualistic concept and will differ from person to person [32]. This important aspect of risk perception becomes a challenge in many risk communication strategies, particularly when risk perception is low.

Source and repetition of information also contribute to decreased risk perception. If the source is not trusted or people do not trust current science or technology in general, there will be lower general risk perception [33]. If information regarding a specific risk has become the social norm and is common knowledge, risk communicators will have a hard time heightening risk perception when necessary [34,35]. A major example of this would be safe driving practices for younger drivers (i.e., college students), which is a very popular risk communication topic. Adults between the age 18–29 are more likely to drive while drowsy compared to other age groups, and drivers in their 20's make up 27% of distracted drivers in fatal crashes; however, college students have grown accustomed to car-related risk information and accept it as common knowledge [36,37]. Other factors contributing to low risk perception include optimistic bias, self-efficacy, perception of benefits, and perceived barriers to preventative action [35,38,39].

In general, higher perceived severity of a risk (understanding the risk has serious consequences) will lead to higher risk perception and cause an individual to engage in more physical protective behavior [40,41]. Higher perceived severity is common in those who have personally experienced certain risks in their area (i.e., tornadoes, earthquakes, wildfires), so it is understandable that heightened risk perception differs from region to region [42,43]. Among college students, local hazards, as well as a risk being uncontrollable or unknown, will contribute to higher risk perceptions [43].

1.6. College Students: Risk Information

The sources of risk information have changed as technology has advanced over the years [44]. For example, in 2010 students were more likely to receive information from family, friends, and television and less likely to receive it from university flyers, professors, or course content [45]. However, in 2012 students preferred to receive emergency preparedness information directly from their school via texts and emails [44]. These results were proven a few years prior when a tornado touched down near the Mississippi State University campus, and most of the students stated they first heard of the tornado through university alert messages [46]. This may be because students indirectly expect their university to "take care of them" in the event of an emergency or disaster [42].

1.7. Service Animals & Disasters

Because there were multiple instances of people being forcefully separated from their animals during Hurricane Katrina [47] there has been growing attention on the issue of animals and disasters. While the PETS Act was passed in 2006, the Act has limitations in terms of breed, disaster phases (e.g., recovery and long-term housing is not considered in the PETS Act, such as rental fees for pets of displaced persons), and refusal of evacuation and reunification issues remain a concern in disasters [48]. Simultaneously, the body of research on pets and disasters is growing– especially in the areas of understanding attachment [49,50] and issues related to grief [51,52]. There is also a growing body of research on the needs of functional-access individuals in disaster planning [53,54]. However, there is a scarcity of research on guide dog trainers and disaster planning and preparedness. The primary research aim for this paper is to identify how guide dog puppy-raisers engage in and perceive preparedness for their guide dog puppies. We focus on college students as a population of puppy raisers because college students are at a unique developmental phase in which perceptions of risk and safety, as well as independence, begin to take on a new role in their lives [55]. Additionally, it is important to understand the ways in which college students prepare for hazards [56]. If being a guide dog puppy volunteer is associated with lower or higher levels of preparedness, this has important implications for higher education institutions in terms of planning and outreach. No studies exist that examine the ways in which college students as puppy-raisers prepare for disasters. Studies so far on dogs on college campuses focus on dogs as stress relief for students [57,58] rather than how having a guide dog puppy in training changes the students' level of preparedness.

2. Method & Materials

2.1. Participants

The researchers gathered data using a Qualtrics survey platform to reach volunteer puppy raisers within a southeastern college town. Qualtrics is an electronic survey platform that collects responses and can be automatically uploaded through the internet or in which responses can be uploaded onto a hard drive. The participants were contacted through an email list serve composed of the guide dog puppy raisers. There were approximately 300 people on the list-serve. The puppy raisers targeted for this survey are volunteers through an organization based in the northeastern United States that recruits' college students throughout the US for puppy-training. Specifically, as part of their puppy-raising program, this school has volunteer groups throughout the east coast, with a lot of groups in southeastern states. One of the researchers for this project knew the leader of this foundation from her personal social network, which enabled the research team to distribute the survey through the email list serve. In other words, the recruitment was through convenience sampling [59] but with the support of the leader of this organization that trains guide dogs. While richness of data is beneficial from open-ended interviews [60], we chose a survey format because it facilitates rapid data collection and is convenient for respondents to take a web-based survey rather than coming to a specific location at a specific time for data collection.

Because of the nature of being a puppy raiser (i.e., taking care of a young puppy, taking it with you everywhere you go), groups usually center around universities, as many of their volunteers are college students. Although the guide dog school is headquartered in the northeast, they have a southeastern coordinator who maintains the southern groups. A reminder email was sent two weeks after the initial survey recruitment and notification.

2.2. Procedure

The survey (Appendix A) was conducted under a university Institutional Review Board in December 2017 and January 2018. The survey link contained a consent form, an introduction to the study and allowing participants to consent and continue on with the questionnaire. The responses were automatically and anonymously uploaded into the Qualtrics system after each submission and transferred to SAS statistical software for data cleaning and analysis.

2.3. Data Analysis

The results from the Qualtrics survey had 84 responses, not including pilot responses. Analyses were conducted on the sample size of $N = 84$ (response rate = 28%), but there were missing data for 31 respondents. Of those 31, 84% did not answer past the first two questions and were not included in the final analyses, and the remaining 16% left partial responses and were included. There were a total of 53 respondents who completed the entire survey. SAS version 9.4 was used to code the survey responses, assess frequencies, and conduct chi-square analyses.

3. Results

Out of the 53 participants who completed the entire survey, the majority were female ($n = 48$, 91%) with a mean age of 28 years ($SD = 14$). The respondents were predominantly White ($n = 47$, 88%). The remaining respondents were Hispanic ($n = 3$, 1%), Asian ($n = 2$, <1%), and African American ($n = 1$, <1%). 40 out of the 58 participants identified as college students. Seniors made up most of the college students ($n = 19$, 48%), followed by graduate and dual-degree students ($n = 9$, 22%), juniors ($n = 8$, 20%), and sophomores ($n = 4$, 10%). It is likely that the high number of responses from college students is because most of the people on the list-serve are students and because they may be more likely to check email frequently because of schoolwork and communication. There were no freshman participants; however, it is less likely for first year students to be puppy raiser volunteers as it takes time to learn about the program, apply, and go through the volunteer process. Among the puppy raisers, 65% of them were repeat raisers, meaning they had already raised one or more dogs, while 35% were first time raisers.

The first items in the survey measured evacuation plans, locations, and logistics (Refer to Appendix A). Among the puppy raisers who said they had an evacuation plan for their puppy in training ($n = 17$), many stated they would just put the dog in the car with them and leave. Some raisers specifically added they would bring food, water, toys, a dog bed, or a crate ($n = 8$). Only two raisers said anything about bringing a preparedness or disaster kit for their dog, and another two mentioned bringing papers and vaccine records. One survey item inquired about where participants would go if needing to evacuate. Many respondents ($n = 33$) stated they would go stay with family, including parents, siblings, or even extended family. Family was also a popular response among college students when asked where they would go if they had to evacuate their parents' house.

Four survey questions specifically asked participants about plans relating to Tropical Storm Irma of 2017 since the storm had hit southeastern states a few months prior to the survey. An analysis of the crude relationship between puppy raiser experience (first time raiser versus repeat raiser) and consideration to evacuate for Irma gave an odds ratio of 15.3. This reflects that the odds of first-time puppy raisers considering evacuation for Irma was 15.3 times the odds of repeat puppy raisers, revealing that first-time raisers were much more likely to consider evacuation, X^2 (1, $N = 39$) = 11.37, $p < 0.05$. Also, of the 33 participants (77%) that said they prepared supplies specifically for their animal before Irma, 32 (74%) mentioned essentials such as food and water, 4 (9%) said they gathered medical supplies or a first aid kit for their animal, 5 (12%) mentioned outside gear for the dogs when going to the bathroom or a way for the dogs to go to the bathroom inside, and a final 2 (5%) mentioned getting together papers or vaccine records. It should be noted that some respondents were included in more than one category as they mentioned more than one of the above answers.

In response to the item inquiring about having a disaster kit at home, over half (58%) of the participants said no. Most participants (82%) said that their companion animal or puppy in training makes them feel safe daily as opposed to the 2% who said their companion animal or puppy in training makes them feel unsafe. Table 1 shows frequencies of the risk perception responses, reflecting what participants view as the highest risk to their companion animal or puppy in training daily versus what they view as the highest risk to themselves.

Table 1. Survey responses to "highest risk to self" and "highest risk to pet".

Risk to Pet			Risk to Self		
Risk	*Frequency*	*Percentage (%)*	*Risk*	*Frequency*	*Percentage (%)*
Auto Accident	32	74.4	Auto Accident	43	74.1
Canine Flu/Illness	5	11.6	Flu/Illness	10	17.3
Severe Weather	2	4.70	Severe Weather	1	1.70
Active Shooter	1	2.30	Active Shooter	1	1.70
Structural Fire	2	4.70	Structural Fire	0	0
Other	1	2.30	Sport/Physical Injury	2	3.50
			Other	1	1.70

4. Discussion

The primary aim of this study was to explore levels of preparedness and risk perception in college students with service dogs (puppies) in training. We expected volunteer raisers who had raised more than one dog to be more prepared for sudden emergencies or disasters (i.e., have a disaster kit, have a plan for their dog). However, over half of the puppy raisers surveyed did not have a disaster kit of any kind and although many prepared food and water for their dog prior to Irma, very few mentioned vaccine records or anything medical-related for the dog. Additionally, first-time puppy-raisers were more prepared than people who had raised multiple puppies. This may be because of the influence of having a "new animal" and potential feelings of protectiveness, like findings in which families with children are more likely to have higher levels of household preparedness [61].

Vaccine records are extremely important to have on hand in case of an emergency as they would be needed when taking an animal to a boarding facility, vet office, or even a shelter [62]. Being turned away at any of those facilities due to lack of records can also play a role in people not wanting to evacuate, especially when they are unable to locate a pet-friendly shelter to begin with [48].

In response to the item inquiring about having an evacuation plan for their puppy in training, over half stated not having a plan at all. Among those that did have a plan, the majority just mentioned evacuating with the dog in the car. Two participants mentioned bringing a crate, and although research suggests transportation challenges as the main cause of evacuation failure for companion animal guardians, it should be noted that these dogs are trained to ride in the footwell of cars (not in crates) and that every puppy raiser should be issued a crate when they receive their puppy in training. Again, few respondents mentioned vaccine records or a disaster kit, with only one participant stating they always kept a kit in the car for their dog. These results support existing literature on preparedness among college students in general, with the most common preparedness activities found to be storing water or food, having a working flashlight, and taking a Cardiopulmonary Resuscitation (CPR) class [63]. In a similar way that students take few protective actions because they assume the university will take care of them in the event of a disaster, puppy raisers may expect the same resources and guidance from the organization for which they volunteer [42].

Many participants said in the event of an evacuation, they would evacuate to a family or relative's place of residence. No specific logistical issues were mentioned; however, that may be due to the socioeconomic status of the survey sample. Most respondents have access to cars—the dogs have been trained and accustomed to riding in a car—and therefore most participants would not need the use of local shelters. Previous hurricane evacuation data reflects that most evacuees stay with relatives or friends [64]. Many college students mentioned evacuating to where their parents live and is important to note since students may in some cases rely on parents for risk-related information more so than media or other interpersonal sources [44].

There was a difference in likelihood of evacuating or considering evacuating prior to Irma between first time raisers and repeat raisers, with first time raisers more likely to do so. This could be due to repeat raisers being more confident about their responsibility towards their puppy in training as they have been through the raising process before. Of the college student participants, many reported being enrolled in majors relating to animals or other involved sciences. The confidence among puppy raisers concerning the safety of their dog could also stem from their area of study being closely related to the risk, a reason many college students tend to have low risk perception [32,33]. It should also be noted, again, that the samples for this study were not diverse in terms of sex and ethnicity, which creates limitations for the implications regarding additional analyses to compare risk perception and preparedness among respondents. Additional research in the area of guide dog puppy raisers should include a sample that has more ethnic and gender diversity, especially since minority groups more susceptible to adverse outcomes in disasters [65].

Many of the participants stated that their companion animal or puppy in training made them feel safe. These results are consistent with other findings that suggest pets help enable a sense of security and serve as a haven and secure base for their guardians [66,67]. This sense of safety from the human-animal bond also suggests that companion animal guardianship can serve as a protective factor during evacuations where guardians are more likely to get to safety for the sake of their companion animal [68]. College students also directly benefit from pets, where just a simple twenty-minute canine therapy session reduced stress and even homesickness [69].

However, despite the sense of security the puppies in training and pets bring to their guardians, there are still daily risk perceptions that differ from person to person. Table 1 reflects that most participants viewed the highest risk to themselves and their companion animal to be an auto accident. Despite the number of young drivers involved in car accidents every year, some participants viewed the highest risk to their companion animal to be an event of lower probability, such as severe weather, structural fire, or an active shooter [36]. Because individuals have less control over these types of events, it is possible that fear caused a heightened risk perception despite an auto accident being a more significant risk [70].

5. Conclusions

This study took place in the United States and therefore the results are not generalizable beyond the specific population of college students with guide dog puppies. Also, as we mention in the broader discussion, the homogenous sample (white, able-bodied, and female) is a limitation. Future research should include diverse groups of volunteers. Additionally, the small sample size of survey responses is a limitation of this study. Although there were differences between first time and repeat puppy raisers, the small sample did not yield more than one statistically significant finding. Future research should also focus on the potential difference in preparedness practices and activities between companion animal guardians and puppy raisers since puppy raisers take their dog most places with them as opposed to a companion animal guardian. Future research should also include in-depth qualitative data in which students as puppy-raisers share their thoughts on safety, risk, and preparedness. This would add richness to the data regarding motivations, perceptions, and processes for preparedness. This might also uncover new information about levels of bonding with their guide dog puppies and how levels of bonding relate to preparedness.

Finally, it is also important for future research to explore the issue of branding or the "trendiness" of having a guide dog puppy among college students, especially in the context of understanding how this experience is different for the person training the dog compared to the person who is ultimately paired with and requires the dog because of a functional-access need. The college students in this sample were white and able-bodied [71] whereas this may not be the case for functional access individuals in which having a service dog may be another form of labeling the person as "disabled" [72,73] and in which preparedness may entail different meanings and processes. In other words, it may be important to understand how the accompaniment of a service-dog elicit stigma from others and the ways in which this impacts functional and access needs individuals in hazards events.

The results from this study can be used as a foundation in future research for service dog organizations into hazard and disaster preparedness among their puppy raiser volunteers. Because service dog volunteer groups are sometimes spread several states away from the headquarters of the guide dog organization, accountability and tracking the dogs could become difficult in the event of an emergency or disaster. For some volunteer groups, there are hierarchical leadership roles in place to help keep track of the raisers and all the dogs. These leadership positions could incorporate potential preparedness protocols into their volunteer group meetings or even just going over what to include in a pet-related disaster kit. Since most groups are required to attend monthly meetings, those are already designated times when many puppy raisers and volunteers come together where preparedness practices could easily be discussed among the group.

Additionally, since having a service dog in training made students feel safer, this information could be used in developing recruitment messages for new volunteers. Having service puppies in training makes students feel safer and possibly less stressed. A key message for college student volunteers could be something along the lines of, "Your service puppy in training helps you to feel safe. Learn how to keep your puppy in training safe during disasters". In general, puppy raisers are a distinct and dynamic group of companion animal guardians that are most likely not accounted for in many companion animal management plans, so it is important to practice individual and community-level preparedness for the safety of both people and animals. Finally, it is also important to consider that preparedness strategies and meanings will likely be different for trainers than they are for people who are later paired with the dogs and this may impact implications for experiences of both the dogs and humans. This discrepancy should be accounted for when organizations are creating training protocols for raising guide dog puppies (i.e., considering a more inclusive and aggressive outreach of puppy-raisers who have functional-access needs, people of color, and more diverse groups of people). This is also reflected as a guideline for best practices in preparedness and disaster planning [74].

Author Contributions: Conceptualization, S.E.D. and S.N.; Formal analysis, S.N.; Investigation, S.N.; Methodology, S.N.; Supervision, S.E.D.; Writing—original draft, S.N.; Writing—review & editing, S.E.D., A.K.F. and Z.C. All authors have read and agreed to the published version of the manuscript.

Funding: This research received no external funding.

Acknowledgments: We thank Curt Harris at the University of Georgia for his comments on the early draft of this paper.

Conflicts of Interest: The authors declare no conflict of interest.

Appendix A Survey Instrument

Preparedness among Puppy Raisers

1. Are you currently a volunteer puppy raiser for the (organization redacted for peer review)?
    a. Yes
    b. No
2. Do you have a disaster preparedness kit in your home/apartment/dorm? ________
3. Do you currently have any pets living with you? If so, please list them.
    a. Yes ____________
    b. No (Skip to Question 6)
4. Do you have special plans for evacuation related to managing the logistics of transporting your pet(s)? _______

5. In your opinion, what do you think is the most serious risk to your pet's or pets' health or safety on a daily basis?

    a. Car/auto accident
    b. Disease or illness
    c. Severe weather
    d. Structural fire
    e. Other

6. In your opinion, what do you think is the most serious risk to your own health or safety on a daily basis?

    a. Car/auto accident
    b. Flu, illness, or communicable disease
    c. Severe weather
    d. Active shooter
    e. Structural fire
    f. Sport or other physical injury
    g. Other

7. What news sources do you use on a daily basis? (Circle all that apply.)

    a. Internet news
    b. Television news stations
    c. Mobile news apps
    d. Social media
    e. Word of mouth on campus

8. If you (and your pet(s)) need to evacuate from Athens, where would you go? ____________________

9. If you (and your pet(s)) need to evacuate from your parents' house, what would be your evacuation destination? ____________________

10. Does having your pet(s) make you feel:

    a. Very unsafe
    b. Moderately unsafe
    c. Slightly unsafe
    d. Neutral
    e. Slightly safe
    f. Moderately safe
    g. Very safe
    h. Do not know
    i. N/A (No pets)

11. When Tropical Storm Irma impacted the Georgia area and Athens, please select the news sources you were using to gather information about the storm:

    a. Internet news
    b. Television news stations
    c. Mobile news apps
    d. Social media

e. Word of mouth on campus

12. When Tropical Storm Irma impacted the Georgia area and Athens, did you prepare any supplies specifically for your pet(s), and if so, please describe:

    a. Yes ____________
    b. No
    c. N/A (No pets)

13. When Tropical Storm Irma impacted the Georgia area and Athens, did you evacuate or consider making plans to evacuate?

    a. Yes
    b. No

14. If yes, what was your evacuation destination and why did you choose that destination?

    ____

15. What else would you like to share with us about Irma or any other hazard events (and your pet(s))? Please describe.
16. What is your age? ________
17. What is your race?

    a. Black or African American
    b. White or Caucasian
    c. American Indian
    d. Asian
    e. Pacific Islander
    f. Hispanic or Latinx
    g. Other _______

18. What is your gender?

    a. Female
    b. Male
    c. Transgender
    d. Other
    e. Prefer not to answer

19. What year are you in college?

    a. Freshman
    b. Sophomore
    c. Junior
    d. Senior
    e. Grad/professional student
    f. Dual degree
    g. N/A (Skip to question 21)

20. What is your major? _____________
21. What is your occupation? ___________

References

1. American Veterinary Medical Association. PETS Act (FAQ). Available online: https://www.avma.org/KB/Resources/Reference/disaster/Pages/PETS-Act-FAQ.aspx (accessed on 3 May 2018).

2. Department of Homeland Security. Pets and Animals. Make a Plan. Available online: https://www.ready.gov/animals (accessed on 3 May 2018).
3. Wilson, S.A.; Temple, B.J.; Milliron, M.E.; Vazquez, C.; Packard, M.D.; Rudy, B.S. The Lack of Disaster Preparedness by the Public and its Affect on Communities. *Internet J. Rescue Disaster Med.* **2008**, *7*, 1–8.
4. Day, A.M. Companion animals and natural disasters: A systematic review of literature. *Int. J. Disaster Risk Reduct.* **2017**, *24*, 81–90. [CrossRef]
5. Australian Veterinary Association. Pet Ownership Statistics. 2016. Available online: http://www.ava.com.au/news/media-centre/hot-topics-4 (accessed on 3 May 2018).
6. American Pet Products Association. Pet Industry Market Size and Ownership Statistics. 2017. Available online: http://www.americanpetproducts.org/press_industrytrends.asp (accessed on 10 September 2017).
7. Hesterberg, U.W.; Huertas, G.; Appleby, M.C. Perceptions of pet owners in urban Latin America on protection of their animals during disasters. *Disaster Prev. Manag. Int. J.* **2012**, *21*, 37–50. [CrossRef]
8. Abrams, L. My Dog is my Co-Therapist. *Reflect. Narrat. Prof. Help.* **2009**, *15*, 52–58.
9. Clancy, E. Animals as Community Stakeholders: Inclusion of Pets in Social Policy and Practice (Occasional Essay). *Fam. Soc.* **2014**, *95*, 285–289. [CrossRef]
10. Irvine, L. Animals as Lifechangers and Lifesavers. *J. Contemp. Ethnogr.* **2013**, *42*, 3–30. [CrossRef]
11. Walsh, F. Human-Animal Bonds II: The Role of Pets in Family Systems and Family Therapy. *Fam. Process.* **2009**, *48*, 481–499. [CrossRef]
12. Scoular, A. What Jobs Are Working Dogs Used For? The Telegraph. Available online: https://www.telegraph.co.uk/pets/essentials/jobs-for-working-dogs/ (accessed on 24 November 2017).
13. Siniscalchi, M.; Stipo, C.; Quaranta, A. "Like Owner, Like Dog": Correlation between the Owner's Attachment Profile and the Owner-Dog Bond. *PLoS ONE* **2013**, *8*, e78455. [CrossRef]
14. Mariti, C.; Ricci, E.; Carlone, B.; Moore, J.L.; Sighieri, C.; Gazzano, A. Dog attachment to man: A comparison between pet and working dogs. *J. Vet. Behav. Clin. Appl. Res.* **2013**, *8*, 135–145. [CrossRef]
15. Baár, M. Prosthesis for the body and for the soul: the origins of guide dog provision for blind veterans in interwar Germany. *First World War Stud.* **2015**, *6*, 81. [CrossRef]
16. Whitmarsh, L. The Benefits of Guide Dog Ownership. *Vis. Impair. Res.* **2005**, *7*, 27–42. [CrossRef]
17. Wiggett-Barnard, C.; Steel, H. The experience of owning a guide dog. *Disabil. Rehabil.* **2008**, *30*, 1014–1026. [CrossRef] [PubMed]
18. Matsunaka, K.; Koda, N. Acceptance of dog guides and daily stress levels of guide users and nonusers. *J. Vis. Impair. Blind.* **2008**, *5*, 295–304. [CrossRef]
19. Walther, S.; Yamamoto, M.; Thigpen, A.P.; Garcia, A.; Willits, N.H.; Hart, L.A. Assistance dogs: historic Patterns and roles of dogs placed by aDi or igDF accredited facilities and by non-accredited Us facilities. *Front. Vet. Sci.* **2017**, *4*, 1. [CrossRef]
20. Fallani, G.; Previde, E.P.; Valsecchi, P. Do disrupted early attachments affect the relationship between guide dogs and blind owners? *Appl. Anim. Behav. Sci.* **2017**, *100*, 241–257. [CrossRef]
21. Valsecchi, P.; Previde, E.P.; Accorsi, P.A.; Fallani, G. Development of the attachment bond in guide dogs. *Appl. Anim. Behav. Sci.* **2010**, *123*, 43–50. [CrossRef]
22. Batt, L.; Batt, M.; Baguley, J.; McGreevy, P. The effects of structured sessions for juvenile training and socialization on guide dog success and puppy-raiser participation. *J. Vet. Behav. Clin. Appl. Res.* **2008**, *3*, 199–206. [CrossRef]
23. Batt, L.; Batt, M.; Baguley, J.; McGreevy, P. Relationships between puppy management practices and reported measures of success in guide dog training. *J. Vet. Behav. Clin. Appl. Res.* **2010**, *5*, 240–246. [CrossRef]
24. Batt, L.S.; Batt, M.S.; Baguley, J.A.; McGreevy, P.D. The Value of Puppy Raisers' Assessments of Potential Guide Dogs' Behavioral Tendencies and Ability to Graduate. *Anthrozoös* **2009**, *22*, 71–76. [CrossRef]
25. Chur-Hansen, A.; Werner, L.-K.; McGuiness, E.C.; Hazel, S. The Experience of Being a Guide Dog Puppy Raiser Volunteer: A Longitudinal Qualitative Collective Case Study. *Animals* **2015**, *5*, 1–12. [CrossRef]
26. Lefevre, K. Dawgs Training Dogs: Guide Dog Training from Start to Finish. The Red and Black. 2017. Available online: https://www.redandblack.com/uganews/dawgs-training-dogs-guide-dog-training-from-start-to-finish/article_b855aebc-1420-11e7-960e-5b0f5b880e8e.html (accessed on 29 March 2017).
27. Katz, J.E.; Sugiyama, S. Mobile phones as fashion statements: evidence from student surveys in the US and Japan. *New Media Soc.* **2006**, *8*, 321–337. [CrossRef]

28. Lloyd, J.K.F. Exploring the Match between People and Their Guide Dogs. Ph.D. Thesis, Veterinary Science at Massey University, Aotearoa, New Zealand, 2004.
29. Case, K. Intersectionality and paradigms of privilege: Teaching for social change. In *Deconstructing Privilege;* Routledge: New York, NY, USA, 2013; pp. 122–134.
30. Brown, V.J. Risk Perception: It's Personal. *Environ. Health Perspect.* **2014**, *122*, A276–A279. [CrossRef] [PubMed]
31. Sullivan-Wiley, K.A.; Short Gianotti, A.G. Risk Perception in a Multi-Hazard Environment. *World Dev.* **2017**, *97*, 138–152. [CrossRef]
32. Kim, S.; Kim, S. Impact of the Fukushima Nuclear Accident on Belief in Rumors: The Role of Risk Perception and Communication. *Sustainability* **2017**, *9*, 2188. [CrossRef]
33. Ma, H.; Zhai, G.; Tsuchida, S. Risk Literacy and Risk Perception among Undergraduates in Mainland China: The Case of BSE. *Hum. Ecol. Risk Assess. Int. J.* **2013**, *19*, 526–537. [CrossRef]
34. Ieong, M.C.F.; Ho, J.-C.; Lee, P.C.-T.; Hokama, T.; Gima, T.; Luo, L.; Chang, P.W.-S. Risk Perception of Nuclear Power Plants Among University Students in Northeast Asia After the Fukushima Nuclear Disaster. *Asia Pac. J. Public Health* **2014**, *26*, 631–641. [CrossRef]
35. Terry, C.P.; Terry, D.L. Distracted driving among college students: Perceived risk versus reality. *Curr. Psychol.* **2016**, *35*, 115–120. [CrossRef]
36. National Highway Traffic Safety Administration. Distracted Driving. 2015. Available online: https://www.nhtsa.gov/risky-driving/distracted-driving (accessed on 5 October 2017).
37. National Sleep Foundation. Drowsy Driving Prevention Week. Facts and Stats. 2005. Available online: http://drowsydriving.org/about/facts-and-stats/ (accessed on 3 May 2018).
38. Haile, Z.; Kingori, C.; Darlington, K.-A.; Basta, T.; Chavan, B. HIV Risk Perception among College Students at a University in the Midwest. *Sex. Cult.* **2017**, *21*, 62–73. [CrossRef]
39. Saling Untied, A.; Dulaney, C.L. College Students' Perceived Risk of Sexual Victimization and the Role of Optimistic Bias. *J. Interpers. Violence* **2014**, *30*, 1417–1431. [CrossRef]
40. Bernat, J.K.; Anderson, L.B.; Parrish-Sprowl, J.; Sparks, G.G. Exploring the Association between Dispositional Cancer Worry, Perceived Risk, and Physical Activity Among College Women. *J. Am. Coll. Health* **2015**, *63*, 216–220. [CrossRef]
41. DeYoung, S.E. Disaster Preparedness: Psychosocial Predictors for Hazard Readiness. Ph.D. Thesis, North Carolina State University, Raleigh, NC, USA, 2014.
42. Lovekamp, W.E.; Tate, M.L. College Student Disaster Risk, Fear and Preparedness. *Int. J. Mass Emerg. Disasters* **2008**, *26*, 70–90.
43. Usuzawa, M.; O'Telan, E.; Kawano, R.; S'Dizon, C.; Alisjahbana, B.; Ashino, Y.; Hattori, T. Awareness of disaster reduction frameworks and risk perception of natural disaster: a questionnaire survey among Philippine and Indonesian health care personnel and public health students. *Tohoku J. Exp. Med.* **2014**, *233*, 43–48. [CrossRef]
44. Koskan, A.; Foster, C.; Karlis, J.; Rose, I.; Tanner, A. Characteristics and influences of H1N1 communication on college students. *Disaster Prev. Manag.* **2012**, *21*, 418–432. [CrossRef]
45. Wilson, S.L.; Huttlinger, K. Pandemic flu knowledge among dormitory housed university students: a need for informal social support and social networking strategies. *Rural Remote Health* **2010**, *10*, 1526.
46. Sherman-Morris, K. Tornado warning dissemination and response at a university campus. *Nat. Hazards* **2010**, *52*, 623–638. [CrossRef]
47. Irvine, L. *Filling the Ark*; Temple University Press: Philadelphia, PA, USA, 2009.
48. Farmer, A.K.; DeYoung, S.E.; Wachtendorf, T. Pets and Evacuation: An Ongoing Challenge in Disasters. *J. Homel. Secur. Emerg. Manag.* **2016**, 13. [CrossRef]
49. Glassey, S.; Wilson, T.M. Animal Welfare Impact Following the 4 September 2010 Canterbury (Darfield) Earthquake. *Australas. J. Disaster Trauma Stud.* **2011**, *2*, 49–59.
50. Hall, M.J.; Ng, A.; Ursano, R.J.; Holloway, H.; Fullerton, C.; Casper, J. Psychological impact of the animal-human bond in disaster preparedness and response. *J. Psychiatr. Pract.* **2004**, *10*, 368–374. [CrossRef]
51. Brackenridge, S.; Zottarelli, L.K.; Rider, E.; Carlsen-Landy, B. Dimensions of the human–animal bond and evacuation decisions among pet owners during Hurricane Ike. *Anthrozoös* **2012**, *25*, 229–238. [CrossRef]
52. Schaffer, C.B. *Human-Animal Bond Considerations during Disasters;* Center for the Study of Human-Animal Interdependent Relationships School of Veterinary Medicine: Tuskegee, AL, USA, 2011.

53. Ivey, S.L.; Tseng, W.; Dahrouge, D.; Engelman, A.; Neuhauser, L.; Huang, D.; Gurung, S. Assessment of state-and territorial-level preparedness capacity for serving deaf and hard-of-hearing populations in disasters. *Public Health Rep.* **2014**, *129*, 148–155. [CrossRef]
54. McDermott, S.; Martin, K.; Gardner, J.D. Disaster response for people with disability. *Disabiity Health J.* **2016**, *9*, 183–185. [CrossRef] [PubMed]
55. Han, M. A Study on Learning Satisfaction's Relations to Career Adaptation and Psychological Independence in College Students. *Indian J. Public Health Res. Dev.* **2018**, *9*, 2241–2247. [CrossRef]
56. Tkachuck, M.A.; Schulenberg, S.E.; Lair, E.C. Natural disaster preparedness in college students: implications for institutions of higher learning. *J. Am. Coll. Health* **2018**, *66*, 269–279. [CrossRef] [PubMed]
57. Crossman, M.K.; Kazdin, A.E. Animal visitation programs in colleges and universities: An efficient model for reducing student stress. In *Handbook on Animal-Assisted Therapy*; Academic Press: London, UK, 2015; pp. 333–337.
58. Jalongo, M.R.; McDevitt, T. Therapy dogs in academic libraries: A way to foster student engagement and mitigate self-reported stress during finals. *Public Serv. Q.* **2015**, *11*, 254–269. [CrossRef]
59. Emerson, R.W. Convenience sampling, random sampling, and snowball sampling: How does sampling affect the validity of research? *J. Vis. Impairment Blind.* **2015**, *109*, 164–168. [CrossRef]
60. Harrell, M.C.; Bradley, M.A. *Data Collection Methods. Semi-Structured Interviews and Focus Groups*; Rand National Defense Research Inst.: Santa Monica, CA, USA, 2009.
61. Ronan, K.R.; Alisic, E.; Towers, B.; Johnson, V.A.; Johnston, D.M. Disaster preparedness for children and families: a critical review. *Curr. Psychiatry Rep.* **2015**, *17*, 58. [CrossRef]
62. Federal Emergency Management Agency. Helping Pets. 2015. Available online: https://www.fema.gov/helping-pets (accessed on 10 November 2017).
63. Lovekamp, W.E.; McMahon, S.K. I have a Snickers Bar in the Trunk of My Car: Student Narratives of Disaster Risk, Fear, Preparedness, and Reflections on Union University. *Int. J. Mass Emerg. Disasters* **2011**, *29*, 132–148.
64. Thompson, C.N.; Brommer, D.M.; Sherman-Morris, K. Pet Ownership and the Spatial and Temporal Dimensions of Evacuation Decisions. *Southeast. Geogr.* **2012**, *52*, 253. [CrossRef]
65. Gaillard, J.C.; Gorman-Murray, A.; Fordham, M. Sexual and gender minorities in disaster. *Gender Place Cult.* **2017**, *24*, 18–26. Available online: http://proxy-remote.galib.uga.edu/login?url=http://search.ebscohost.com/login.aspx?direct=true&db=edsbl&AN=vdc.100048551427.0x000001&site=eds-live (accessed on 10 May 2018). [CrossRef]
66. Brooks, H.; Rushton, K.; Walker, S.; Lovell, K.; Rogers, A. Ontological security and connectivity provided by pets: a study in the self-management of the everyday lives of people diagnosed with a long-term mental health condition. *BMC Psychiatry* **2016**, *16*, 409. [CrossRef]
67. Zilcha-Mano, S.; Mikulincer, M.; Shaver, P.R. Pets as safe havens and secure bases: The moderating role of pet attachment orientations. *J. Res. Personal.* **2012**, *46*, 571–580. [CrossRef]
68. Thompson, K. Save me, save my dog: Increasing natural disaster preparedness and survival by addressing human-animal relationships. *Aust. J. Commun.* **2013**, *40*, 123–136.
69. Binfet, J.-T. The Effects of Group-Administered Canine Therapy on University Students' Wellbeing: A Randomized Controlled Trial. *Anthrozoös* **2017**, *30*, 397–414. [CrossRef]
70. Sarathchandra, D.K.; Haltinner, K.; Lichtenberg, N. College Students' Cybersecurity Risk Perceptions, Awareness, and Practices. In Proceedings of the 2016 Cybersecurity Symposium, Coeur d'Alene, ID, USA, 18–20 April 2016; pp. 68–73.
71. Dunlap, M.; Scoggin, J.; Green, P.; Davi, A. White students' experiences of privilege and socioeconomic disparities: Toward a theoretical model. *Mich. J. Community Serv. Learn.* **2007**, *13*, 19–30.
72. Clare, E. Stolen bodies, reclaimed bodies: Disability and queerness. *Public Cult.* **2001**, *13*, 359–365. [CrossRef]
73. Samuels, E. *Fantasies of Identification: Disability, Gender, Race*; NYU Press: New York, NY, USA, 2014.
74. Bolin, B.; Kurtz, L.C. Race, class, ethnicity, and disaster vulnerability. In *Handbook of Disaster Research*; Springer: Cham, Switzerland, 2018; pp. 181–203.

animals

MDPI

Review

Back to the Future: A Glance Over Wolf Social Behavior to Understand Dog–Human Relationship

Giada Cordoni [1] and Elisabetta Palagi [1,2,*]

1 Natural History Museum, University of Pisa, Via Roma 79, 56011 Calci, Pisa, Italy; giada.cordoni@unipi.it
2 Ethology Unit, Department of Biology, University of Pisa, Via Volta 6, 56126 Pisa, Italy
* Correspondence: elisabetta.palagi@unipi.it

Received: 25 October 2019; Accepted: 11 November 2019; Published: 18 November 2019

Simple Summary: Wolves, the ancestors of dogs, are one of the most cooperative canine species. This cooperative propensity derives from the fact that each subject needs other group members to obtain resources and increase survival. The pack functions as a unit in which each individual collaborates in territory defence, hunting, and rearing of offspring. For this reason, even though a clear hierarchy exists among wolves, subordinates can provide help to dominants to obtain social tolerance in a sort of commodity exchange. Wolves can make peace after aggression, console victims of a conflict, and calm down the aggressors. This set of behaviors, also called post-conflict strategies, requires a social attentiveness towards others' emotional state and the ability to coordinate appropriate reactions. Adult wolves also play. They engage in play fighting, which strongly resembles real fighting, by finely modulating their motor actions and quickly interpreting playmates' *intentions*, thus maintaining the non-serious playful mood. All these cognitive and social skills were a fertile ground for the artificial selection operated by humans to redirect the cooperative propensity of wolves towards dog–human affective relationship.

Abstract: This review focuses on wolf sociobiology to delineate the traits of *cooperative baggage* driven by natural selection (wolf-wolf cooperation) and better understand the changes obtained by artificial selection (dog-human cooperation). We selected some behaviors of the dog's ancestors that provide the basis for the expression of a cooperative society, such as dominance relationships, leverage power, post-aggressive strategies, and playful dynamics between pack members. When possible, we tried to compare the data on wolves with those coming from the dog literature. Wolves can negotiate commodities when the interacting subjects occupy different ranking positions by bargaining social tolerance with helping and support. They are able to manage group disruption by engaging in sophisticated post-conflict maneuvers, thus restoring the relationship between the opponents and reducing the spreading of aggression in the group. Wolves engage in social play also as adults to manipulate social relationships. They are able to flexibly adjust their playful interactions to minimize the risk of escalation. Complex cognitive abilities and communicative skills are probably the main proximate causes for the evolution of inter-specific cooperation in wolves.

Keywords: social tolerance; social attentiveness; reconciliation; consolation; appeasement; play fighting; leverage; behavioural plasticity; *Canis lupus lupus*

---

1. Introduction

> *"Domestication is a sustained multigenerational, mutualistic relationship in which one organism assumes a significant degree of influence over the reproduction and care of another organism in order to secure a more predictable supply of a resource of interest, and through which the partner organism gains advantage over individuals that remain outside this relationship, thereby benefitting and often increasing the fitness of both the domesticator and the target domesticate".* [1]

Even though we intuitively understand what domestication is, there is a surprising lack of consensus on its definition. Beyond the agreement that domestication involves a relationship between a *domesticator* and a *domesticated organism*, there are many debates on what this relationship entails and how and when it occurs. Many definitions of this process take into account only the perspective of the *domesticator*, emphasizing the impact of humans in this role. Historically, humans deliberately and opportunistically select a wild species for creating its "domesticated counterpart" of whom they control all aspects of its life cycle [1]. According to "*domestication syndrome*" [2], a set of morphological, physiological, reproductive, and behavioral traits can be observed in domesticated species but not in their wild ancestors. In animals, these traits can include, for example, increased fecundity, altered coat color, reduced body size, facial neoteny, increased docility, and hypersociability [3–5].

Dog, the oldest domesticated animal by humans, certainly shows many traits of the "*domestication syndrome*" such as, reduced body size and snout length and increased docility, tameness, and playfulness [5,6]. Many researchers suggest that the beginning of dog domestication took place in the Early Upper Paleolithic (~30,000 years ago), when people still lived in small groups as hunter–gatherers and agriculture was not yet practiced [7–10]. Nevertheless, the fossil remains confidently indicate the appearance of dog in Europe ~15,000 years ago [11]. Basing on morphological and genetic analyses, wolves (*Canis lupus lupus*) are undoubtedly the ancestors of modern dogs [9,11–13]; while the wild wolf phenotype changed markedly, the genotype changed only minimally, leaving domestic dogs, genetically speaking, still as wolves [10]. The domestic relationship between people and dogs is the result of a wolf ecological strategy to cope successfully with the Late Pleistocene environmental changes due to the increased human population. The plasticity characterizing wolves permitted them modifying their ecological niche by joining the human niche; people possibly facilitated this change by incorporating some young wolves into their groups and by selecting over time the more docile and tameness subjects [10,13,14].

According to the *Domestication Hypothesis*, it has been proposed that through an evolutionary and ontogenetic positive feedback processes, dogs have become more socially tolerant and attentive than wolves, two characteristics that are crucial for cooperation to occur [15,16]. However, the studies supporting the *Domestication Hypothesis* were mainly based on wolf–dog behavioral difference in relation to their interactions with humans [15–21]. Range & Virányi [20,21] proposed an alternative (but not exclusive) theory, the *Canine Cooperation Hypothesis*, according to which these differences may reflect only an improved capacity of dogs to accept humans as social partners instead of an increased in their general tolerance, attention, and cooperation degree. Some cognitive studies have shown that wolves can attentively use the information provided by a familiar human to solve a task [22], can follow human gaze as readily as conspecific gaze [23] and are more successful than dogs in copying the actions of conspecifics [24,25]. Moreover, when wolves and dogs (reared under the same conditions) were faced with a series of object-choice tasks, wolves showed similar results to dogs in responding to communicative and behavioral cues, but they outperformed dogs in their ability to follow causal cues [26]. All these authors suggested that high level of cooperativeness characterizing wolf society may come together with a high propensity to pay close attention to others' actions. Also, by moving the focus from human–wolf to wolf–wolf interactions, researchers have highlighted the high cooperativeness and cohesiveness characterizing wolf packs [20,26–28].

This review aims at delineating the possible pathways of the behavioral changes that, over the time, have led from wolves to dogs and, consequently, to the strong dog-human relationship. To address this issue, we deal with different aspects of wolf sociality. In particular, we "take a glance" to dominance, post-aggressive and playful dynamics between pack members by comparing the findings on wolves with those on dogs [29]. Can the wolf-dog behavioral difference be credited only to the domestication process? Can the domestication process have induced a shift of social tolerance and attentiveness from conspecifics to humans, thus leading dogs towards an exclusive inter-specific cooperation?

2. Social Tolerance by Dominants, Leverage Power by Subordinates, and Peaceful Strategies by All Group Members

Wolf pack is defined as a cohesive family group, including a long-term bond breeding pair, mature offspring, and pups; occasionally, an unrelated individual may join the group [30,31]. All wolves participate in pack life by creating a system of division of *labour* in which individuals cooperatively hunt and defend their territories and collectively rear the pups [30,32]. The socialization begins at around four months of age when cubs start to follow the adults on hunting trips ("hunting school"). During this period, puppies improve their motor and perception skills and perfect mutual interaction and coordination with conspecifics. The affinitive relationships develop during puberty, when maturing individuals are slowly integrated into the daily life of the group [30–32]. The strict social association between pack members finds support in a study of Cassidy and McIntyre [33]. The authors recorded 121 territorial inter-pack conflicts in Yellowstone National Park, and in 17.6% of cases, wolves engaged in aggression to defend their pack fellows.

Within the pack, puppies generally occupy lower ranking positions compared to their parents and older siblings. When wolves reach sexual maturity (~2 years), they disperse from their natal group, attempt to pair with other dispersed wolves and start their own packs, thus avoiding competing for dominant-breeder *status* with natal group members [30,32]. However, under some conditions, both in the wild and in captivity, mature individuals delay dispersal or do not disperse at all; in these cases, competition for dominant-rank may be stronger [32,34,35]. In captive packs, wolves often have a linear hierarchy in which all males are dominant over females [32,36,37]. Nevertheless, the more appropriate term to define dominance relationships in a typical wolf pack (nuclear, extended or complex families) is "age-graded dominance hierarchy" [31,36]. Moreover, the subordinate individuals can sometimes oppose their leader's actions; for this reason, Zimen [38] defined the leadership in wolf packs as a "qualified democracy", in which no subject decides alone to carry out activities that are crucial for the group survival. Recently, Range and colleagues [21] found that, under feeding conditions, captive wolves are more tolerant compared to dogs. Indeed, high-ranking dogs monopolized the resources while low-ranking individuals showed deference by staying apart without trying to obtain food from the dominant subjects. Conversely, subordinate wolves overtly challenged the dominant ones to subtract food from them. Wolves are cooperative hunters [34] and, in term of mutual beneficial exchanges, all members of the pack have the possibility to access food, independently of their ranking position. On the contrary, domestic dogs rely on humans for food and feral dogs are solitary scavengers, hence, both do not depend on conspecifics' support for obtaining the resource [39]. These findings suggest that, in the wolf society, the power is not entirely "in the hands" of the physically stronger subjects. The subordinates can exert leverage power [40,41] because of their support to the pack life and their cooperation is gained by high-ranking individuals through peaceful sharing instead of aggressive coercion [42].

Nevertheless, despite this cooperating social system, the presence of aggression is the other inevitable side of the coin that leads to a temporary interruption of the inter-individual relationships [29,43]. To cope with aggression and the consequent social damage, as it occurs in many social mammals (human primates [44]; non-human primates, [45–49]; dolphins [50]; spotted hyenas, [51]; red-necked wallaby [52], wolves engage in post-conflict contacts such as reconciliation (i.e., the first affinitive contact exchanged by the former opponents relatively shortly after a conflict [53]).

By analyzing 3344 conflicts, Cordoni and Palagi [37] provided the first evidence for the occurrence of reconciliation in wolves by observing the pack hosted at the Pistoia Zoo (Italy), which was categorized as a "disrupted family" due to the absence of the alpha female. The high level of conciliatory contacts was uniformly distributed across the different sex–class combinations. Interestingly, reconciliation was not linked with rank distance between opponents but it positively correlated with coalitionary support (defined as a third party joining an ongoing conflict by attacking one of the opponents in support of the other, [54]). Generally, in social mammals, the high level of support can unveil high level of cooperation. In wolves, alliances [55] and reconciliation [37] act as diffuse non-dispersive mechanisms that concur in strengthening group cohesiveness. The occurrence of reconciliation was confirmed by other studies both in wild [56] and captive wolves [57]. Also, in the wild condition, wolves showed high level of conciliatory contacts and, once more, the finding was explained by the authors in the light of the strong cohesion between pack members [56].

Even though few doubts remain for the occurrence of reconciliation in wolves [37,56,57], contrasting results derive from canine reconciliation. Cools and co-workers [58] demonstrated reconciliation in small groups of dogs sharing a pen. Recently, Cafazzo and colleagues [57] by studying four captive small packs of dogs and wolves provided evidence for reconciliation in wolves, but not in dogs. Indeed, in this study, the dogs avoided affiliating with their opponents after conflicts. The social repairing function ascribed to the reconciliation mechanism [53] is probably useless for dogs. The difference in socio-ecological habits between dogs and wolves may cause the difference in their conciliatory tendency. In dogs, the absence of cooperative hunting and collective rearing of offspring limits the need to maintain friendly and peaceful social relationships with conspecifics [57].

Beyond reconciliation, other types of post-conflict interactions can occur. Group members not involved in aggression (bystanders) can spontaneously offer friendly contacts to both victims ("consolation"; [53]) and aggressors ("appeasement"; [59]). Even though the cognitive and emotional skills underpinning these contacts are still under debate, "consolation" and "appeasement" seem to serve different functions.

In wild and captive wolves both "consolation" and "appeasement" are present [43,54,56]. In the Pistoia pack [43,54], the two post-conflict behavioral strategies occur within two minutes after the end of the aggression and are performed with comparable levels. Despite these similarities, "consolation" and "appeasement" seem to play different roles. In the Pistoia wolves, in about 45% of cases, bystanders offer affinitive contacts to aggressors, which generally occupy high-ranking positions. Such calming interactions have the immediate effect to reduce the likelihood of renewed aggression toward other group members by the previous aggressor. On the other hand, affiliation that bystanders direct towards the victims follows the relationship quality linking the subjects more than their hierarchical positions: the stronger the bonding, the higher the frequency of affinitive contacts. Furthermore, these contacts protect the victim against the reiterated attacks from the previous aggressor. In sum, "consolation" may represent a "victim protection" strategy and "appeasement" a "bystander protection" strategy, thus highlighting the functional dichotomy characterizing these two behaviors [43,54]. The Figure 1 illustrates a reciprocal muzzle licking during a post-conflict triadic affiliation also named "consolation".

(a) (b)

Figure 1. Reciprocal muzzle licking during a post-conflict triadic affiliation ("consolation") between an adult (the consoler) (**a**) and an immature subject (the victim) (**b**) Photos by Elisabetta Palagi.

Although bystander post-conflict affiliation has been demonstrated in dogs [58], it does not seem that familiarity between interacting subjects significantly affects the behaviour. Cools and colleagues [58] showed that bystander affinitive contacts were more frequently directed to victims than to aggressors. Moreover, victim-directed affiliation was markedly higher than affinitive interactions between the former opponents. The scarcity of data did not permit the authors to investigate the possible functions of dog "consolation" and "appeasement" leaving the question still open.

3. The Playful Wolf: Tactics and Plasticity for a Deeper Knowledge of Others

Beyond dominance, affiliation, and conflict management, social play is another type of interaction that can affect the social dynamics within a group [60]. Immature individuals of many mammalian species engage in different types of playful activities [61]. The pervasive distribution of play suggests that the core neural circuitry underpinning the modulation of this behavior may have evolved early in mammals and it may be shared by different species [62].

During play, animals execute motor action patterns that are recruited from "serious" contexts such as agonistic, anti-predatory and sexual [61]. However, these patterns are performed in an exaggerated, incomplete, repeated, mixed and unexpected manner [63–65].

The first playful experience occurs between mothers and infants [66]; these sessions represent a good training ground for preparing infants to the future interactions with peers [67,68]. During ontogeny, play concurs in implementing physical, cognitive, emotional, and social skills of individuals by guaranteeing a safe and fruitful "environment" in which making training [45,63,69–73]. Held and Špinka [74] suggested that play could provide individuals with psychological benefits by promoting opioid-mediated pleasurable experiences: the better individuals feel, the more individuals play. The resultant positive feedback between play and animal well-being becomes an important tool for improving and ameliorating the management of animals in captivity.

Although benefits provided by social play can vary depending on context (e.g., pre-feeding, mating period), habitat, playmate characteristics (e.g., sex, age, and rank) and species sociality (e.g., tolerant versus despotic), the intrinsic nature of play requires an implicit agreement, cooperation, and negotiation between players, who have to trust one another [65,75–77]. Playmates reach this goal by fine-tuning their contact interactions and by avoiding the performance of behaviors that might be misinterpreted. By playing fairly, animals may acquire the social competence and rules that are at the basis of a peaceful coexistence [60].

Peaceful cohesiveness is a feature of wolf society that is guaranteed also by playful activity [54,60,78]. Although most studies have focused on play in domestic dogs [6,79–81], this behavior does not appear as an artefact of domestication since also wolves play even during adulthood [78,82].

It has been suggested that, through play, adult subjects can evaluate and manipulate the social relationships with group-members [71]. Cafazzo and colleagues [82] investigated play behavior in four captive wolf packs, two composed by immature peers and the other two by mixed-age subjects (puppies and adults). A positive linkage between play frequency and relationship quality was found; indeed, those dyads that spent more time in relaxed play (defined as play sessions involving a limited number of offensive patterns) engaged in more interactions belonging to the affinitive domain. Moreover, in mixed-age groups, but not in peer groups, the frequencies of aggression are negatively correlated with play levels. Interpreting this result, the authors suggested that play can limit aggressiveness between group-members only after the establishment of clear hierarchical positions.

The observation of the wolf pack hosted at the Pistoia Zoo (Italy) led to contrasting results compare to those of Cafazzo and colleagues [82], although direct comparison can be misleading due to the different conditions under which the wolves were reared. In the Pistoia group, composed by adult related individuals, playful activity was not affected by both social relationship quality and aggression level, but it was strictly linked to dominance. Play was negatively correlated to the rank distance between fellows; in other words, subjects with closer ranking position played more than subjects differing greatly in rank [78]. In another study on the same pack, Cordoni and Palagi [54] compared the level of aggressiveness and steepness of hierarchy in two different periods (Sample 1 and Sample 2). During Sample 1, two adult females died and this event probably provoked delayed social effects which were manifest during the Sample 2 characterized by higher aggressiveness and steeper linear hierarchy. The re-arrangement of the dominance relationships within a pack occurs gradually and it is usually manifest later in time respect to the occurrence of the perturbing event (e.g., the removal of particular subjects; [83]). Intriguingly, in the Pistoia pack, the hierarchical difference between the two Samples had a significant effect on playful but not on affiliative dynamics. During the riskier period (Sample 2), wolves consistently reduced their playful activity, avoided playing during high-tension contexts (e.g., pre-feeding time) and limited the number of players per session by preferring dyadic (only two players) at polyadic interactions [54]. As a whole, these findings suggest that "rank rules dictate play rules". By playing, wolves can acquire information about physical and cognitive skills of fellows with whom they can compete in the future for dominant position. Nevertheless, in order to maintain a "not serious" mood, wolves had to manage play in a flexible manner by place the interaction in the right time, in the right context, and with the right players [45,54,72,73,78].

Pal [84] observed 24 free-ranging dog puppies from birth to 13 weeks of age. In each litter, there were particular subjects that initiated offensive play more often than the other siblings did. Moreover, the same-sex offensive play was the predominant interaction between puppies. These findings suggest that, starting from five weeks of age, free-ranging dogs can employ playful activity to acquire information about individuals with whom they may compete in the future for reaching dominance positions, as it occurs in wolves [84].

In domestic dogs, adult-adult play does not follow the 50–50 rule that is a balance between offensive ("dominant") and defensive ("subordinate") patterns exchanged by playmates during the interaction. In play, adult dogs seem to hold the same dominance position they have outside the playful context [79]. Also, dog puppies do not adhere to the 50–50 rule by performing offensive playful patterns much more frequently than defensive ones [80]. Play becomes more asymmetrical as the puppies matured. Along with dog developmental pathway, the winning and the losing positions during playful interactions mirror the dominance relationships between individuals. Intriguingly, Cordoni and colleagues [6], by studying 49 domestic dogs in an off-leash dog-park, evidenced that the level of play asymmetry did not differ between "friends" (i.e., dogs that lived together or regularly interacted) and "strangers" (i.e., dogs that have never interacted before the observation). In the light of this result, dog play can be not only predictive of the dominance relationships between playmates, but it may serve an important function in maintaining good social bonds with specific partners. Nevertheless, independently of the inter-individual relationship, the playful sessions characterized by high asymmetry, and, consequently, by high competition, had a shorter duration compared to the more balanced sessions. The decrease in play duration and the use of clear communication can represent strategies used by dogs for overcoming the risk of escalation during very asymmetric playful interactions [6].

In wolf puppies, social play seems to be well-balanced with both immature partners performing a similar amount of self-handicapping behaviors, this reciprocity decreases when one of the players is an adult. When play involves two mismatched wolves, the session generally becomes more asymmetric [85]. The flexibility in managing the playful arousal is evident in the study by Cordoni & Palagi [54]. The adult wolves of the Pistoia group changed their play modality according to the period of observation. In Sample 1, the period characterized by low level of hierarchical steepness and aggressiveness, subjects engaged in more self-handicapping and role reversal manoeuvres, thus making their playful sessions more symmetric. This suggests that animals are able to flexibly adjust their playful tactics according to the social circumstances and that the play asymmetry is not always predictive of the dominance status of the players. In this view, the analysis of some social factors such as the exact quantification of (i) the hierarchical steepness, (ii) the level of affiliation and (iii) the bidirectionality of agonistic conflicts is mandatory before beginning a study on play in adult wolves.

4. Conclusion

"Can the wolf-dog behavioral difference be credited only to the domestication process? Can the domestication process have induced a shift of social tolerance and attentiveness from conspecifics to humans thus leading dogs towards an exclusive inter-specific cooperation?"

Through a systematic comparison between data coming from the available studies on wolves and dogs we propose possible answers to these questions.

A group of wolves moves as a unique entity, with each subject relying on others' support to gain benefits and increase survival and fitness. In this network of cooperating individuals, some are dominants and some subordinates. These hierarchical relationships are not based on a mere deference of subordinates towards the dominants but on an exchange of services that both counterparts seem to put in act. In the wolf cooperative system, the social tolerance provided by dominants to subordinates can be repaid by subordinates with their help and support in the group maintenance activities. This does not necessarily mean that individuals cannot use their high-ranking position to obtain priority in certain domains. Indeed, a dominant can exert a strong control over resources to the detriment of subordinates. However, keeping everything under control through threats and overt aggression is energetically demanding, and, for this reason, trading can be also present when coercion by dominants is possible. In these situations, also the subordinates can exert their little amount of power over the dominants, that is called leverage. This delicate equilibrium permits and sustains the development of cooperation in social groups.

A further prerequisite for the development of a cooperative society is social attentiveness, which occurs when a subject is sensitive to others' behaviors and needs. The level of attentiveness can change according to different factors such as contexts, kinship, relationship quality shared by the interacting subjects. When an individual is attentive to fellows, he/she can adjust and coordinate his/her competitive or cooperative reactions. The link between attention and action coordination appears clear in the "consolation" and "appeasement" dynamics occurring after a conflict. The ability to perceive the emotional mood of the victim (anxiety) or the aggressor (arousal) is highly beneficial to the subject who can reduce, through the post conflict affiliative interaction, the probability of further aggression thus concurring in the maintenance of group cohesion.

An important mix of cooperation and social attentiveness is mandatory during play fighting, an activity which is complex to manage and that can lead easily to misinterpretation. Animals have to finely read and rapidly interpret each single pattern of the playmate in order to react in an appropriate manner. This needs a profound knowledge of the partner, because the reaction can change as a function of the relationship linking the two players. This is a highly cognitive demanding behavior which comes into place when animals are strongly motivated to cooperate.

In conclusion, the domestication process has acted over *cooperative baggage* already present in the sociobiology of the dog's ancestors. In wolves, such *baggage* has been driven by natural selection towards conspecifics, while in dogs, this *baggage* has been redirected by the artificial selection towards humans.

Author Contributions: Conceptualization, G.C. and E.P.; Writing—original draft preparation, G.C. and E.P.; Writing—review and editing, G.C. and E.P.

Acknowledgments: Thanks are due to the Pistoia Zoo (Italy), its Director Paolo Cavicchio, the wolf keepers for allowing and facilitating the work; Maria Vittoria Ulivi, Iolanda Ferranti, Chiara Stefani and Federica Pacini for their collaboration in the data collection.

Conflicts of Interest: The authors declare no conflicts of interest.

References

1. Zeder, M.A. Core questions in domestication research. *Proc. Natl. Acad. Sci. USA* **2015**, *112*, 3191–3198. [CrossRef]
2. Hammer, K. Das domestikationssyndrom. *Kurturpflanze* **1984**, *32*, 11–34. [CrossRef]
3. Larson, G.; Piperno, D.R.; Allabyd, R.G.; Purugganan, M.D.; Andersson, L.; Arroyo-Kalin, M.; Barton, L.; Vigueira, C.C.; Denham, T.; Dobney, K.; et al. Current perspectives and the future of domestication studies. *Proc. Natl. Acad. Sci. USA* **2014**, *111*, 6139–6146. [CrossRef] [PubMed]
4. Sánchez-Villagra, M.R.; Geiger, M.; Schneider, R.A. The taming of the neural crest: A developmental perspective on the origins of morphological covariation in domesticated mammals. *R. Soc. Open Sci.* **2016**, *3*, 160107. [CrossRef] [PubMed]

5. von Holdt, B.M.; Shuldiner, E.; Janowitz Koch, I.; Kartzinel, R.J.; Hogan, A.; Brubaker, L.; Wanser, S.; Stahler, D.; Ostrander, E.A.; Sinsheimer, J.S.; et al. Structural variants in genes associated with human Williams-Beuren syndrome underlie stereotypical hypersociability in domestic dogs. *Sci. Adv.* **2017**, *3*, e1700398. [CrossRef] [PubMed]
6. Cordoni, G.; Nicotra, V.; Palagi, E. Unveiling the "secret" of play in dogs (*Canis lupus familiaris*): Asymmetry and signals. *J. Comp. Psychol.* **2016**, *130*, 278–287. [CrossRef] [PubMed]
7. Ovodov, N.D.; Crockford, S.J.; Kuzmin, Y.V.; Higham, T.F.G.; Hodgins, G.W.L.; van der Plicht, J. A 33,000-year-old incipient dog from the Altai mountains of Siberia: Evidence of the earliest domestication disrupted by the Last Glacial Maximum. *PLoS ONE* **2011**, *6*, e22821. [CrossRef]
8. Germonpré, M.; Lázničková-Galetová, M.; Sablin, M.V. Palaeolithic dog skulls at the Gravettian Předmostí site, the Czech Republic. *J. Archaeol. Sci.* **2012**, *39*, 184–202. [CrossRef]
9. Thalmann, O.; Shapiro, B.; Cui, P.; Schuenemann, V.J.; Sawyer, D.K.; Greenfield, D.L.; Germonpré, M.B.; Sablin, M.V.; López-Giráldez, F.; Domingo-Roura, X.; et al. Complete mitochondrial genomes of ancient canids suggest a European origin of domestic dogs. *Science* **2013**, *342*, 871–874. [CrossRef]
10. Morey, D.F.; Jeger, R. Paleolithic dogs: Why sustained domestication then? *J. Archaeol. Sci.* **2015**, *3*, 420–428. [CrossRef]
11. Frantz, L.A.F.; Mullin, V.E.; Pionnier-Capitan, M.; Lebrasseur, O.; Ollivier, M.; Perri, A.; Linderholm, A.; Mattiangeli, V.; Teasdale, M.D.; Dimopoulos, E.A.; et al. Genomic and archaeological evidence suggest a dual origin of domestic dogs. *Science* **2016**, *352*, 1228–1231. [CrossRef] [PubMed]
12. Skoglund, P.; Ersmark, E.; Palkopoulou, E.; Dalén, L. Ancient wolf genome reveals an early divergence of domestic dog ancestors and admixture into high-latitude breeds. *Curr. Biol.* **2015**, *25*, 1–5. [CrossRef]
13. Germonpré, M.; Lázničková-Galetová, M.; Sablin, M.V.; Bocherens, H. Self-domestication or human control? The Upper Palaeolithic domestication of the wolf. In *Hybrid Communities. Biosocial Approaches to Domestication and Other Trans-Species Relationships*; Stepanoff, C., Vigne, J.D., Eds.; Routledge: London, UK, 2018; pp. 39–64.
14. Perri, A.A. Wolf in dog's clothing: Initial dog domestication and Pleistocene wolf variation. *J. Archaeol. Sci.* **2016**, *68*, 1–4. [CrossRef]
15. Miklósi, Á.; Kubinyi, E.; Topál, J.; Gácsi, M.; Virányi, Z.; Csányi, V. A simple reason for a big difference: Wolves do not look back at humans, but dogs do. *Curr. Biol.* **2003**, *13*, 763–766. [CrossRef]
16. Virányi, Z.; Gácsi, M.; Kubinyi, E.; Topál, J.; Belényi, B.; Ujfalussy, D.; Miklósi, Á. Comprehension of human pointing gestures in young human-reared wolves and dogs. *Anim. Cogn.* **2008**, *11*, 373–387. [CrossRef] [PubMed]
17. Gácsi, M.; Gyõri, B.; Virányi, Z.; Kubinyi, E.; Range, F.; Belényi, B.; Miklósi, Á. Explaining dog wolf differences in utilizing human pointing gestures: Selection for synergistic shifts in the development of some social skills. *PLoS ONE* **2009**, *4*, e6584. [CrossRef]
18. Udell, M.A.; Dorey, N.R.; Wynne, C.D. Can your dog ready our mind? Understanding the causes of canine perspective taking. *Lear. Behav.* **2011**, *39*, 289–302. [CrossRef]
19. Gácsi, M.; Vas, J.; Topál, J.; Miklósi, Á. Wolves do not join the dance: Sophisticated aggression control by adjusting to human social signals in dogs. *App. Anim. Behav. Sci.* **2013**, *145*, 109–122. [CrossRef]
20. Range, F.; Virányi, Z. Tracking the evolutionary origins of dog-human cooperation: The Canine Cooperation Hypothesis. *Front. Psychol.* **2015**, *5*, 1582. [CrossRef]
21. Range, F.; Ritter, C.; Virányi, Z. Testing the myth: Tolerant dogs and aggressive wolves. *Proc. R. Soc. B* **2015**, *282*, 20150220. [CrossRef]
22. Range, F.; Virányi, Z. Social learning from humans or conspecifics: Differences and similarities between wolves and dogs. *Front. Psychol.* **2013**, *4*, 868. [CrossRef] [PubMed]
23. Range, F.; Virányi, Z. Development of gaze following abilities in wolves (*Canis lupus*). *PLoS ONE* **2011**, *6*, e16888. [CrossRef] [PubMed]
24. Range, F.; Virányi, Z. Wolves are better imitators of conspecifics than dogs. *PLoS ONE* **2014**, *9*, e86559. [CrossRef] [PubMed]
25. Maglieri, V.; Prato-Previde, E.; Tomasi, E.; Palagi, E. Wolf-like or dog-like? A comparison of gazing behaviour across three dog breeds tested in their familiar environments. *R. Soc. Open Sci.* **2019**, *6*, 190946. [CrossRef]
26. Lampe, M.; Bräuer, J.; Kaminski, J.; Virányi, Z. The effects of domestication and ontogeny on cognition in dogs and wolves. *Sci. Rep.* **2017**, *7*, 11690. [CrossRef]

27. Marshall-Pescini, S.; Virányi, Z.; Range, F. The effect of domestication on inhibitory control: Wolves and dogs compared. *PLoS ONE* **2015**, *10*, e0118469. [CrossRef]
28. Hansen, W.C.; Fitzpatrick, J.; Tapper, I.; Temrin, H. Wolf (*Canis lupus*) hybrids highlight the importance of human-directed play behavior during domestication of dogs (*Canis familiaris*). *J. Comp. Psychol.* **2018**, *132*, 373–381. [CrossRef]
29. Cordoni, G.; Palagi, E. Play and post-conflict mechanisms in wolves: A comparison with human and non-human primates. In *Wolves: Biology, Behavior and Conservation*; Gotsiridze-Columbus, N., Ed.; Nova Science Publishers: Hauppauge, NY, USA, 2012; pp. 133–156.
30. Mech, L.D.; Boitani, L. Wolf social ecology. In *Wolves: Behaviour, Ecology, and Conservation*; Mech, L.D., Boitani, L., Eds.; The University of Chicago Press: Chicago, IL, USA, 2003; pp. 1–34.
31. Packard, J.M. Wolf behavior: Reproductive, social and intelligent. In *Wolves: Behaviour, Ecology, and Conservation*; Mech, L.D., Boitani, L., Eds.; The University of Chicago Press: Chicago, IL, USA, 2003; pp. 35–65.
32. Mech, L.D. *Alpha status*, dominance, and division of labor in wolf packs. *Can. J. Zool.* **1999**, *77*, 1196–1203. [CrossRef]
33. Cassidy, K.A.; McIntyre, R.T. Do gray wolves (*Canis lupus*) support pack mates during aggressive inter-pack interactions? *Anim. Cogn.* **2016**, *19*, 939–947. [CrossRef]
34. MacNulty, D.R.; Smith, D.W.; Mech, L.D.; Vucetich, J.A.; Packer, C. Nonlinear effects of group size on the success of wolves hunting elk. *Behav. Ecol.* **2012**, *23*, 75–82. [CrossRef]
35. Stahler, D.R.; MacNulty, D.R.; Wayne, R.K.; vonHoldt, B.; Smith, D.W. The adaptive value of morphological, behavioural and life-history traits in reproductive female wolves. *J. Anim. Ecol.* **2013**, *82*, 222–234. [CrossRef] [PubMed]
36. Zimen, E. A wolf pack sociogram. In *Wolves of the World*; Harrington, F.H., Paquet, P.C., Eds.; Noyes Publishers: Park Ridge, IL, USA, 1982; pp. 282–322.
37. Cordoni, G.; Palagi, E. Reconciliation in wolves (*Canis lupus*): New evidence for a comparative perspective. *Ethology* **2008**, *114*, 298–308. [CrossRef]
38. Zimen, E. *The Wolf: A Species in Danger*; Delatorre Press: New York, NY, USA, 1981.
39. Vanak, A.T.; Gompper, M.E. Dogs *Canis familiaris* as carnivores: Their role and function in intraguild competition. *Mammal Rev.* **2009**, *39*, 265–283. [CrossRef]
40. Lewis, R.J. Beyond dominance: The importance of leverage. *Q. Rev. Biol.* **2002**, *77*, 149–164. [CrossRef] [PubMed]
41. Norscia, I.; Palagi, E. *The Missing Lemur Link: An Ancestral Step in the Evolution of Human Behaviour*; Cambridge University Press: Cambridge, UK, 2016; pp. 1–289.
42. Dale, R.; Range, F.; Stott, L.; Kotrschal, K.; Marshall-Pescini, S. The influence of social relationship on food tolerance in wolves and dogs. *Behav. Ecol. Sociobiol.* **2017**, *71*, 107. [CrossRef]
43. Palagi, E.; Cordoni, G. Postconflict third-party affiliation in *Canis lupus*: Do wolves share similarities with the great apes? *Anim. Behav.* **2009**, *78*, 979–986. [CrossRef]
44. Cordoni, G.; Demuru, E.; Ceccarelli, E.; Palagi, E. Play, aggressive conflict and reconciliation in pre-school children: What matters? *Behaviour* **2016**, *153*, 1075–1102. [CrossRef]
45. Palagi, E.; Antonacci, D.; Norscia, I. Peacemaking on treetops: First evidence of reconciliation from a wild prosimian (*Propithecus verreauxi*). *Anim. Behav.* **2008**, *76*, 737–747. [CrossRef]
46. Palagi, E.; Cordoni, G.; Borgognini Tarli, S. Possible roles of consolation in captive chimpanzees (*Pan troglodytes*). *Am. J. Phys. Anthropol.* **2006**, *127*, 105–111. [CrossRef]
47. Palagi, E.; Leone, A.; Demuru, E.; Ferrari, P.F. High-ranking geladas protect and comfort others after conflicts. *Sci. Rep.* **2018**, *8*, 15291. [CrossRef]
48. Cordoni, G.; Palagi, E.; Borgognini Tarli, S. Reconciliation and consolation in captive western gorillas. *Int. J. Primatol.* **2006**, *27*, 1365–1382. [CrossRef]
49. Palagi, E.; Norscia, I. Bonobos protect and console friends and kin. *PLoS ONE* **2013**, *8*, e79290. [CrossRef] [PubMed]
50. Weaver, A. Conflict and reconciliation in captive bottlenose dolphins, *Tursiops truncatus*. *Mar. Mam. Sci.* **2003**, *19*, 836–846. [CrossRef]
51. Wahaj, S.A.; Guse, K.R.; Holekamp, K.E. Reconciliation in spotted hyenaa (*Crocuta crocuta*). *Ethology* **2001**, *107*, 1057–1074. [CrossRef]

52. Cordoni, G.; Norscia, I. Peace-making in marsupials: The first study in the red-necked wallaby (*Macropus rufogriseus*). *PLoS ONE* **2014**, *9*, e86859. [CrossRef] [PubMed]
53. de Waal, F.B.M.; van Roosmalen, A. Reconciliation and consolation among chimpanzees. *Behav. Ecol. Sociobiol.* **1979**, *5*, 55–66. [CrossRef]
54. Palagi, E.; Cordoni, G. Being a victim or an aggressor: Different functions of triadic post-conflict interactions in wolves (*Canis lupus lupus*). *Aggress. Behav.* **2015**, *41*, 526–536.
55. Fox, M.V. *The Soul of the Wolf*; Little Brown: Boston, MA, USA, 1980.
56. Baan, C.; Bergmüller, R.; Smith, D.W.; Molnar, B. Conflict management in free-ranging wolves, *Canis lupus*. *Anim. Behav.* **2014**, *90*, 327–334. [CrossRef]
57. Cafazzo, S.; Marshall-Pescini, S.; Lazzaroni, M.; Virányi, Z.; Range, F. The effect of domestication on post-conflict management: Wolves reconcile while dogs avoid each other. *R. Soc. Open Sci.* **2018**, *5*, 171553. [CrossRef]
58. Cools, A.K.A.; van Hout, A.J.M.; Nelissen, M.H.J. Canine reconciliation and third-party-initiated postconflict affiliation: Do peacemaking social mechanisms in dogs rival those of higher primates? *Ethology* **2008**, *113*, 53–63. [CrossRef]
59. Das, M. Conflict management via third parties. In *Natural Conflict Resolution*; Aureli, F., de Waal, F.B.M., Eds.; University of California Press: Berkeley, CA, USA, 2000; pp. 263–280.
60. Palagi, E.; Cordoni, G.; Demuru, E.; Bekoff, M. Fair play and its connection with social tolerance, reciprocity and the ethology of peace. *Behaviour* **2016**, *153*, 1195–1216. [CrossRef]
61. Fagen, R. *Animal Play Behavior*; Oxford University Press: New York, NY, USA, 1981.
62. Siviy, S.M. A brain motivated to play: Insights into the neurobiology of playfulness. *Behaviour* **2016**, *153*, 819–844. [CrossRef] [PubMed]
63. Špinka, M.; Newberry, R.C.; Bekoff, M. Mammalian play: Training for the unexpected. *Q. Rev. Biol.* **2001**, *76*, 141–168. [CrossRef] [PubMed]
64. Burghardt, G.M. Defining and recognizing play. In *Oxford Handbook of the Development of Play*; Pellegrini, A.D., Ed.; Oxford University Press: New York, NY, USA, 2011; pp. 9–18.
65. Burghardt, G.M. A brief glimpse at the long evolutionary history of play. *Anim. Behav. Cogn.* **2014**, *1*, 90–98. [CrossRef]
66. Palagi, E. Not just for fun! social play as a springboard for adult social competence in human and non-human primates. *Behav. Ecol. Sociobiol.* **2018**, *72*, 90. [CrossRef]
67. Tomasello, M.; Gust, D.A.; Evans, A. Peer interaction in infant chimpanzees. *Folia Primatol.* **1990**, *55*, 33–40. [CrossRef]
68. Pellis, S.M.; Pellis, V.C. *The Playful Brain: Venturing to the Limits of Neuroscience*; Oneworld Publications: Oxford, UK, 2009.
69. Byers, J.A.; Walker, C. Refining the motor training hypothesis for the evolution of play. *Am. Nat.* **1995**, *146*, 25–40. [CrossRef]
70. Dolhinow, P. Play: A critical process in the developmental system. In *The Nonhuman Primates*; Dolhinow, P., Fuentes, A., Eds.; Mayfield Publishing Company: Mountain View, CA, USA, 1999; pp. 231–236.
71. Pellis, S.M.; Iwaniuk, A.N. Brain system size and adult-adult play in primates: A comparative analysis of the roles of the non-visual neocortex and the amygdala. *Behav. Brain Res.* **2000**, *134*, 31–39. [CrossRef]
72. Cordoni, G.; Palagi, E. Ontogenetic trajectories of chimpanzee social play: Similarities with humans. *PLoS ONE* **2011**, *6*, e27344. [CrossRef]
73. Palagi, E.; Cordoni, G. The right time to happen: Play developmental divergence in the two *Pan* species. *PLoS ONE* **2012**, *7*, e52767. [CrossRef]
74. Held, S.D.E.; Špinka, M. Animal play and animal welfare. *Anim. Behav.* **2011**, *81*, 891–899. [CrossRef]
75. Bekoff, M. Social play behaviour. Cooperation, fairness, trust, and the evolution of morality. *J. Conscious. Stud.* **2001**, *8*, 81–90.
76. Bekoff, M. The significance of ethological studies: Playing and peeing. In *Dog Behavior and Cognition*; Horowitz, A., Ed.; Springer: New York, NY, USA, 2014; pp. 59–75.
77. Palagi, E.; Burghardt, G.M.; Smuts, B.; Cordoni, G.; Dall'Olio, S.; Fouts, H.N.; Řeháková-Petrů, M.; Siviy, S.M.; Pellis, S.M. Rough-and-tumble play as a window on animal communication. *Biol. Rev.* **2016**, *91*, 311–327. [CrossRef] [PubMed]

78. Cordoni, G. Social play in captive wolves (*Canis lupus*): Not only an immature affair. *Behaviour* **2009**, *146*, 1363–1385. [CrossRef]
79. Bauer, E.B.; Smuts, B.B. Cooperation and competition during dyadic play in domestic dogs, *Canis familiaris*. *Anim. Behav.* **2007**, *73*, 489–499. [CrossRef]
80. Ward, C.; Bauer, E.B.; Smuts, B.B. Partner preferences and asymmetries in social play among domestic dog, *Canis lupus familiaris*, littermates. *Anim. Behav.* **2008**, *76*, 1187–1199. [CrossRef]
81. Palagi, E.; Nicotra, V.; Cordoni, G. Rapid mimicry and emotional contagion in domestic dogs. *R. Soc. Open Sci.* **2015**, *2*, 150505. [CrossRef]
82. Cafazzo, S.; Marshall-Pescini, S.; Essler, J.L.; Virányi, Z.; Kotrschal, K.; Range, F. In wolves, play behaviour reflects the partners' affiliative and dominance relationship. *Anim. Behav.* **2018**, *141*, 137–150. [CrossRef]
83. Yachmennikova, A.A. Agonistic behavior inter-connections in hidden patterns in wolf pups group during the juvenile hierarchy stabilizing period. In *Wolves: Biology, Behavior and Conservation*; Maya, A.P., Crussi, H.F., Eds.; Nova Science Publishers: Hauppauge, NY, USA, 2012; pp. 109–132.
84. Pal, S.K. Play behaviour during early ontogeny in free-ranging dogs (*Canis familiaris*). *App. Anim. Behav. Sci.* **2010**, *126*, 140–153. [CrossRef]
85. Essler, J.L.; Cafazzo, S.; Marshall-Pescini, S.; Virányi, Z.; Kotrschal, K.; Range, F. Play behavior in wolves: Using the '50:50' rule to test for egalitarian play styles. *PLoS ONE* **2016**, *11*, e0154150. [CrossRef]

animals

MDPI

Erratum

Erratum: Clay, L.; Paterson, M.; Bennett, P.; Perry, G.; Phillips, C. Early Recognition of Behaviour Problems in Shelter Dogs by Monitoring Them in Their Kennels after Admission to a Shelter. *Animals* 2019, *9*, 875

Liam Clay [1,*], Mandy Paterson [1,2], Pauleen Bennett [3], Gaille Perry [4] and Clive Phillips [1]

1 Centre for Animal Welfare and Ethics, University of Queensland, Gatton, Queensland 4343, Australia; mpaterson@rspcaqld.org.au (M.P.); c.phillips@uq.edu.au (C.P.)
2 Royal Society for the Prevention of Cruelty to Animals Queensland, Brisbane, Queensland 4076, Australia
3 School of Psychology and Public Health, La Trobe University, Bendigo, Victoria 3552, Australia; pauleen.bennett@latrobe.edu.au
4 Delta Society, Summer Hill, New South Wales 2130, Australia; perrygaille@gmail.com
* Correspondence: liam.clay@uqconnect.edu.au; Tel.: +61-422-706-076

Received: 4 December 2019; Accepted: 10 December 2019; Published: 15 December 2019

The authors wish to make the following corrections to their paper [1]:

In Table 3, arousal was repeated as a row heading at the end—it should have read "friendliness" (see Table 3 below).

In Table 4, there were errors in the behaviours, their method of recording, the interactions and p-values. The new Table 4 is given below.

Scientific results were correctly stated in the text but incorrect in these two tables. The authors would like to apologize for any inconvenience caused.

Table 3. The behaviours contributing to the emotional states of fear, anxiety, aggression, arousal, and friendliness.

Behaviour Number															
1	2	3	4	5	6	7	8	9	10	11	12	13	14	15	16
Fear															
Diverting	Ears Back	Lip Licking	Lowered Body	Lowered Head	Shiver	Stiff Tail	Tail Low	Tail Tucked	Tense Body Posture	Weight back	Yawn				
Anxiety															
Fast tail	High tail	Jumping	Licking	Lip licking	Medium	Pacing	Panting	Stiff Tail	Tense body	Weight Back	Weight Forward	Whining			
Aggression															
Biting	Ears Forward	Growling	High Tail	Lip Licking	Lowered Head	Medium Tail	Snapping	Standing	Stiff tail	Still tail	Targeting	Vertical Lip Raise			
Arousal															
Barking	Diverting Gaze	Fast Tail	High Tail	Jumping Up	Jump Off	Licking	Medium Tail	Mouthing	Pacing	Panting	Weight Forward	Whining			
Friendliness															
Balanced	Body Curve	Direct eye	Ears Forward	Ears Open	Fast Tail	Handler Interaction	Jump	Medium Tail	Play	Relaxed body	slow	Sniff	Soft	Tail Loose	Walking

Table 4. Differences in kennel behaviour between dogs that were euthanased and adopted, either overall or on certain days.

F/D	Behaviours	Interaction	*p*-Value
D	Tense body	Day 1	0.001
F	Balance/relaxed	Overall	0.004
F	Jumping in kennel	Overall	0.03

D = Duration, F = Frequency.

Reference

1. Clay, L.; Paterson, M.; Bennett, P.; Perry, G.; Phillips, C. Early Recognition of Behaviour Problems in Shelter Dogs by Monitoring them in their Kennels after Admission to a Shelter. *Animals* **2019**, *9*, 875. [CrossRef] [PubMed]

MDPI
St. Alban-Anlage 66
4052 Basel
Switzerland
Tel. +41 61 683 77 34
Fax +41 61 302 89 18
www.mdpi.com

Animals Editorial Office
E-mail: animals@mdpi.com
www.mdpi.com/journal/animals

www.ingramcontent.com/pod-product-compliance
Lightning Source LLC
Chambersburg PA
CBHW061827210926
51850CB00006B/825

* 9 7 8 3 0 3 9 4 3 0 4 4 4 *